LET THE WARMTH & MYSTIQUE

OF

AFRICA

BURROW INTO YOUR HEART!

AFRICA

Breathe My Air

Stir My Dust

Lose Your Heart

RONALD AND LOUDELL POSEIN

XULON PRESS

Xulon Press
2301 Lucien Way #415
Maitland, FL 32751
407.339.4217
www.xulonpress.com

Cover Design by Harold Schnell and Xulon Press.

Unless otherwise indicated, Scripture quotations taken from the English Standard Version (ESV). Copyright © 2001 by Crossway, a publishing ministry of Good News Publishers. Used by permission. All rights reserved.

Printed in the United States of America.

ISBN-13: 9781545606049

ACKNOWLEGEMENTS

Years before I ever thought of putting some life experiences on paper, I told my wife that if I ever wrote a book it would be dedicated to myself. None of this, "I dedicate this book to my dear wife who made a huge sacrifice as I spent months working on the manuscript." Truth be known, my wife was happy to have me out of the kitchen re-arranging things for her.

Seriously, I dedicate this book to all the not so wonderful people I have met in my life. They probably taught me more about myself than my great friends did.

LouDell and I dedicate the book to Africa where we often commented that we will never laugh as much at home as in Africa.

Our grown children, Rhonda and Stephen, dedicate the book to their fellow friends who grew up in a third world culture and never could figure out where they belonged. They are referred to as TCK's (Third Culture Kids).

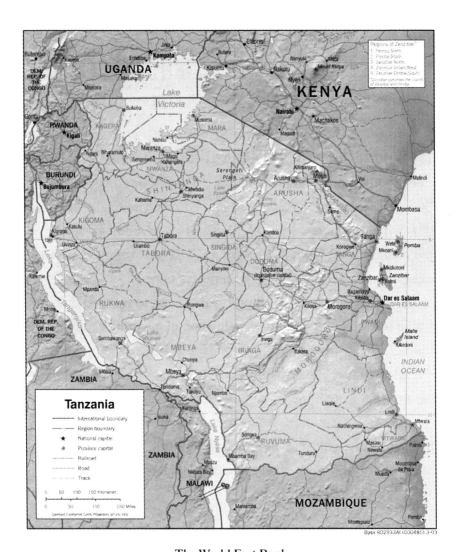

The World Fact Book

INTRODUCTION

Each person has unique God-given talents. In our possessive society we think those innate giftings or talents are our property, developed by self-initiative, self-assertion, perseverance, and hard work. Business owners go to a lot of expense for their employees to attend courses on how to be assertive. Unless God had given us all of these things, we would be as unproductive as a solitary stalk of corn standing in the middle of a barren, drought-ravaged field. The saying, "The Power of One," hinges on the understanding that "One" refers to God alone, whereas the "powerlessness of one" alludes to each one of us. We have the awesome privilege of being God's conduit to accomplish His will in our own lives and in the lives of others. Scripture puts it this way: "For while we were still weak, at the right time Christ died for the ungodly". (Romans 5:6 ESV). Powerlessness is something we are born with; power is something we think we can create for ourselves.

I will mention things about African culture which, as Westerners, we find absurd or unbelievable in this 21st century. Most things are based on a value system; until that value system is understood, many things will not make sense. Be content to know that many things in our Western culture make absolutely no sense to the African mind. When I mention things about African culture, it is with respect to their value system. I love the diversity in the African culture. In my travels I have found no other people who are so across-the-board genuinely friendly, accepting, and open to receive and share.

CHAPTER 1

God's Call

"Ron, I want you in Africa." The voice in my ear was crystal clear. It was 9:30 p.m., dark, and I was standing outside of our small country church on the wooden walkway, which was so common for country churches in those days. Rural Black Mud, Alberta, was the place—who knows how these names came into being? After all, is mud any other colour? The year was 1956; I was 11 years old and had not yet committed my life to Christ.

A missionary service had just concluded. Being the first missionaries sent to Africa by the German Branch of The Pentecostal Assemblies of Canada, the couple were visiting churches to raise awareness of potential church work in Africa.

Thus started my own dream of Africa, although arriving there would take another 15 years, with many experiences before then. One of those experiences was marrying LouDell Lang. As our dating became serious, I shared my calling to Africa. I am a private person, and she was the first person with whom I had shared it. She did not have a calling and that presented a dilemma. She was hoping that time would dim that desire in me. A stint in Germany, two children, and three different pastorates came along, but my ultimate purpose in life remained focused on Africa. In spite of LouDell's lack of clear direction from God, she took my calling as her own. When a calling is clear, it is easier to tough out the tough times. In LouDell's case, she struggled with lonesomeness and the general extreme differences in culture.

Boarding School

For both of us, sending our daughter, Rhonda, and our son, Stephen, to boarding school was a traumatic experience and one we still prefer not to dwell on. When we do think about it, memories surface and tear at our hearts'. I will say that many times we stood shedding tears in their empty bedrooms and praying for them. Rhonda was seven years old, and Stephen was six when they left home. They spent three months at school and one month at home. From the outset, let me say that the options of home schooling and boarding school both have their strong pros and cons. Each family must make their own decision about it.

A number of things made boarding school necessary for us. Our work involved church planting almost exclusively in rural areas, which meant we were away from our home every week. LouDell had tried home-schooling Rhonda the first year, but it was not a positive experience—for her or for Rhonda. No local competent schools existed where we lived. Rhonda was especially outgoing and needed the interaction of peers. We were the only Europeans in the towns we lived in for most of the first four years. Our children also needed to be enrolled in a school which had a curriculum in line with Canadian schools since that is the country they would eventually again call home.

Africa's Mystique

Africa, the Dark Continent, holds a mystique like no other. Anyone who has spent time there acknowledges it is forever in their blood, constantly drawing them back. Many have debated why that might be. Is it the people of thousands of tribes and customs? Most surely this has to be a big part of the answer. Is it the wide-open spaces? That tranquility has to be part of it as well. Is it the abundance of endless species of wild animals? Early childhood readers certainly contributed to that. Is it the uncounted thousands of types of bugs, insects and snakes? What about the landscape of deserts to rainforest to the highest freestanding mountain in the world? Snow-capped Mount Kilimanjaro, standing within only a few degrees of the equator, is a total mystery as wild animals graze at its tropical base. Or possibly it is the night sky, where stars from both the northern and southern hemispheres are visible.

It is a land of threes:

Three "S's"—Sights, Sounds and Smells

Three "B's"—Bugs, Birds and Beasts

Three "D's"—Discomfort, Dilapidation and Disease

Three "I's"—Inconvenience, Innovation and Inconsistency

Three "W's" (Wonders)—Serengeti Park, Ngorongoro Crater, Mount Kilimanjaro

Famous missionaries and explorers spent years exploring the country. Their reports sent waves of fear into the hearts of the readers and, in some, sparked desires to become part of the experience. They were men like John Hanning Speke, David Livingstone and Henry Morton Stanley. To discover the source of the mighty Nile River was a quest which took decades to unravel. After years of enduring extreme sickness, hardship and even death, explorers discovered that its origin lay in Tanzania. Running through hundreds of kilometers of dry land, it becomes a giant of six kilometers wide emptying into the Mediterranean Sea. Tanzania is home to snow-capped Mount Kilimanjaro, the highest in all of Africa at 19,341 feet. Thousands of people each year are drawn to the challenge of reaching the peak without oxygen. Professional climbers of other world mountains usually begin using oxygen at 13,000 feet.

Lake Victoria is the second largest fresh water lake in the world, bordering the three East African countries of Tanzania, Kenya and Uganda. Drawing tens of thousands of visitors annually are world-famous parks like the Serengeti, the Selous, Lake Manyara and Tarangire, plus many smaller ones.

Paleontologists flock to Olduvai Gorge to uncover the remains of the earliest man and animals. Topography can change rapidly. Plains turn into hills and then into brush, and even rainforest. Each area is home to different species of wildlife. Although not mentioned in some encyclopedias, the Ngorongoro Crater is a wonder of the world. In some publications it is referred to as the 8th wonder of the world. Its size is impressive: 10 miles across and 2,000 feet deep. It has its own ecological system of rainforest, plains, hills, a lake and rivers. Within its boundaries lies a shallow alkaline lake, home to thousands of flamingoes. Feeding on the rich plankton, they turn the shallow lake into a mass of pink as they return to lay eggs and rear their young. A small forest grows as well as grassy plains. All of this sustains Cape buffalo, lions, elephants, rhinos, Thomson's gazelle, jackals, hyenas, and many species of birds. Some animals never migrate out of it.

Watching the massive twice-yearly Serengeti migration of a million wildebeests is a sight one never forgets. As far as the eye can see, for 360 degrees, the plains shimmer with moving wildebeest. Ever present to prey on the sick, weak and young are lions, hyenas, cheetahs, foxes and vultures.

In the late '80s, the World Health Organization officially classified the AIDS virus as a disease. It had taken hold in East Africa long before that. First-hand reports by local people said that it had actually been around since the late 1950s. By then it had already become so common that people recognized it as a special disease. People would get sick, develop running sores, lose their appetite, and lose weight. This last symptom gave it its local designation of "slims" disease. Death was not far behind. Biblical time is followed: our 7 p.m. is their 1 o'clock nighttime. Women's right to property, upon the husband's death, is something they still strive for. We do not understand the custom, but when a husband dies, his brothers may come in and take everything, including the house. A person wonders under what circumstance it actually originated. This African proverb underlines the insignificance placed on women: "A person sent by a woman is not afraid of death."

Infrastructure, especially in the rural areas, is non-existent. The introduction of the cell phone in the '90s was a saviour for most people in terms of communication. Roads during our time there were seldom maintained and left to fend for themselves. The ravages of each rainy season altered their course.

Society is open due to the fact that people live in such close proximity. "There is no secret in Africa" is a common and true expression. Shattering the early morning stillness is the loud Muslim call to prayer. Emanating from huge loudspeakers, it echoes over the towns and cities. The calls to prayer by multiple mosques are very audible. The life expectancy of a Tanzanian in the '70s was 40 to 45 years of age. Many street-side stores exist, but hawkers bartering their goods are a common sight on every sidewalk. They even come right into banks and barter their shoes or milk or vegetables with bank employees—while you wait in line.

In the midst of less than comfortable living conditions, according to Westerners, Africans are happy people, content with what life is serving them. I suppose it will take some serious change in our conditions to get back to realizing what produces true happiness.

One particular sight still stands out in my mind. A man with useless, dangling, shriveled legs, but with a healthy upper body, is hand-pedaling his three-wheel bike over the rough streets. Pedals are mounted in front of him where handlebars would normally be. With this construction he can pedal with his hands. On the back of his tricycle hangs a hand-written sign, "God Is Faithful." Wow!

The British took over colonization from the Germans in the very late 1800s and left their mark. They had an obsession with record keeping. Their love of paperwork resulted in numerous offices of all kinds scattered throughout the town, many of them in minimally renovated houses. The second mark left was vehicles driving on the left side of the road. Having the steering wheel on the right and the gas pedal near the door takes some getting used to.

Preparing for the Mission Field

When LouDell and I applied to go to the mission field, we were informed of the requirements and procedures.

First, write a letter of request to Head Office for appointment.

You have to be a senior pastor for at least two years.

You must be ordained (this takes at least two years after your first full-time pastorate).

Take a St. John's Ambulance First Aid Course.

Don't have more than two children.

In case of abduction we will not pay any ransom.

Then we received "The Letter":

"We wish to inform you that at our most recent meeting of the General Executive, your names have been approved for overseas mission work in Tanzania. God bless you."

Since then we have heard "GOD BLESS YOU" a few times. What is not said becomes the critical thing. We were young and pastoring a church and asked, *Now what do we do?* The office wrote back and said, "We don't know. You have to figure it out yourself." A missionary advised us to begin by writing the field director of the mission in Tanzania. Good advice. That was our introduction to fending and thinking for ourselves, all within the confines of our organization. Many people take "thinking for yourself" or "finding your way" as indications that they can do whatever they want. That is never the case; there are always parameters. When we work within them, we are protected from unforeseen possible negative consequences. These negative things can affect our personal lives as well as impact the work we are involved in.

Although we didn't know it at the time, we would face many situations and be compelled to make decisions without assistance from anyone, mostly because there was no one nearby to help us. Another important aspect was that people 6,000 miles away could not possibly fully understand the situation. There were times when decisions had to be made quickly. The communication time factor played a huge part in demanding self-reliance.

Living in a foreign country brings culture, remoteness and isolation into play to a pronounced degree. Overnight, the normal Christian support network no longer exists. Communicating with local Christians is impossible due to language differences. There are no Christian magazines and no Christian radio broadcasts. There is no T.V.; thus, no Christian T.V. programs. There are no mid-week church services, no Christian bookstores or books, no familiar hymns in the church. It is as though you have suddenly been catapulted onto an island. And indeed you have been! Your Christian walk had better be something you carried within you, and not something based on external sources. An African rugby team had a motto which aptly describes what I am talking about: "We don't build character; we test it." When going to a mission field, the character of the missionary already needs to be in place or disaster is not far away.

Beggars

Anyone who has travelled to Africa knows there are things that characterize its mystique. Among these are beggars on the narrow sidewalk blocking your direct route to anywhere. I use the term "beggar" in a kind and sensitive way. In a society without social assistance, health care or pensions, it is the only available method of survival. Birth defects, accidents or disease have disabled these people in some way. Some are totally immobile, so they just lie on the sidewalk. Each morning they are brought there by a relative or friend who, later in the day, picks them up again. In essence, this is not much different from the accounts of people's lives that we find described during biblical times.

Mwanza city designated Thursdays as "Beggars Day." Many did not adhere to just one day, however. On Thursdays beggars were to be found on every street. Some were just plain poor people; some were lepers with stubs where fingers and toes used to be. Some had holes eaten into their cheeks, making their teeth visible; some had their noses completely eaten away. Certainly not a pleasant close-up sight for most Westerners! Visiting a leper colony quickly makes one's own personal, incidental troubles fade in importance.

Remarkably, as I soon discovered, these people on the street could fend for themselves. There was a certain fingerless and toe-less leper whose regular station was the post office. I saw him on a regular basis. A typical construction site fence is erected with corrugated iron sheets—standing three meters on end—around the site. These crowd onto half of the already narrow sidewalk. Early one morning (I love this about Africa: businesses open at 7 or 8 a.m. at the latest), I was walking on the side of the street opposite one of these construction (obstruction) sites. Glancing ahead a bit, I noticed the leper was on the same sidewalk and was acting a bit irritated. Then my ears tuned in to the fact that across the street a man was exchanging some unkind words with him. Before I knew it, and to my astonishment, the leper somehow picked up a fair-sized rock and hurled it at the fellow. It had some force behind it, judging by the way it rattled off the iron sheets and the speed with which the other fellow bolted down the street. That day I gained a new respect for the spirit of self-preservation.

Some beggars were lame and used a long, thick pole as a cane. The extreme cases had one leg shriveled and dangling, slapping the pole as they propelled themselves along. They grabbed the pole with both hands, and in one motion planted the pole and swung around it. Some were blind, so were being led by children. One lady stands out in my mind. She had totally lost mobility in her legs. Laboriously she shuffled along the filthy, sometimes muddy sidewalk by lifting her body with her arms and pushing forward. Others had no legs and used their hands as legs, wearing rubber sandals like shoes.

These beggars were extremely shrewd as well. On Thursdays they would arrive in droves from outlying suburbs. Some, as I said, had escorts, but a regular practice was to hook up with street children and adopt them for a day—a kind of profit-sharing arrangement. The beggars with children received handouts in proportion to the number of children they had. On their own, the street children are seen as urchins. Due to their stealing and bad manners, they would not get near the handout if they were alone. Everyone benefited from the collaborative arrangement.

This is the way it worked. Beggars, with children in tow, went from storefront to storefront asking for alms. Most stores in Mwanza are to be compared to a Wild West setting. Each street has ground-level stores, approximately 10 to 12 feet wide by 10 feet deep. A counter in front allows you only to look in and see what is displayed hanging on a wall. These stores are side by side with a single wall serving two stores. There is no public access into the store. On Thursdays the merchants would send a worker to the bank to get a bag full of five-cent pieces (equivalent at that time to about .001 cents Canadian) to hand out to each beggar. Some businesses on side streets that were into manufacturing would hand out a meal at their premises. In large numbers beggars sat in front of the place and ate rice and spinach. Some places gave out bread or bananas.

Sidewalk Markets

In the smaller towns and cities like Mwanza and Musoma, if you were driving through the middle of town at 6:30 p.m. or later, you wouldn't know where the stores were because everything is shuttered up. Every city or town has its open fresh fruit and vegetable market as well. Mostly these markets consist of a long cement platform with a roof and many people selling their wares. In smaller centers, each table is homemade by the person leasing the selling space. The tables were rarely square and it did not matter that every leg had it's own unique length. A variety of fruit is sold such as pineapple, papaya, mango, bananas, lemons, coconut, avocado, limes and, once in a blue moon, oranges. Peanuts (locally known as ground nuts) are already shelled; you just have to roast them in the oven to make them edible. They sure are good! In rural areas each village has a market day that is rich in colour and in the variety of products available. In our own yard we grew coconuts, papayas, oranges, avocados, lemons, limes and pomegranates.

Sidewalks have their own culture. They are always teeming with people. Merchants display goods on the already narrow space. Many sidewalks are lined with people sewing on treadle machines. Sewing clothes is a big business as every school student needs two uniforms. Ordering a dress or pants is much cheaper than going to a business that does sewing professionally. Many of these sewers do a remarkable job. They sew made-to-measure clothing as well as any professional tailor shop.

The majority of the people walk while carrying purchases on their heads. This ranges from lumber, to iron sheets, to reinforcing bars, to stalks of bananas and all kinds of fruit. Young girls are masters at stacking fruit on large round aluminum pans and then carrying it on their heads. Taking it down to sell without dumping the whole thing is another of their perfected arts and one which always made me marvel.

Hawkers are everywhere trying to sell trinkets, African art, tools, aluminum pots and pans, peeled oranges, mango, pieces of pineapple, squeezed sugar cane juice, newspapers, ladies cloth wraparounds, and wrenches. Always to be found are the squatting sellers of Arabic coffee. The cups are tiny but hold potent black coffee. Very hygienic too. After every use the cup is given a quick rinse in a pail of water. I have never seen the water in the pail changed. No doubt it is done once a day at least.

Driving in Africa

Why would any country spend a lot of money building a nicely paved road and then allow only vehicles to use it? What does it matter that the speed limit is not posted? Is this just another thing you are expected to absorb by some mystical mental telepathy? Ask any policeman, and he will tell you that drivers should know those things.

The reason has always baffled me, but Africans love nighttime. Driving on a highway at night is not recommended due to the dangers. These include: grey donkeys aimlessly meandering; upwards of 75 cows being herded over the road by young boys; the same with goats and sheep; people dressed in dark clothes riding bicycles (no lights, no reflectors); carts loaded with firewood being pulled by muscled men; there are always staggering drunks filled to the brim with local home brew; lumbering, black smoke belching trucks, overloaded buses and vehicles without headlights or taillights (it's dark, remember).

Vehicles with headlights are no less a hazard. Invariably both headlights are on full beam and shining directly into your eyes, blinding you to anything that may be on the road ahead. The close calls I have had are too scary to even recount. Once a long safari took me into the night. Coming along at a nice clip, I saw bright headlights on the road ahead. There was nothing unusual about that, except there were four headlights which seemed to span both lanes. Coming closer and a lot slower, I saw two big trucks standing side by side while the drivers were having a chat. I simply waited until they were done, and then everyone happily moved on.

Bus drivers are always in a hurry. It would probably be more accurate to say they are enjoying the exhilaration of just touching the edge of losing control. Careening past at 140 kilometers an hour, they create some serious side wind. Traffic lanes in Africa are only slightly wider than a bus. Only rarely do roads sport centre lines. This translates into *very* close encounters. Together with all of these challenges, the police find it convenient to occasionally set up roadblocks at night with heavy iron pipes spanning both lanes. Sometimes they hang little kerosene lanterns on the pipes, but even travelling at a breakneck speed—considering the conditions—of 70 or 80 kph, these little lamps appear without warning. Other times the lights have long since burned down to nothing. Crashing into the barrier is not something the gun toting, groggy policemen see as humorous at five o'clock in the morning. Being awakened from a nice nap is irritating.

Following a friend of mine one early morning, I suddenly saw sparks flying. Yup, he had unceremoniously awakened some policemen, and they were not impressed to see a pipe embedded in their makeshift office wall down in the ditch. The kerosene lamps had long since gone out, but that was no excuse for hitting the barrier. When everything was said and done, my friend paid handsomely for the expected repairs to the mangled pipes and added three hours to his safari. Months later I encountered the same roadblock, and the pipes still sported their new bent design. The repair money had found a better use.

African Manpower

Driving through rural areas, you see many people standing in a line and swinging their Chinese-made hoes in unison. Whatever we may think of Chinese products, their steel products for Third World countries are made for serious work. The Flying Pigeon bicycles are unbelievable as far as durability goes and the weight they can carry. I saw a man pushing his bicycle with a 220-pound bag of corn lodged in the frame, another 220-pound bag on the back carrier, and yet another 220-pound bag on the top part of the frame. (Tanzanians bag everything in 100-kilogram sizes; they see anything less as a bother to handle.) No wonder he was pushing it! I have also seen a person riding such a bicycle and carrying four passengers. That takes some leg muscles, not to mention ingenuity for seating space.

For us, that is a mighty heavy bag to throw around. In the interest of making things easier for our farm manager, I bought some bags at the local market that held only 50 kg. It wasn't long before he was complaining that they were too small. In the absence of machinery to move things, strong backs are still the order of the day. The industrial process of moving bags involves three people. Two people lift the bag two feet off the ground and the third, "the carrier," bends down and positions himself so the bag lines up with his shoulders. Down comes the bag, he straightens up and *sprints*—yes, you read that right—to the drop-off point. This involves running up a steep ramp made of filled bags piled up. Contrary to our thinking that such a journey might include moaning and groaning, it resounds with laughter and singing.

I needed to buy some corn for our orphanage, so I went to a local government warehouse. What hit my eye, as I walked in, was unbelievable. The warehouse was the size of half a football field—55 meters long by 30 meters wide. There were thousands and thousands of bags of corn, piled in sections 30 feet high. All of them got there by manpower.

Typically, all land is "hoed" into long straight mounds about a foot and a half high. Beans, corn, millet and sweet potatoes are all planted on top of and on the sides of these long, elevated rows. It took me a while to figure out the reason behind it. When the heavy rains come, the plants do not wash away. The long rows also prevent wholesale erosion by the water; after, the rainwater would remain in the trench. When we say "rain" in Africa, we mean rain big-time. I also learned that these crops complement each other, plus it makes good use of the land surface.

People love to do things in a group. Forget our puny, flimsy, lightweight toys; these Chinese hoes have heavy steel blades that are 10 inches long by six inches wide. We are familiar with jokes that illustrate the absurd—like selling ice to Eskimos. In Tanzania there is an equivalent: try selling a hoe handle—or any other handle, for that matter—when there are numerous trees around with limbs that make perfect handles. My point is that hoes do not come with handles. The hoeing process looks like this: four, five or more (women usually) stand side by side and raise the hoe, with its seven-foot handle, high and back over their heads. Then, with force, they bring it down so the blade is buried completely in the soil. They give it a tug to bring the soil back toward themselves, forming a mound. Doing it together, they form mounds that are one continuous line. They step over it and repeat the process. When done, the whole field has rows and rows of furrows and mounds—all in a line. This planting on the top of the mounds confused me. I didn't see the sense of planting a seed on a mound, where the soil would quickly dry out. It all made sense, though, once I experienced the first torrents of the monsoon rains. The huge runoff does not create washouts in a furrowed field. Lesson learned.

In the areas around Lake Victoria where we lived, it seemed that 100 per cent of all farm cultivation was done by hand. Our tractor on the orphanage farm was the only one in at least a 15-mile radius. Africans are early risers and, when driving by the fields in the morning, you can see large portions that have already been hoed. Basically the hard work has to be done by 10 a.m. After that, the tropical sun becomes too hot. It is common to see women hoeing with a child tied on her back. Now that is true bonding.

Early morning is also the time for the young boys to take the cows to pasture; only a few tribes use girls as shepherds. Cows and boys distracted by interesting things don't worry about checking for traffic. Tanzania has no private land ownership, so cattle can be grazed anywhere. In the front yard of huts, women sweep with a broom—branches tied together on a handle made from a tree branch. Taking the fine dust off every day and lightly sprinkling the yard with water leaves the area as hard as pavement.

Around 10 a.m., massive tarps are spread out and corn, rice, millet or cassava is spread on it to dry. Clothes are spread out to dry on any available grass or bush. Don't waste your time trying to sell clotheslines in this country. Who needs them? Children are everywhere amusing themselves by throwing rocks at passing vehicles, playing soccer with rolled-up plastic tied with a string, shooting birds with their slingshots, or steering their homemade cars with a long wire. They have amazingly creative minds.

In the afternoon two girls can be seen grinding corn in the yard using a large mortar and pestle. Each wields a three-inch diameter pestle. The first girl forcefully thrusts the end into the mortar holding the corn; as she withdraws her pestle, the other girl thrusts her pestle in. Poetry in motion. In the end, the corn becomes fine cornmeal flour.

A common sight is a child sitting on the roadside selling whatever is in season. It may be mangoes, oranges, pineapples, bananas, peanuts, beans, rice, cassava, European potatoes, sweet potatoes or roasted corn. A rather nice convenience—something like a fast food drive-through. If you're looking for a pastime while you are on a longer trip, buy some roasted corn on the cob. It has been left on the stalk until it is rock hard. Chewing on a cob of that will keep you busy albeit, at the end, your jaws are aching from the workout. Nothing beats the taste, however.

CHAPTER 2

Some Tanzanian History

Tanzania was originally called Tanganyika and was ruled by the Germans in the late 1800s to early 1900s. It was known as German East Africa and actually covered the present countries of Tanzania, Kenya, Rwanda, Burundi, and part of Congo. A good book on the history is *Battle for the Bundu: The First World War in East Africa* by Charles Miller. When the British defeated the Germans, things changed. The Germans were very strict, if not ruthless. When a thief was caught, his left hand was cut off. If he repeated the offence, the right hand went. The stump of a tree still exists in the city of Mwanza where these public deterrents were administered. It is said that during the Germans' rule, there was literally no thieving.

The evidence of their rule is still evident in many places today. For example, many of the German government office buildings, known as German *bomas*, still stand and are used. They made some sort of cement from a dirt and clay mixture. Walls were built two feet thick—virtually indestructible. Driving into Serengeti Park from Mwanza, you bypass small bridges over streams. I say *bypass* because today the soil around them has eroded so they stand a meter or higher above the surrounding areas. At some of these bridges, the dirt has been brought back in the front and back of them, so you actually do drive over them.

The British introduced their own style of record keeping that consisted of multiple copies of the same document. I remember filling out up to six copies of a given document. This was especially prevalent in banks and in the customs and immigration areas and still is to some extent today. Searching for usable carbon paper was a challenge. This obsession with administration led to a proliferation of offices. Many offices, or sub-offices of main offices, are found in renovated houses or other buildings even to this day. This means that the Land Office, for example, might be spread all over town, with up to five or six offices housing different departments. They are usually dark, with 12- to 14-foot ceilings and sweltering hot, especially at the coast. Tired-looking fans lazily turned, endeavoring to move some air. Window coverings consisted of pieces of cloth customarily tied in a knot about three feet up to keep them open.

With the British administration came the notorious rubber stamp. The British loved rubber stamps—the more the merrier. During our time there, 1973–2010, the rubber stamp still trumped your signature. It didn't matter that you could go to a street corner vendor and find a fellow sitting behind a small shoeshine type box cutting rubber stamps by hand with a dull razor blade. They were masters at their craft. So, technically, anyone could have a rubber stamp made with any name on it. If you asked for a stamp with a company name, proof of your authority for having it made was never required. I always travelled with rubber stamps for the mission, Starehe Children's Home, and our personal one. On safari you never knew when an official letter might have to be written. To make it look "official," for sure it needed one or two stamps on it. Tanzanians are masters at reading character in a very short period of time. They can sit with a total stranger and very quickly accurately describe the person's character.

Early, Late or On Time?

When we first started our ministry, it bugged me to no end that people would show up a half-hour to one hour later than I had arranged to be at a church. My decision was firm that if it happened again, I would turn around and go home. It was obvious to me that they were not very committed.

On one occasion we drove out to a rural church about an hour from Musoma. I had arranged with the pastor that we would be there at 10 a.m. As usual, I was early and when 10 a.m. arrived, not a single person had come to the church. As gracious as I always was, I added 15 minutes onto my grace period. Pretty generous of me! Truth be known, the 15 minutes were at my wife's insistence of mercy. Then I left. My wife wasn't exactly in favor, but I was the driver and knew my way home across country.

The next day the pastor came to my office, very upset that I had left. I reminded him that our arrangement was for 10 a.m. I did not want my time wasted sitting in the hot African sun waiting when people did not care enough to be on time. He graciously explained that the former missionary always came an hour later than what he said he would, so the people adjusted to his time schedule. That same missionary had "taught"'me that people are always late, so he went an hour later than what he had arranged with the pastor. Hilarious! Each was adjusting to what they knew the other would be doing.

I informed the pastor that I was not interested in playing games. If I said 10 a.m., I would be there at 10 a.m. From that day forward, I never had any people in any church come late. Word spreads very quickly.

Dealing with leadership in nearly a hundred churches was a challenge I relished. Whether a pastor was actually doing his job was difficult to figure out. One pastor gave me some doubt, so I devised a simple plan of verification. Due to the number of churches, we made a schedule of visitation six months in advance. Naturally, the pastor would be there along with his congregation. The actual number of the congregation could be suspect since a white-faced visitor draws a crowd, out of curiosity if nothing else. It was a waste of breath to ask if all present were regular attenders. The answer would always be a resounding yes.

As scheduled, I visited the pastor's church and we had a fine service. Out of schedule, I returned to the same church the next Sunday. The congregation numbered about half of the attendance on the previous Sunday. The pastor was nowhere to be seen. When it was time to begin the service, the deacon asked me to wait a while as the pastor was just a bit late. I knew what that meant and took over leading the service. About halfway through my sermon a bewildered, sweating, puffing, rubber-booted pastor rushed into the church. I politely asked him to take a seat while we finished the service. By his attire I knew he had been out herding his cattle while his real sheep were without a shepherd. After that all the pastors were on edge because they never knew where I would actually be going on a certain Sunday.

Some customs are more difficult than others for us to accept. As a sign of friendship and confidence, men hold hands in public. There would be times when I was walking along with a pastor and he would grab my hand. I had to grit my teeth to translate the idea of what it suggests in the Western culture and convert it into the African cultural meaning. Now, a man holding a woman's hand also has a meaning. It means they are on their way to make love, not only that they are in love. "In love" affections are not openly displayed.

In some tribes a woman does not make eye contact with a man. Doing so is translated as a come-on. For many years, a woman wearing slacks was a dead giveaway that she was a loose woman. This stemmed from a tribe whose women were known as "loose" because they customarily wore slacks. It became a trademark.

We're Too Young To Contemplate These Things

LouDell and I spent two years as assistant pastors at Immanuel Pentecostal Church in Winnipeg. This was a church of over 200 people. Pastor Arthur and his wife, Adina, were so kind and mentored us as true spiritual parents. Our national leadership had been informed of our desire to serve in Africa. That was gratefully accepted, but there were some stipulations. I needed to be ordained. In order to achieve that, I had to pastor a church on my own. Serving as an assistant in Winnipeg was wonderful on a resumé, but having to administrate and navigate issues that arise in a church setting requires wise decision-making, tact, and an eye to maintaining good public relations. Recently I heard a good definition of tact: addressing a tough issue and coming away still being friends with that person.

In 1970 we accepted an invitation to pastor the congregation in Ponoka. It was while pastoring there that the Missions Committee of our denomination approved us as the next missionaries to Tanzania. In August 1972 we were officially appointed as missionaries. Uganda and Tanzania were involved in cross- border skirmishes, so our departure was delayed until September 1973. It would not be wise to send a new missionary to a country on the brink of war. The possibility existed that all expatriates might have to escape to a neighboring country if things got worse. In 1975 war did break out. The famous words of President Nyerere declared that Tanzania was going to war with Uganda. I remember his saying, "We have a snake in our house."

Pastoring in Ponoka, Alberta, I learned a valuable lesson, which I never forgot and made part of my life. It was also incorporated in the lessons that I later taught pastors. In August 1972 our district superintendent approached us with the decision of their committee. A larger church was without a pastor, and they wanted us to relocate.

This did not make any sense to us at all. I explained to him that we had been approved to go to Tanzania. Why should we go to the trouble of resigning from our present church, and packing up our things and moving, for possibly a very short length of time? Being a member of the national committee, which had approved us, he was well aware of our missionary approval. We also had two small children. He agreed with our assessment of the situation. Within a month he called us again and said that the district committee still felt we should move. LouDell and I discussed it at length. We came to the conclusion that there was not one positive thing in favor of accepting the recommendation of the committee. However, since they were our leaders, we accepted their decision without any bad feelings. We relied on the fact that God was using them to lead us. It is good to follow the biblical principle of respecting godly leadership.

In October 1972 we moved to Barrhead, Alberta, and pastored there until July 1973. Without hesitation, we can say that God blessed us beyond anything we can record here. It was God's favor on us for respecting our leadership in spite of the physical difficulties connected with making the move. God honors us when we honor those in authority over us.

Our understanding was that the congregation was aware of our temporary assignment. Either they had forgotten that little fact or were hoping for another outcome. Whatever the case may be, they were shocked when we were fully appointed as missionaries and resigned the pastorate. Leaving was not easy.

Being appointed as missionaries involved the normal process of applications, various committee approvals, interviews, and finally appointment. There was a myriad of things to arrange at home, the trip itself, and then our four-year term in Tanzania. The most sobering thing for us was the fact that we, at this young age, had to make a will and designate a guardian for our children should they be left without parents. Rhonda was five, and Stephen was three. Wow! To face that reality was tough. Another guardian had to be appointed in Kenya when the children attended school there while we lived in Tanzania. Rev. Jack Lynn took on that responsibility.

Luggage and freight allowance for overseas was limited, so we had a home auction to sell off our things. Watching our children's toys being sold was particularly difficult. At their age, their understanding of God's call was not relevant. It was a stark realization that our children would be called upon to make just as many sacrifices as we ourselves. That proved to be a harsh reality during our years in Africa. At the end of the day, we stood with only as many possessions as we could load into a vehicle.

On September 1, 1973, we bid a tearful goodbye to parents and relatives and boarded the plane to begin our adventure of being career missionaries. That experience had many highs and lows, but what an adventure it was for 35 years—one we will never forget nor ever regret.

LouDell's first impression was seeing a bulldozer operator wearing a suit. Forever embedded in our memory is the day we left Nairobi, Kenya, for Musoma, Tanzania. Only a few days before that, we had stepped off a 767 jet. Now, two days later, with our children Rhonda (5) and Stephen (3) in tow, plus eight suitcases, we headed for the Wilson Airport for our final destination flight. Apparently, the pilot slated for the flight did not show up, so a substitute was called. He was as angry as a bull at having to fly. He wasted no time in a slow takeoff. Taxiing to the runway takeoff area, he whirled the plane around and in one continuous motion gunned the engine to full throttle. He careened off the runway, and within minutes the ground fell away below us as we passed over the Rift Valley. All five first-time passengers, crammed shoulder to shoulder, in that small six-seater single-engine plane feared for our lives, the blood draining from our faces as the plane jumped and bucked in the severe turbulence. It is easily within the realm of possibility that the man flying our plane was an experienced World War II fighter pilot.

Rainy season was approaching; the skies held ominous black thunderclouds, and the pilot's attitude matched them exactly. Apparently, another pilot had booked off at short notice and he was conscripted to do the flight. He was not happy! With the buffeting winds tossing our plane around and the air pockets dropping us unexpectedly, it was not long until a few of the five passengers began making use of the "air sickness" bags. Just to be clear, there is not much fresh air in that washtub cabin. If your stomach managed the tumbling and sloshing from side to side, then it still had to deal with an odor that it wanted to imitate.

Imagine your shoulder being squeezed against the side, which is an inch from the outside. Looking out, you looked straight down and began pressing against your fellow passenger, who was already shoulder to shoulder and trying to distance himself from the window. Amid the assault on our bodies, stomachs and senses, other things still managed to worm their way in and make an impression. Forever seared into my mind's eye is looking down and seeing my first grass-roofed mud hut. The dark red soil belonged in a paint- by-number book. Then there was the utter lack of definition of order. No geometry of nicely squared off parcels of land. No north, south, east, or west roads. Meandering cattle paths joined groups of huts. Seemingly without design of any kind, it was a vast open expanse leading nowhere.

LouDell was seated behind me. She reached around and grabbed my leg. I think the scars are still visible. Getting into Tanzania airspace, the weather cleared, but not the spirit of the pilot. Musoma had a dirt airstrip with a two-room airport. At that time very few planes landed there. The pilot dive-bombed the building and did a low flyover "to wake up the workers," he said. That two-hour flight took a few years off our young lives—that is for sure!

Within a few minutes, we had exited. The pilot had the plane at the end of the airstrip and was revving for takeoff. Again, forever imprinted in our minds is the sight of the plane roaring past us in a billow of dust. There went our last connection to the Western world, and we were in Africa for four years. It was a hollow feeling. It had all happened so quickly. There was no easing into anything.

Rental housing was basically unavailable, so for a few months we stayed with missionaries stationed in Musoma. Later we cleaned and renovated the servant's quarters, which measured 14 feet by 26 feet. Some noteworthy things about these quarters: no hot water; shower water running straight outside through a hole which rats loved to enter; bat droppings on the kitchen cupboards every morning; and a neighbor brushing his teeth with all kinds of guttural utterings, all no further than three feet from our wooden shuttered bathroom window. No African scene would be complete without the high-pitched yapping of dogs all over the neighborhood, plus crowing roosters at five in the morning.

Many things are shocking in a new country and culture. Especially sobering for us was the fact that the average life expectancy was 40 to 45 years of age. Many children died before their second birthday. The rate of mothers dying in childbirth, or shortly afterward, was numbing.

Starting Over

It was our first Christmas in Africa. Lonesomeness hits hardest at holiday times. Missionaries stationed in Mwanza invited us to spend Christmas with them. It was a brutal three-hour drive from Musoma, where we lived. The roads were sandy, dusty, full of pot-holes, and amply supplied with detours into the ditch to circumvent particularly bad stretches on the main road. One day into our visit, our children, Rhonda and Stephen, plus two of the fellow mission-aries' children, all got sick. A local clinic prescribed some medicine. Within a few hours, Stephen was hallucinating. This was our first lesson about African medicine. Generally, the people's immune system is so strong that it is common practice to double the dosage of medication. Stephen's hallucinations lasted for four hours.

The ladies stayed home with the sick children while I accom-panied the missionary to a pre-arranged service two hours into a rural area. I was asked to preach and was translated by the veteran missionary. My language school course still had three months to go to completion, so preaching in Kiswahili was out of the question. Still being in the Western pastoral frame of thinking, I began by saying, "We live in tremendously exciting, momentous days." My interpreter hesitated, looked at me and said, "I think, just start over."

Drain the Spirit of Epilepsy

I met Justina when she came to the Pentecostal Assemblies of God Bible College located in Mwanza. The life stories of stu-dents give me a mental picture of the life which lies behind a face. Justina gave me a quick glimpse into her life, and I later spoke with her to fill in all of the details. She was born into a heathen family. Justina was born in a small rural African home near Serengeti Park. At what age her memory recall starts she does not know. What she does know is that the first memories she has are of a witch doctor treating her. The bark of a tree was boiled. She was forced to drink the bitter concoction, resulting in bringing on vomiting for extended periods of time. When the "medicine" didn't work, her body was cut so "the sickness demons" could escape. Her skin was cut and blood was sucked out in an attempt to drain her body of the spirit of epilepsy. Her father believed that the epilepsy was a demonic curse that had to be driven out.

Justina was born with two strikes against her, and a third and fourth would count her out in her teen-age years. The first strike was that she was born as a girl. In her society that sealed her fate as never being seen worthy of attending school. Who would waste money to send a girl to school? Who needs education to carry water, chop firewood and, least of all, bear children? As Justina got older, she had only one wish: to attend school. She expressed this wish to her father many times but already knew the answer by observing other girls in her tribe. When the age of puberty arrived, they were married off to whomever could come up with the highest number of cows. Unfortunately, by the time a man acquired enough cows, he was 35 or 40 years of age. Marriage could also mean becoming a second or third wife. Justina knew that no one would pay a bride price for her.

The second strike came: as a twin sister at her birth, she was epileptic. Epilepsy is not a disease confined to the Western world. Her father knew that no man would ever want to marry her, and thus his chances of getting a bride price for her were nil.

Justina could not believe her ears when, contrary to the custom of the Iraqw tribe, her father agreed that she could attend school in a neighboring village. Boarding school is a common part of education in Tanzania. Justina was fortunate in that she could walk to school, even though it was four kilometers away. Children are often nine or 10 years old before they begin school.

One strike seemed to have been taken away, but another now took its place. At school a young boy, Daniel Awe, shared with her about this man, Jesus, who was alive. If you believed in Him and repented of your sins, He would make you His child. He could also heal. Justina believed and was immediately healed of her epilepsy. Daniel was a young boy in the church at Mbulu, which I visited as part of my regular schedule. Later I also built a lovely church building in the town. Still later he became the leader of our National Church.

Justina's decision to become a Christian began a series of more terrible events for her. Unacceptable to her father and the witch doctor was Justina's resolve never to allow the witch doctor to touch her again. African culture dictated that somehow he must get credit for the cessation of epileptic fits. This failure to cure could ruin his whole life's credibility over a wide area. Word spreads fast in Africa.

For years Justina's father beat her and made life miserable. A group of Christians met in a church within two kilometers of Justina's home. She would do anything to attend Wednesday services, which took place around 4 p.m. After coming home from school, she attended to her regular "girl duties" of gathering firewood, pounding corn into fine flour, and going to the nearby stream

31

and bringing water for cooking and bathing. Washing clothes was done once a week on the rocks at the river.

Justina wanted to look presentable, which is an innate trait of Africans, so she would hide her "good" clothes in the nearby cornfield. That way she could disappear from the hut compound without anyone really noticing. Justina knew for certain that when she got home, her father would berate her and give her a beating in front of five brothers and two sisters. This went on for a number of years, but Justina kept attending church faithfully. What she got at church far outweighed any physical pain she would suffer each time she returned home.

One morning her father instructed her not to leave the family compound as something important was going to happen. The something special turned out to be a visit from the witch doctor. Upon seeing him, she bolted from the hut and ran into the cornfield. Her brothers were sent after her. She struggled as hard as she could to resist going back to the hut. In the desperate struggle, her clothes were ripped off and she ran naked and hid in the bush. After some time, the witch doctor and the father realized that Justina was not going to give in. Her sisters were sent to her with some clothes. Adding to the embarrassment of the father was the fact that Justina was the only Christian in the whole surrounding village, and even the witch doctor had no power over her.

Strike number four against her came at a time which should have been the happiest of her life. A young man proposed marriage to her. To negotiate a dowry price he would have to sit with a committee of elders, which included her father. This presented an impossible situation because the young man was a Barbaig by tribe. Justina's Iraqw tribe despised the Barbaig tribe. She was accused of being bought by the young man. This stemmed from the father's admittance that he had not helped her in any way to this point in life. The thinking was that if the young man offered her a better future, she would accept that.

She assured her parents that their fears were unfounded, and she was going to get married. Upon hearing this, they formally disowned her as their daughter. Further, she was also stripped of her tribal birth. In spite of pleading with her parents and family to attend her wedding, not one member of her family came.

I will let her relate the story in her own words. She can also fill in some other details. For much of the time she cried as she recalled those frightful and hurtful days.

My very first childhood memories are of the witch doctor cutting my skin and sucking blood out of me. Growing older, I learned that he was attempting to extricate the "spirit of epilepsy" with which I had been born. Numerous times, a drink prepared from a special tree bark made me vomit violently. For 14 years I was subjected to every conceivable, and inconceivable, form of attempted treatment. No witch doctor likes to be defeated. It is bad for business.

Two powerful realities assured me that I would never attend school. I was a girl, and sick at that. Beyond school, I also realized that no bride price would ever be paid for me. Then the first miracle happened. Inexplicably, Father sent me to a neighboring village to school! Within a short while, a boy (he is now my district superintendent) told me about Jesus. Together with salvation, I received healing! Spiritual freedom and a healthy body brought me much persecution, but it was easy to bear after the spiritual darkness I had endured.

33

Now our home, the village, and the witch doctor were disrupted. I was the only Christian! Everyone disowned me: my parents, my seven brothers and sisters, the tribal clan and the village. Constant parental beatings and expulsion from home became my life.

The witch doctor was consumed with rage because none of his medicines worked in Father's home any longer. Contrary to my culture, I defied Father's strict instructions not to attend church (my church was just being built from Living Memorial money donated when the grandfather of my missionary, Ron Posein, died and left money to build a church). During the day I hid my good clothes in the bush so that in the evening I could more easily slip out of the hut undetected. Upon my return I could expect severe beatings and berating.

After four years, my parents totally banished me from home. Waiting on tables in a small roadside eating place helped me live from day to day, but the witch doctor still sought revenge! Young men were sent to threaten me and to kill me, just as they had killed another girl who had dared not denying her faith. One evening, eight strong young men informed me that they had come to kill me. They waited outside until I finished work. A single door led out of my place of work, so I had no way of escape. As they lunged toward me with outstretched arms holding machetes, I cried out, "Jesus, help me!" Just as suddenly, they turned and began beating each other. From my room, a tiny board shed behind the hotel, I heard the commotion carry on all night.

Later they confessed, "We have seen powerful witchcraft, but nothing compares to this!" Jesus had fought for me again. Today I am 29 years old with a husband from a tribe my parents despised. On my wedding day I sadly stood alone with him. In a few days, I am graduating from our three-year Bible school. Each day I thank God for two special things: one is salvation—I could never deny Jesus!—and also for my little church, which became a place of refuge. A number of years later, my parents came to see me. They said, "If God helps us and forgives us for how we have treated you, maybe even we can be saved." On his deathbed, Father accepted Jesus. Two of my sisters are saved, but Mother has not yet committed her life to Christ.

CHAPTER 3

Only One Key

Theft and the "borrowing of documents" were always rampant. In order to protect workers from passing the blame for missing items onto another, a procedure was developed throughout the country. That procedure was to have only one person in possession of a key to an office, cupboard or door. Even we used that procedure at our home and at the orphanage. It was the only foolproof way of ensuring security.

I made a trip from Mwanza to Dar es Salaam to clear a vehicle through customs and port. By road the trip is 1,000 km, and by air two hours and $350. Business is never accomplished in one day, so that necessitates taking a room in a hotel, at another $120 per night. Arriving at the government office dealing with vehicle clearance, I was informed that the man who had access to the clearance forms was not in. The forms are locked in a cupboard, and he is the only man with a key.

"When will he be back?"

"We are not sure."

"What do you mean, you are not sure? Where has he gone?"

"He has gone to England, and we are not sure when he will be back."

Can you imagine? An important matter like this, and one man holds the only key — and he has left the country! Talk about a waste of people's time and money, not to mention the inefficiency. Home wasn't just across the street, but I had no choice except to come back another day.

We Don't Sell Firewood

The Arusha area, lying at a higher altitude, could get pretty cold in the months of July and August. Naturally, in the tropics no house has built-in heating. Our house had a wonderful floor-to-ceiling fireplace which opened into the living room as well as the dining room. I decided to buy eucalyptus firewood and dry it for the next year. Driving past some bakeries, I saw cords of firewood piled up. I found out that they bought it from the forestry department at a forest about 20 km out of town.

Taking our yard worker with me, I drove up there with our mission's seven-ton Isuzu truck. We arrived at the gate, which every establishment has, and were asked what we wanted. I said I wanted to buy some firewood.

"We do not sell firewood" was the curt reply.

"But wait a minute: I see that every bakery in town has cords of eucalyptus piled up outside that they use to bake bread."

"We don't sell firewood" was the retort to my observation.

"But they told me they purchased it from right here."

Now I was confused. My immediate thought was that there might be an extra hidden cost of doing business (known as bribery in the West). I began to wonder how much that was going to cost me over and above the actual price.

"So where do the bakeries get the firewood?"

"We sell fence posts, not firewood!"

"OK, I want 150 fence posts."

"No problem."

The next year I burned fence posts, not firewood.

Smoking Toenail

In some things, electricians possess next to no fear. Electricity, although 220 volts, is not an issue. At times I had so-called tradesmen at the house working on the myriad of things which go wrong in a Third World electrical maze. A fellow was working on an electrical plug, and suddenly I smelled something. The fellow had touched a live wire, and his toenail was smoking! He said that is the reason they wear rubber flip-flops. Shutting off the power mains would never occur to them. More about that later.

It seems that at every turn in a new culture, something unfamiliar hits you in the face. A man once asked if our "electric city" was working. It took me a minute to connect the dots. He was talking about *electricity*!

No Ordination to Ministry for You, Sir

Getting ordained to ministry is a huge accomplishment for a national pastor, who has most probably taken six years to complete the required studies. Having done that, he still needs to pass other qualifying requirements and appear before a committee for oral assessment. Most pastors do not begin studies until well into their 30s because they either do not have the funds or cannot find a replacement to carry on the work at their churches while they are studying. At that age, the majority of national pastors have families and small farms to look after. So it is a huge accomplishment to have cleared all of those hurdles and finished Bible school studies.

Marwa was both excited and nervous as his turn came to be grilled by the ordination committee. He had crippled legs and walked by swinging his body around a post, which served as a crutch. Everything was going well, and the interview seemed to be drawing to a positive conclusion. Then came the final question concerning his family life: "There is a rumor that your wife beats you. Is that true?"

No use trying to hide the truth—yes, it was true. Interview over—no pastor could be ordained when he allowed this to happen. Marwa had persevered to become self-sufficient, obtain a reasonable education, and raise a family. All of that, however, paled in the face of the African custom of male dominance.

Bosch Wheel Alignment Tool

I took my Volkswagen Westfalia Camper into the Arusha German Technical College for service. Being a training college, they had every piece of equipment Bosch ever made. That way the students could learn first-hand alongside a German mechanic. The German director and I were standing outside and looking down the road a bit at seven students walking around a truck. Never knowing what people may be up to, he asked, "What do you think they are doing?"

I said, "It looks like they have some string, and I think they are doing a wheel alignment on the truck." He started swearing and saying that couldn't be as the entire top-notch Bosch equipment is in the shop at their disposal. I said, "That may be true, but what are they going to do when they are back living in the rural area?"

It was beyond his comprehension that state-of-the-art tools were hanging on the wall and these students were using string made out of sisal. He didn't believe me and called a fellow over. Sure enough! It was a wheel alignment they were doing. When you have been in Africa for a few years, you learn to read signs that are not necessarily posted on a signpost.

Forget The Heat; Keep Those Windows Up

Road hazards are commonplace, and one has to be vigilant constantly. Roads in the Mara Region were narrow and sandy, offering little opportunity to pass slow-moving buses. On one such occasion, I saw an opening to get past a bus. Just as we were passing, a fellow decided it was a good time to relieve himself—out of the bus window. Moral of the story: always pass a bus with your car windows tightly rolled up. As a rule, bus drivers do not see the need to make pit stops.

Luggage or Life

Fear is not a common characteristic of many people. We were following a bus going 120 kilometers an hour on a narrow and roughly paved road. In order to accommodate the luggage of travelers, which may consist of live goats, chickens, pots and pans, bedding, etc., buses have roof carriers covering the full length of the bus. These normally have sides on them to contain the luggage. As we were following, the "tan boy" (who we in Canada would call the conductor) crawled out of the front passenger window and climbed up on top of the bus to check the luggage. Satisfied that all was still there, he climbed back down. All the while, the bus is zooming along at 120 kilometers an hour. Do I need to tell you that I kept my distance until I saw the man disappear back through the window and into the bus?

The Cruelest Gift Ever

As missionaries connected with the German Branch of The Pentecostal Assemblies of Canada, we were blessed with parcels sent from home. Two parties collaborated to make this possible. First of all, the Tanzania Revenue and Customs Department allowed them into the country duty free. Secondly, the Women's Ministry, along with sending used clothing, spent much money in buying special food and hygiene items that were not readily available in Tanzania—things like soup mixes, toothpaste, cooked ham and spices. These were all packed between used clothing, which we distributed to our people mostly in the rural areas. The ladies also made small things that we could give out to children in the churches.

In one such parcel we received a box of liqueur-filled chocolates. Now that was a treat—even though I am a teetotaler—but I was only going to concentrate on the chocolate, right? Chocolate bars were not available in Tanzania at the time. With some anticipation we were tempted to caress the box before gingerly opening it. *Surprise, surprise!* The box had been used only to protect what was inside:

sparkly crosses made from plastic bread closers for distribution to Sunday school children. At least the children would be happy.

The disappointment was tempered a bit when I used the box to trick visitors who came our way. Germans were especially good targets and registered the greatest disappointment upon opening the box.

Even God Is Probably in Doubt

Tanzania and Kenya were occasionally at odds after the East African Community (Tanzania, Kenya and Uganda) broke apart. Frequently, on the spur of the moment, some political spat would result in the countries closing their borders with one another. Usually it was announced on BBC radio, which had more African news than the local radio station. On one such occasion, a Canadian pastor happened to be visiting us in Bukoba. Living in Bukoba already gave you the feeling that civilization was far away. Flights in or out were sporadic. In the span of a week, flight cancellations could exceed the number of actual flights. Travelling by road was a 16-hour undertaking, with no rest stops or eating places on the way. Naturally, the pastor was totally stressed when we received the news that the borders had again been closed indefinitely. In such situations there was literally no way out of Tanzania. The Western reaction is always "THEY CAN'T DO THAT!"

That evening we went over to our Baptist missionary friends, Dave and Betty Ann. We were the only two white couples in the town. Over coffee, everyone is looking for good news, or at least some comfort from those who live in the country. With this heavy on his mind, the visiting pastor asked Dave about the border closure and when he thought it might be opened. Naturally, he had already asked me the same question a dozen times or more since we had heard the news. My answer, indicating we had to wait and see, was not exactly a life jacket being thrown to a drowning man. It was more like "Tread water; we have sent for help." Dave could keep a very solemn and straight face at times. He looked at the pastor who was so eager to hear good news and said, "Oh, I am sure the borders will be opened, but God Himself is probably in doubt as to when that may happen." Not exactly the comfort our visitor was looking for! The night was long and restless for him.

The borders did open for expatriates and missionaries in about a week. The pastor got out as quickly as possible. Visitors like to choose their missionary experiences. That way, they fit into the imaginary romanticism of missionary life in a foreign country.

Second Cup, Please

During the dry season in Tanzania, water is always at a premium. People need it for drinking, washing dishes, making tea, washing clothes and bathing. However, cows, goats and sheep also need water. They have a tendency to disregard signs not to "dirty" in the water. When there is only one pond in the area, what are you going to do?

A Canadian pastor was visiting and went to a village church with another missionary and myself. When it came time for tea, it was unmistakable where the water came from. The pastor had a very hard time getting it down but was totally shocked when we accepted refills. The water is usually boiled so, other than the aroma, it is OK to drink. While sipping, think tea, not general-use pond.

Get Her Out of Here

Trent and Rhonda came to join us in the work at Starehe Children's Home. Rhonda became sick and was admitted to the Hindu Union hospital in Mwanza. After numerous tests and days on medication, it became clear that she was not responding to treatment. In fact, she was getting a lot worse—and fast. Dr. Kocher advised us to get her to Nairobi as quickly as possible. I contacted AIM Air (African Inland Mission Air). They had a plane available, so they immediately dispatched it to Mwanza—about a one-hour trip. When they arrived, we (Trent, LouDell and I) had Rhonda at the airport and checked through immigration. They allowed her to remain in the car. As soon as the plane landed, we were given permission to drive right onto the tarmac beside the plane. Rhonda had to be helped into the plane. Later in Nairobi, she did not remember anything about the trip.

The insurance company in Canada gave them a lot of hassle when they applied to have the airfare reimbursed. They felt they should have been contacted first (in Canada—at 1:00 a.m.). Then they would have made all the arrangements. Being the Field Director for our mission, I took responsibility for the decisions. They contacted me to explain their approved procedures. Rather bluntly I told them that, had we waited, the only arrangements they would have made would have been for a funeral. So many times in an emergency, God gave us calm hearts and went before us step by step in getting things arranged.

Medicine in the Third World

Medicine in Third World countries can be rather interesting. In a pharmacy you can literally buy any medicine over the counter— no prescription is needed. Just explain what your ailment is, and there you go. I was standing in line to get some "prescribed" medicine and overheard a conversation between the pharmacist and the person ahead of me. Looking at the prescription, he asked what their problem actually was. Once the person had described it, the pharmacist recommended a much better medicine than what the doctor had prescribed.

As Europeans, we had to be very careful since the local people have built up some serious immunity to common diseases. As a normal course of action, the doctor prescribes a far higher dosage than that which is outlined on the bottle. On another occasion I heard parents berating the pharmacist because their child had suffered severe hallucinations due to the dosage written on the bottle by him. As could be expected, the pharmacist could not find the doctor's prescription note, which was issued just the evening before. After the parents left, I am sure it was located and immediately shredded. No use leaving hot evidence lying around.

In Mwanza there was a wonderful elderly Asian doctor couple who had practiced medicine for many years. Dr. and Mrs. Kochar were very familiar with all of the local ailments of the tropics. He could be abrupt at times. One of two questions were the first we were greeted with. "Now what happened?" or "What kind of medicine do you want?" *Ah, might be a good idea to check us first? We* were amused by this part of his character. It did show us that he was always willing to help. The Kochars were sorely missed when they retired and moved back to India.

Local newspapers reported that critical studies verified that up to 90 per cent of medicines coming into Tanzania from other countries were highly unreliable. Most of them have some critical ingredients missing. Over time, people build up immunity and tend to have a future relapse. The disease is never totally cured; rather, it suffers only a temporary setback.

Many children we received at our orphanage came to us with tuberculosis. LouDell would take them into the infectious disease part of the hospital for treatment. However, as she sat in the waiting room, there were infected people coughing and spitting on a regular basis. It's no wonder that she and her helper became infected and had to endure a seven-month period of treatment.

At times it took some persuasion to convince the doctor to prescribe medicine. Two things came into play. First of all, it was only a child; secondly, the suspicion that the child was HIV positive was the cause of the sickness. They didn't want to waste TB medicine on an HIV positive person, even if they had TB.

Double the Sentence

Pius Mkonge (*mkonge* means "sisal" in Kiswahili) was a pastor in the coastal region of Tanga. The adage, "No Hurry in Africa," is not a good character trait when it comes to pastoring. Pius was given his annual holiday leave but decided that "deadline" did not actually mean deadline. As the work was still young in the region, a missionary was the regional superintendent. Endeavoring to instill the importance of sticking to the rules and developing good pastoral work ethics, the missionary deducted days not worked from the PEP (Pentecostal Evangelism Program) money that Pius received each month. He was of the mind that this was grossly unfair treatment.

That fall, Pius came to Bible college and, during a class, stood up and asked, "What do you think of a missionary who deducts money—a partial month—just because a pastor has taken a few days longer than approved holidays?"

I answered, "That is very bad!"

Pius beamed with the realization that he might able to reopen his case. Then I continued, "He should have been deducted a whole month's allowance!"

Joyous facial expressions slowly withered, and Pastor Pius did a slow-motion slide back down into his seat. I knew the rest of the story because the resident missionary had shared it with me. Exchanging stories of experiences is a highlight when missionaries get together. Most of us lived 600 to 1,000 km from each other, so the times when they did occur were full of fun.

What Kind Of Meat Is This?

On one occasion, some PAG (Pentecostal Assemblies of God) executive members and I visited a pastor. His church was way out in the bush near a forest and very close to the Kenya border. His family was not well off, and the rains had not been good that year. Even if food has to be borrowed from a neighbor, every guest must eat well. In due course his wife brought us food. The meat had a reddish tinge and did not taste like cow or goat. Pastors began to question him as to what it was. He never admitted or denied it, but to this day we are sure it was monkey.

A Good Convert

During a church service, the pastor gave glory to God because a person had newly become a Christian. The pastor was so happy because now he had much meat to eat. The convert was a poacher. He probably became a Christian, but leaving a lucrative profession is not easy.

Their Eyes Were Sunk Into Their Heads

Mwanza Region was suffering a severe drought. Hot tropical winds relentlessly stirred up the fine dust, which should have been soaked by the long rains and supporting a bountiful corn crop. Children were starving, and adults were without energy. Water ponds turned to mud, and women walked 10 km or more to fetch drinking water hauled in by the district government. The national government declared Shinyanga as a disaster area and appealed to foreign nations to donate food.

Feeling the plight of pastors, the superintendent, Rev. Francis, visited Pastor John. Francis saw how Pastor John's children were emaciated, their eyes sunk into their heads, their hair turning red, and their stomachs bulging. Severe malnutrition bordering on starvation was obvious. Rev. Francis had seen a lot of suffering in his ministry duties. As far as starvation was concerned, this was the worst he could remember. Tears literally came to his eyes as he observed the destitute and starving family. He emptied his pockets of all the money he had on him and gave it to Pastor John.

Sometime later Francis became aware of the fact that Pastor John had over 100 head of cattle being looked after by a herder in an area where grass was available. Francis was furious, and at the next council meeting asked John why he had not butchered some cattle so his children could eat. John replied, "What would my children who may still be born eat?" That philosophy could not be comprehended, even by a national.

Healed After a Stroke

A pastor in the Arusha Hanang District suffered a major stroke. While in the area, we visited him in his mud hut. He was lying in a dark room, a window of only about 12 inches by 12 inches letting in light. His bed was made of tree trunks for legs and woven cow hide as a mattress. He was totally immobile. After a time of conversation, we prayed for him—with little to no faith, I must admit. Within a year we heard that he was fully recovered and pastoring again.

You Want Him to Die

Tanzania has a tropical climate. This means there are two rainy seasons and a number of months of dryness in between. The pictures we see of lush tropical gardens and landscapes harbor something just out of sight. If vegetation is not severely and regularly cut back, a forest of vegetation soon overtakes everything. That being the case, plus the fact that we were sent to be missionaries and not use most of our time tending yards, we always employed people to keep things in the yard looking presentable. Short grass also drastically reduces hiding spots for snakes.

Hand tools are the things Africans feel most comfortable with and do the best job with as well. Charles was a wonderful yard worker. We relied on him to cut (slash) the grass, cook food for the dogs, guard our home while we were on safari, bring in the fruit from our trees, and keep our vehicle clean after a long dusty safari. Eventually he married a lovely young lady, and children soon followed.

The time had come for us to return home to Canada and do deputation work to raise awareness of needs within the area in which we worked. It would only be a short four-month stay.

Excitement again welled in our hearts as we returned to our home in Tanzania. I believe there is no equal anywhere in the world for African greetings. They are accompanied by much laughter, shaking of hands, and general, unabashed, prolonged jubilation. Always the "great" compliment is shared: "You have gotten so fat!" Really? The compliment stems from the harsh reality of a society that knows hunger and starvation. Getting fat expresses that God has blessed you with much food. This is very humbling!

Charles' wife, Lois, was there, but Charles was not among the greeters. Responding to our inquiry, the joy in Lois's face was replaced by sorrow. Just up the hill, Charles was lying in his bed dying of full-blown AIDS. As quickly as time allowed, we followed the rock-strewn path to his house.

Stepping from the bright tropical sunshine through a mud hut door and into the darkness of the hut, we were not prepared for what met our eyes. First we heard his raspy, weak voice, and as our eyes adjusted to the darkness, we saw the mere skeleton of a man trying desperately to bravely sit up on the bedside. Barely lifting his head was all he could accomplish. Not being able to greet us, as was so culturally important, caused him nearly as much pain as his disease. It was ravaging his body and visibly sucking his strength with every breath. Totally unable to get up, his wife fed him by spoon in small amounts as he was able to tolerate. On the mud floor at the side of his bed was a small, rusted, old tin can, which served as his bedpan. A few short months earlier, we had left a strong, healthy, hearty man; now we looked into the face of a human skeleton, his chest heaving and fighting for every breath.

Unfortunately, the struggle in and around the home was not only physical, but spiritual as well. Sitting at the mud hut door, day and night, were relatives of Charles. They were not just sitting there to commiserate with a sick person; they were badgering Lois to take Charles to a witch doctor. They sat there for weeks trying to force her to comply. They knew of a very powerful doctor just across the border in Kenya. Lois was a believer in Jesus, and witch doctors were a spiritual dark power of the past for her. She understood only too well the bondage of darkness from which she had been released.

It goes without saying that, should Lois agree to take Charles to the witch doctor, a whole entourage of family members would accompany them—all expecting to be royally cared for. Travelling would be done by taxi. Riding in a bus would be too degrading when you are going to see an important man. As well, you must be dropped off at his door. The expectation of food and lodging mirrored the class of the taxi—high class. Then there would be the "fees," which never ever had any hint of mercy or compassion connected to them. Because of the Christian factor, the witch doctor would declare this to be a VERY hard case, and it would drag on for weeks. Naturally, the expenses would all be covered with borrowed money. Many years of debt repayment would lie ahead.

Each day, as Lois stood firm in her resolve, she endured the verbal and psychological abuse. Over and over the accusation was hurled at her, "You want Charles to die! Otherwise, you would agree to have him treated by the medicine man." This is very near to having a curse put upon you. Finally, in what I believe was wisdom from God in answer to her cry, she asked Charles in front of all of his relatives whether he wanted to go to the witch doctor. He said no. After that they dispersed, and Charles died a few weeks later.

We accompanied the wooden bier down a dusty path to Charles' final resting place. A shallow, narrow grave, as though hungry for this moment, received a companion of death. Decay would not take long as the body, shrouded in a thin white cotton sheet, would soon be part of the earth itself. As an act of respect, bystanders threw in handfuls of dirt before turns were taken with the knarled, short-handled shovel to fill in the rest. A few rocks placed around the spot would mark the grave, but within a short while they, too, would disappear, even as life had disappeared and no one would remember that a grave even existed. "Dust to dust" is not just a figure of speech in Africa; it is reality.

For Lois, an agonizing vigil ensued. AIDS has an insatiable appetite. It is a hungry soul looking for its next victim, and that is usually the partner of the deceased. It can lie menacingly dormant and then suddenly jump up into your face.

During this period in Africa, it was common practice for pastors performing marriages to demand that each partner be tested for the

HIV virus prior to the wedding. They became tired of performing a wedding and then, only a few months later, presiding at the funeral of either the bride or the groom.

Lois prayed and waited for the inevitable telltale signs of rapid weight loss, running sores, aching bones, and a body devoid of any energy. Days, weeks, and then months went by, and still she remained healthy and strong. A suitor came into her life and eventually a proposal of marriage. Who would be so unthinking as to marry a woman whose husband had died of AIDS?

The pastor who was approached to do the wedding flatly refused: a wife whose husband had died of AIDS was a walking apparition within an invisible casket. Inevitably the "slims" disease would wrap its arms around her. Did she now want to pass on the dreaded disease? Time after time, Lois attended the AIDS testing clinic, and each time her blood showed no virus. Even well-meaning Christians had the notion cross their minds that the negative result was dressed in a "bribe robe."

Facing the reality that God had protected Lois from being infected, the pastor agreed to perform the wedding ceremony. LouDell and I were privileged to witness it and take part in the cutting of the cake—a roasted goat head with unblinking, staring eyes. What a celebration for and with Lois!

A Shotgun Diagnosis

For a number of months, LouDell had been very sick. Her stomach could not tolerate any fatty food. Even the miniscule amount of fat in an egg sent her into excruciating stomach cramps. In spite of that, she continued to cope. However, as her body lacked proper nourishment, weakness became more apparent. We were living in Arusha at the time and travelled by plane to Mwanza to teach some courses at the Bible college. While there, as we became more desperate, we visited a Canadian Aid doctor. Competent medical help in those days in Tanzania was basically non-existent. His diagnosis? "It could be hepatitis, worms, kidney stones, pancreatitis or cancer." Most encouraging!

A fellow missionary, Rainer, had also come to Mwanza to teach. Within two weeks, they were expecting his wife's sister to arrive at the Kilimanjaro Airport near Arusha. Due to LouDell's weakened condition, it was decided that she would fly to Arusha; the rest of us would drive there. A ticket was booked, and we went to the airport. The plan was that after LouDell had boarded, we would leave for Musoma, where the fellow missionary lived. The drive takes roughly three and a half hours. The plane was delayed by two hours, then another two hours. Finally it did arrive from Bukoba, but the front wheel had jammed sideways on takeoff. They made an emergency landing right in front of the Mwanza airport terminal building (about 50 meters away), and right beside a loaded plane ready for takeoff! Scary, to say the least. Having seen what we had just witnessed, we did not think it was a good idea for LouDell to get on that plane. Jamming into a small pickup truck and heading for Musoma was the only option open to us. The time of day, plus rough and dusty roads, did not contribute to a pleasant drive.

Early the next day we left for Arusha in Rainer's Peugeot car. The roads were not kind to those French vehicles. Near the Ngorongoro Crater, we hit an unavoidable rock—the road was littered with them. The bolts holding the drive shaft to the gearbox sheared off. Managing to tie it together with some wire, we started out again. Within a few meters, the back wheels locked up totally. The roads were the clear winners. There we were at 6 p.m. (it gets dark at 7), high up on the dusty Ngorongoro Crater escarpment, cold and with two small children and a sick wife in the car. Due to the danger of wild animal attacks, it is illegal to be on the road after dark. Finally, a big truck came by. We talked the driver into taking the women and children to the Crater Lodge. Finding out that he was going to our home city of Arusha, I hitched a ride with him as well. We were accustomed to "jamming," so four adults and two children in the front seat was possible. Rainer was staying with his vehicle to guard it. The ladies and children were just getting out of the truck when Rainer also showed up at the lodge; he had arranged for a Maasai warrior to guard the car overnight.

I arrived in Arusha around 1 a.m. By that time, home was a welcome sight. Early the next morning, I went to the Peugeot garage and was amazed that they actually stocked the part we needed. Together with a mechanic, and driving my own four-wheel-drive Toyota Land Cruiser, we were on our way to fix the vehicle. We arrived back to the broken-down vehicle about 3 p.m. I picked up the ladies and kids from the Crater Lodge, and we headed for Arusha. Rainer arrived around midnight.

The doctors, not being able to identify LouDell's health issue, recommended that she return to Canada. When she arrived home, the doctors had just gone on strike. It took three weeks until she was able to secure an appointment. Even after a battery of tests, doctors could not pinpoint the cause of her illness. She began to feel better and returned to Tanzania. After a few months, she again began to suffer. Doctors in Nairobi ran many more tests, which all returned inconclusive results. Our elderly doctor decided the only thing left to do was to perform an exploratory surgery. Her gall bladder was totally diseased. There was no explanation why this did not show up on her many tests.

CHAPTER 4

A Cell Phone Before There Were Cell Phones

Most of our churches were located far from Arusha.

Arusha city was central, so we lived there. For the most part, we booked church visits six months in advance. All of this needed to be co-ordinated at a committee meeting or by using a very unreliable mail service. That was only the beginning. Mailboxes existed only in larger towns, which pastors seldom visited. Added to that was the reality that few postboxes actually existed, plus a yearly rental fee had to be paid. This meant that up to a dozen or more people used the actual renter's post office box number. Further, the renter of the postbox number took the mail and either kept it until the addressee came, or he would send it with someone in a neighboring village who knew the person to whom the letter was addressed. Many times that was like putting a note in a bottle and throwing it into the ocean, hoping it would hit the right current and person in time.

Churches were up to 150 miles away over terrible roads and located in villages where no phone service existed. Unexpectedly, I had an open day and wanted to visit a particular church in the Hanang Region, 125 kilometers away. Not having had the opportunity to inform them of my coming, I didn't know whether anyone would be there. Reaching the church, I was surprised to see a lot of people. The mystery was solved when I spoke to an elderly lady. The night before, she had a dream that I was speaking to her in English from a little box (telephone booth). She saw my car too. So they were expecting my arrival. God had a cell phone long before they were invented on planet Earth.

Beer Wisdom

Julius, a caregiver at Starehe Children's Home, was getting married. According to African custom, a dowry negotiating committee comprised of elders from both sides was set up. They also doubled as the main wedding planners for food and drink. The subject of food and drink is always a vital one. Many of Julius's relatives are Muslim, and their demand was that beer be served at the wedding. Julius adamantly refused, but he knew it would continue to be a contentious issue. He needed guidance from the Lord.

At the next meeting of the elders, Julius announced that beer would be served. The elders were elated that "this young buck" had finally seen it their way. After the jubilation and congratulations had run their course, Julius went on to say that for meat, pork would be on the menu. A great groan of lament went up, but Julius stuck to his guns. It was either soda pop and goat, or beer and pork. This young fellow, with the Lord's help, had defeated them at their own game.

Send Her To The Island Of Death

Many tribes in Tanzania had strict rules to safeguard the purity of their culture. The Mangyati tribe in the Arusha Region are known as the fiercest tribe in Tanzania. The Maasai are erroneously given that distinction, but if people from these tribes meet on a road or pathway, the Maasai always gives way and leaves the path.

In order to ensure continued moral purity, tribal elders enforced the strictest rule. Should a girl become pregnant out of wedlock, two things were the resulting consequences. First, by dugout canoe, she was ferried across to an uninhabited island. Left to fend for herself, she seldom survived — witnessed by the array of human skulls on the island. If she survived and gave birth, she was allowed to return to her family.

Secondly, every hut that she had entered for the last three months had to be burned to the ground. In every way, the whole family suffered the pain, tragedy and expense of her moral failure. As usual, the man was not included in the picture of retribution.

Paulo, pastor and respected elder in the community and member of the village council, had been part of deciding the fate of wayward youth in their tribe on a number of occasions. With his consent, the sentence, without qualms, was meted out in order to uphold and preserve tribal morality.

The elders were summoned to discuss another reported indiscretion. It would be an open and closed case, as usual. Mercy was left to the gods of the island. To Pastor Paulo's great surprise and distress, his daughter was the "case" this time. Suddenly he was faced with a dilemma.

59

He related to us the agony of his soul. His tribe's demand for moral purity; his own flesh and blood—his daughter; the mercy he had been afforded by Jesus' forgiveness of his sins. Which should take preeminence? Each had its list of unarguable merits. After days and sleepless nights, Paulo made the decision that God's grace deserved the highest merit. He and his wife, Esther, sent their daughter to live with an uncle in another town some distance away. After the baby was born, they accepted their daughter back, showing mercy even as they themselves had received it.

They also broke custom in that they did not burn down every hut that their daughter had most recently stepped foot into. Today she is a pastor's wife with a beautiful family.

In Danger Of Losing A Tribal Mark

In the early 1970s, a Bible training program took place at the Nyasirori Mission Station. This was on an old gold mine site about 25 miles from the nearest town of Musoma. The station was four miles off the main road into the "bush," as we referred to it.

Although we lived in Musoma town, we stayed at the station while we took our turn teaching Bible school students. An early memory we have is of a lady coming to the station with her infant baby. An older missionary informed LouDell that the lady wanted her to name the child. Perplexed, we saw the same lady come back four days later, again requesting that LouDell name the baby. Apparently the babies were actually twins, but one was born later than the other.

Tribes have different ideas about twins. Some see them as a blessing. In some rural tribes, the mother allows only one of the babies to nurse—an idea that is repugnant to us. The thinking is that a mother most likely cannot feed both babies, so both will end up being weak. Better to remain with one healthy baby. That practice is happening less often these days, thanks to the availability of baby formula.

Being a rural area, old customs still prevailed. The Wazanaki tribe called this area home. As a rite of passage, the front teeth of the young men were filed to a point. If a young man flinched or exhibited signs that the procedure was painful, it was a great shame to his manhood. As can be imagined, these files were not the sharpest, nor did they have the finest filing protrusions. Although I never saw one, I imagine them to be what we would know as a rasp. The first president of Tanzania, Julius Nyerere, came from this small tribe. In photos his pointed teeth are clearly visible.

Intertribal mixing was rare in those days; thus, students coming for Bible training at the mission station from other parts of Tanzania—and being from different tribes—were very interested in the Wazanaki tribal customs. One of their customs was circumcision, which a number of other tribes do not practice. Most tribes in Tanzania know of one another's customs in these main areas. Students of other tribes were naturally very curious about the circumcision rite. One was announced while students were at the school. Taking place on a Saturday, it gave the students an opportunity to attend. It is always accompanied by much frenzy and the presence of elders, witch doctors, relatives and onlookers. There is a beating of drums; dressing in skins; wearing ostrich feather headdresses and satanic masks; waving machetes; banging any old objects of metal; making threatening gestures with spears; colorfully dressed, the tribes people are totally involved. Dust becomes

AFRICA

a main component due to the commotion of all the people milling about. As the day wears on, the local brew begins to take effect. The whole ceremony's concoction can lead to spontaneous actions.

In this instance, after the candidates had been circumcised, the leaders began scanning the bystanders and immediately recognized that one of our students was from a tribe that does not practice male circumcision. In a wild frenzy they went to grab him and bestow on him "the mark." Charles was a hyper fellow to begin with and ran headlong to protect his honor. Running through the brush and thorn bushes, he twisted his ankle. With eyes as big as saucers, heart pounding, fear written all over his face, and in pain, he hobbled back onto the mission compound, having long since outdistanced his pursuers.

Fingerprinting

Shortly after coming to Tanzania in September 1973, the superintendent, Rev. Stephen, suggested we drive to a place called Mugeta in Musoma Region near Serengeti Park. We were renting a building to use as a church, and the agreement needed to be renewed. Being new, I understood very little Kiswahili. After a two-hour drive, we reached the small group of buildings along both sides of the road.

Pushing aside the cloth hanging in the doorway, we entered the house and sat in a typical small entrance room called a *sebleni*. Rev. Stephen negotiated with the lady who owned the building. I was not yet accustomed to the language, which can sound like a strenuous argument at times. Suddenly Rev. Stephen jumped up, grabbed her hand and began coloring her thumb with a pen. I was

62

dumbfounded. I wondered what in the world was happening! Then he pressed her thumb down on a piece of paper outlining the rental agreement. The document was official.

The work done, we returned to Musoma. To this day a thumbprint is used as absolute proof of a person's identity and is commonly used in "signing" legal documents along with the signature of the person — if they can write.

Orphan in a Large Family

Stephen attended Bible college in Mwanza. In spite of having a full-fledged family, he was an orphan. He shared his story with me.

Having heard the gospel in his village, he gave his heart to the Lord. Witch doctors persuaded his father that Christianity would bring bad luck to the village. Unbelievable as it may seem, Stephen's father had been a military attaché in Ottawa for a few years. As is customary in Africa even today, the father called a *baraza* (a family meeting where only males attend). The father took his place at the head, and all of his sons — 12 in number — sat on the floor in a circle around him. The father addressed a number of issues, including the well-being of his many cattle; the prospect of that year's crop; his respected eldership in the village; the schooling of his many children; and the distribution of the inheritance, should he pass away. Then the main topic was brought up.

"Stephen, you have brought shame upon me and my family, not to mention our village. You have decided to attach yourself to this new religion, which has come into our area. We cannot agree to these new teachings, which want to replace the old and proven beliefs of our ancestors. I give you an opportunity to right your-self and deny this new, contentious teaching. Give us your words."

Stephen answered that he had found great joy and peace in becoming a follower of Jesus Christ. Giving that up was unthink-able. He would not exchange light for darkness and freedom for bondage. Knowing each one intimately made his decision easy.

After a long pause, he heard his father speak to him for the last time in his life. "You have spoken and expressed yourself. Now I speak. You are no longer my son." Each brother sitting around the circle spoke, saying, "Stephen, from this moment, I do not know you as my brother."

Stephen was told to vacate the circle and thus seal his future as having no family.

This meant:

No father.

No mother.

No brothers or sisters.

Shame in the clan and tribe.

No hut to go into.

No welcoming squeals of delight from his younger brothers and sisters as they see him coming up the path to the family hut.

No smiling faces and hands reaching out to carry his bag.

No crowd of neighbors' children entering into a festive wel-coming mood.

No inheritance.

No name to be included in the annals of a family tree.

Then the Vision

Pastor Justin was a powerful evangelist. Scores of people surrendered their lives to Christ in his services. Many were healed as well. In one service, he testified that he had received a vision. His present wife would die (she was in the service, and in African culture it already served as a "fait accompli"). Further, the Lord had shown him who his new (young) wife was to be. He was to take her immediately to care for her until his wife died. Because of Justin's popularity, the whole community supported his "vision."

A short time later, he publicly testified again that the first vision identifying his new wife was a mistake, and that the Lord had now shown him the right one. When asked to meet with a member of the denominational executive, he claimed he was like the Old Testament prophet—dumb. He only wrote notes. The importance of remaining true to God's Word was a lesson to all who heard of this spiritual tragedy. Scripture makes a declaration and asks a sobering question. It declares that the heart is wicked, and who can know where it will go on its own? Allowed to become unbridled, it stampedes like a warhorse.

Giving Up Heaven on Earth

Two things are nirvana in the African culture: prestige and riches. Pastor Paulo had both within his grasp at the age of 16. Paulo's father had everything that life could offer. He had reached the pinnacle: he was a tribal chief and respected village elder; he had hundreds of head of cattle, sheep and goats; he had all the wives he wanted, plus a quiver full of children.

Paulo was the first-born son of his father's first wife, automatically making him the rightful heir to everything. After the evening meal, while sitting around the fire stoked with old corncobs, Paulo's father informed him that tomorrow he was to remain in the yard. Herding the livestock would be the responsibility of his younger brother for that day.

Around 3 a.m. (9 a.m. Western time) the next day, Paulo was asked to accompany his father. This had never happened before, so something very extraordinary must be about to take place, Paulo thought. He was methodically shown all that his father possessed. As they walked the pathways of the village, people bowed in respect and greeted him with *Shikamoo* (an old Arab slave greeting, translated "I grab your feet" or "I fall at your feet in deep respect"). Children playing by their huts were in awe and whispered about this important man; bystanders spoke to each other in hushed tones.

All of Paulo's brothers and sisters and half-brothers and half-sisters were introduced to him. He knew them well, but in African culture, it is normal to present the known and obvious as though to a complete stranger. Their mothers were introduced to him as well.

The tour took them to the fields, where herdsmen gave an accounting of their charges. There had been much increase in the last calving season. Wild animals and thieves had been dealt with, so no animal had perished. Returning home as the sun was setting, they joined family and friends as they (men only) customarily sat outside the hut reminiscing about the day's happenings. Pots and pans clanged as women busied themselves in preparing the evening meal in an adjoining cook hut. Herdsmen (boys) slowly brought in the goats and sheep to be held in a pen, which was part of the kitchen hut.

The evening meal of cooked cornmeal and beans having been consumed around kerosene-burning lamps around 10:00 p.m., the father came to the point of the whole day. It was the time when lamps are taken into the mud-walled rooms and people prepare to go to sleep on a happy stomach of heavy food.

Slowly and deliberately Paulo's father began to speak. "Paulo, you are my first-born from my first wife. You are my pride and testimony to my virility in youth. All that you have seen today is your inheritance. Cattle, fields, servants, brothers, sisters, half-brothers, half-sisters, and even my wives will be under your authority. You will become tribal chief and elder in my place. In order for you to take possession of everything, when the time comes, you must leave your belief in Jesus. Think about it. Isn't the religion of our forefathers of many generations better than this new teaching by these strangers?"

Paulo did not have to think long—heaven on earth could wait. He would strive for the heavenly home and its riches. "My honorable father and wise one, you have shown respect for me today according to my birthright. It is true: I am the first-born son of your first wife. It is my customary right to inherit all you have shown me today. All I have seen humbles me. Becoming tribal chief when you farewell this compound is not a small matter. Becoming head of your family is also not an insignificant responsibility. You have said there is one thing standing in the way, and that is my following Jesus. Today I announce to you that I will not give up following Jesus." Disappointment showed on his father's face, but custom dictated that he could not withdraw the ultimatum. Paulo would never succeed him. Earthly riches and prestige would go to another son. I knew and worked with Pastor Paulo. Until the day he bid goodbye to this world, he was content to wait for his heavenly riches as he enthusiastically served Christ.

A Dead Muslim Muezzin Speaks

Through a powerful mosque loudspeaker, for years Ali called the faithful Muslims to prayer each morning at 5 a.m. He was the muezzin (a Muslim elder who leads in prayers over mosque speakers in the call to prayer). He had never been sick a day in his long life. Upon falling ill, he was taken to the local Muslim hospital. Death being obviously imminent, he called for his wives and all the children. Arriving at his bedside, they were horrified to discover that hospital staff were already preparing to move his body to the morgue.

Friends, relatives and religious leaders soon arrived and loudly wailed in sorrowful moaning cries. Many questions were asked as to how he could pass so suddenly. In the midst of all of the commotion, dead Ali sat up in his bed. That sent some of the weaker of heart, scared out of their minds, to bolt for the hospital courtyard. Others stood wide-eyed.

"I have something to say to my family and hospital staff. In the place where I just came from, I met a man in white robes. The man asked where I was going. I said, 'I did not know. I was not directed to go anywhere. But as I looked down the road, the way before me did not look good.' The man said, 'I am Jesus, whom you have been denying and refusing all these years. I am giving you one last opportunity to confess me to your people. Tell them that I am Jesus, the Son of God. You have been wrong in telling people that I am not the Son of God.'"

Then he began to shout, "Bring me a pastor! Bring me a pastor! Bring me a pastor!" Horrid, sacrilegious and desperate were the cries that echoed through the halls and rooms of the small Muslim hospital in the southern Tanzanian city of Kigoma. (As a side note, just 12 kilometers away lies the small village of Ujiji. This is the exact place where explorers Livingstone and Stanley met). No Christian, by their presence, had ever desecrated that place. Staff wrote the muezzin off as having lost his mind and refused his request.

In "holy" desperation, the doctor on duty ordered an immediate injection of the strongest sedative they had available. Still the man continued to cry out, "Bring me a pastor!" A second injection had no more effect than the first. Finally, in total embarrassment, the staff called for the local Pentecostal pastor.

The muezzin said, "Pray for me. I have seen Jesus. I need to go to heaven." He continued on with his testimony in a room that was by now filled to capacity with doctors, staff, religious leaders, family, friends and mourners— plus onlookers through the open windows.

After his proclamation he fell back again—physically dead, but spiritually alive, I believe. This incident caused a big uproar in the town in spite of the fact that religious leaders tried their best to keep it quiet. Ali was well known in the area, so interest in him was high. He was also the grandfather of Pastor Julius, who became our children's pastor at the orphanage.

Slow Down!

Our son, Stephen, was home from Rift Valley Academy boarding school. He and I were following an overloaded, underserviced (as usual) public transport bus creeping up a very steep and winding hill leading out of Bukoba toward the Uganda border. During my years in Africa, I had been behind such buses many, many times. Waiting for a chance to pass required some patience. On this day I said to Stephen, "I think we should stay back in case the driver misses a gear and the 'tan boy' (the conductor on every bus) is unable to hop out quickly enough to put the block of wood behind the wheels to stop the bus from rolling backwards." Brakes on these buses are not mechanical things that work with any regularity.

We slowed to a crawl and allowed the bus to gain on us. A minute later, as we rounded a corner, we saw the bus on its side, with people trying to clamber out of broken windows. Listening to the still small voice without beginning to analyze is not our strong point as people, but it is God's way of protecting us.

What a Wonderful Bargain

Festus made a very good living by travelling to villages in the Bukoba region selling religious books. Into the New Year and up to March was very lucrative as he sold small pocket diary calendars called *takwimu*. They actually still sell well into June! I ordered them by the thousands from the African Inland Press in Mwanza for him to sell.

On a fine summer day, Festus came to our yard and waited outside. To my query as to the reason behind his beaming face, he gave me the wonderful news: he was getting married!

"Wonderful!" I said. "How much was the bride price?" That is a perfectly legitimate question in Africa.

"Oh," he said, "that is part of the good news. I paid her father 2 kg of sugar and 5 kg of meat. That was a great price!"

"Now that does sound like a very good deal, Festus," I said. "However, I must tell you that sounds like a seven-day bride. "

"No, no," he replied. "She will make a good wife."

About a month later, Festus was again at my place picking up books. I was anxious to hear about his tribal wedding. "It was great," he said and filled me in on the details.

"Where is your wife?"

"Oh, about a week after the wedding, she went to visit her family as is customary."

My premonition was right: she never returned; she was a one-week bride. Festus returned to the life of a bachelor and kept on selling books. There are many life lessons in the story of Festus.

Naked Or Not, Wrestle Her Into Church

Kibara, in the Musoma Region, is a sleepy fishing village on the shores of Lake Victoria. As with all of these small villages, life is normally slow-paced. Fishermen in their dhows, waiting for favorable prevailing winds to set out at daybreak to fish, characterize it. Others sit under mango trees and mend nets. Mangy, slinking, and shrill-barking dogs roam the streets; men sit under trees, whiling away their time along the dusty roadside. Kids play soccer with rolled-up plastic serving as balls. Small front type shops sell soap, lotions, and locally produced aluminum cooking utensils, plastic cups, etc. Cows and goats wander on the loose foraging piles of garbage. Women carry water on their heads. Chickens scratch in the dirt to find whatever chickens look for in the dirt. The local open vegetable and fruit market is covered by dirty, torn plastic sheeting. The ever-present mentally deranged person or persons that such a village may have also form part of the landscape. They, too, wander about unhindered, taking whatever they want or need from anywhere—no one takes much notice of them. They are accepted as a normal part of the village landscape. Whenever they run around naked, a kind lady ties a *kitenge* (a loose-fitting piece of locally produced cloth designed to wrap around the chest or waist) around them.

A deranged woman was well known in the town. Hardly anyone took note of her screaming or throwing dust in the air as she passed. She was known by everyone and for years had been running, mostly naked, up and down the dusty streets—sometimes during the day, at times in the middle of the night. Whether she had been born that way or whether syphilis or some other venereal disease had corrupted her mind, no one seemed to remember. On my trips to the area, I saw her many times.

On bright moonlit nights I also remember her running by my vehicle screaming and carrying on. Every time my heart would go out to her. To avoid taking the children's or some mother's bed in a hut, I slept in my vehicle. It wasn't comfortable, but it was less disruptive to the local pastor's family. It had its advantages in that I could string a mosquito net inside the vehicle and head off to dreamland with the buzz of mosquitoes humming outside the net.

Pastor Eliakim Maguli did take special note of this woman. Pastor Maguli was a man who knew God as the I AM THAT I AM. Whenever the woman wandered close to the church on Sunday mornings while a service was on, he ordered his deacons to go out and wrestle her into the church and to the altar. People gathered around and, while restraining her, they laid hands on her and prayed fervently for healing. Then she was taken outside again to continue her aimless meandering. This went on for a number of years without the pastor or church ever wavering from what was by a now a normal occurrence and a bit of a snicker event in the village. Then one day, in the midst of praying for her at the altar, God totally healed this woman. Her mind was renewed to its normal function. She was set free from the naked wanderings, digging in garbage piles for food, and sleeping under the hot tropical sun or monsoon rains. I cannot even begin to imagine the changes in her life. I wonder how her family felt and reacted to her return.

For most of us, if we had not seen God do anything the first few times, we would have long ago joined the townsfolk in accepting her as she was. Paradoxically, inside the church, we would have continued to preach that with God all things are possible and *Jesus heals*! With Pastor Eliakim, it was a message that held true every day for every person in every situation.

Can you even begin to imagine what that meant to this woman— being clothed and in her right mind?

Let It Rain

Pastor Gordon from Alberta visited while we were working in the Bukoba region known as West Lake, very near the Uganda border. Our Sunday travels took us to a church in an area which had not experienced rain in some time. Normally, rains should have started about three weeks prior, but the sky remained cloudless. Naturally, it was an item of prayer for the rural people, who were dependent on rain for their crops. It was a hot tropical day for which we needed no reminder. Sitting in a mud-walled church, the iron sheet roof, only a few feet above our heads, radiated heat like a sauna.

Pastor Gordon was asked to pray for rain. The hot, searing sun made certain that it was a prayer of pure faith. Two hours later, Pastor Gordon had not yet concluded his message when the pelting rain, bouncing off the tin roof, soon brought an amen. No one's voice could be heard above the thundering noise. The ladies of the church had been cooking to ensure that, as soon as the service ended, we could eat and begin our homeward five-hour journey. We still had to contend with the three miles (five kilometers) of narrow "slippery slope," banana tree encroached back road before we hit the main muddy road. Hurriedly we said our goodbyes and bolted for the vehicle through the ever-increasing torrent of rain.

The secret of operating a four-wheel-drive vehicle is to keep the pedal to the floor to keep the front wheels going in the direction you want. The front wheels co-operated, but the back end of the vehicle insisted on sliding into the path of the banana plants. We were going down the road sideways with mud flying from every wheel. But we were moving. It was obvious that the *thud, thud, thud* of plants hitting the side window would soon turn into the sound of shattered glass. However, the good Japanese glass held. Years later, people from that congregation reminded us of the day that Pastor Gordon prayed for rain.

CHAPTER 5

I Want To Talk To The Real God – He Sat Right Up Here

This is one of the most remarkable stories of God speaking that I ever remember hearing. Belying her age, Marietta jumped, clapped her hands, and danced with the rest of the tightly packed, sweating congregation. As is customary in so many villages in Tanzania, the church measured about 15 feet by 40 feet and 12 feet high to the peak of the corrugated iron roof—translation: "sauna" in the hot tropical sun.

Built out of interwoven sticks, rocks pushed into the openings and plastered inside, it was all bonded together with mud mixed with cow dung. Later, once the termites finished off the sticks, the structure still stood. Windows were two feet by two feet and closed with wooden shutters; some hung precariously from loose hinges.

I was spending three days ministering to local churches very near the Tanzania-Uganda border. Large rolling hills with banana plantations scattered throughout the area contributed to its beauty and sense of peace. Without exception, the after-service custom was that I would be invited to the pastor's hut for food. There would always be much hustle and bustle of people inside and outside the hut. Women would be cooking over rocks outside, with children close by chasing each other. At least always one scrawny, totally lethargic dog was a mainstay lying in the shade along the mud wall. Inside we would be chatting around a small homemade wooden table.

Today was an exception in two ways; first, I was the only one invited to eat at an elderly lady's home. It had never happened before and never happened again in all of my 35 years in Tanzania. Africans are highly social, and a visitor calls for an opportunity to spend time together. Secondly, the day was exceptional because of the story the woman shared. Her hut sat atop the crest of a high hill, with the valley vistas of distant banana plantations below. A little way up the hill were manicured fields of corn, millet and cassava. Grass on the surrounding hills looked like velvet and appeared to have been cut with a lawn mower. Few views I have seen in Africa could compete with this serene place. It was stunningly beautiful in every way.

Marietta was old and illiterate. Illiterate only as Westerners understand it: she could not read or write. In Africa, we Westerners are the illiterate ones because we can read only written signs. To us, the clay pot on the grass hut roof doesn't mean anything, nor does a chair in the shade, nor a rock strategically placed. But to the African it tells an immediate story. Marietta was well advanced in age and already a widow before she heard and accepted the gospel. That meant a radical change in her life: no more taking chickens or goats to a witch doctor; no more participation in heathen rituals; and no more praying and offering to the spirits.

Hers was a typical round, grass-roofed mud hut 12 feet in diameter with eight-foot walls and an open ceiling culminating in the peak of the roof. Inside there was one small room with a bed and just enough space for four homemade wooden chairs. Another small adjoining room was the cooking area, from which smoke seeped into the sitting area where I sat. My eyes revolted against that, but Marietta sat close to the fire as smoke lazily curled around her weathered face. She never made any attempt to avoid the smoke. It was part of her life. Her eyes were immune to its sting.

Approximately half of the sitting room had some sisal poles lying across the walls, making a storage area where she kept clothes and other personal items. This was the focal point of her story. As we finished our leisurely meal of cooked plantains, chicken and beans, she began to tell her story.

"You know I became a follower of Jesus just a few years ago, and you can see that I am old. A pastor came to our area and told us about Jesus. I agreed to be a follower of Jesus. Since then I have attended the place where followers of Jesus meet. My children, who are all grown, were not happy about this. A son of mine came and tried to persuade me to turn back to our old ancestral religion. He was very rude and said to me, 'You old senile woman, why do you allow yourself to be deceived by these strangers? Turn back to the religion of our forefathers, which is the right way.' "

She continued with her story. "I was perplexed and did not know what to do. I am old and all I wanted to do was follow the true God, so I said, 'God, I do not know who You are. You know that I do not know how to read or write or do many other things. I don't know which is the right religion—Jesus or that of my dead forefathers and ancestors. I want to do the right thing. If You are *real*, please show me the right path.' "

Then she became excited and animated as she pointed to the storage area just a few feet above our heads. "He sat right up there!

He spoke to me and told me to read three Scriptures. One was Matthew 5:5

(ESV) "Blessed are the meek: for they shall inherit the earth." John 14:21 (ESV) "Whoever has my commandments and keeps them, he it is who loves me. And he who loves me will be loved by my Father, and I will love him and manifest myself to him." Then this person told me to read John 15:5 (ESV) "I am the vine; you are the branches. Whoever abides in me and I in him, he it is that bears much fruit, for apart from me you can do nothing."

"I can't read or write! I didn't even know what *John* was! On Sunday I asked the pastor to read the three places; I remembered very well what I was told. Each one told me that God loved me, and that I was following the right God revealed through Jesus, His Son. Now my children cannot persuade me to follow our old ways. God heard me! They were promises to me!"

On another occasion I was visiting an old Christian man in his hot mud hut near the town of Moshi in an area called Kahe. As we sat drinking sweet tea which his wife had prepared for us, he said that a man had come to talk with him a few weeks back.

"This man was trying to explain things out of the Bible. He said that I was not fully understanding God's Word. He asked if I had been baptized and when I said I had, he asked how. I replied, "By immersion in water in the name of the Father, the Son, and the Holy Spirit." He went on to tell me that this was wrong and that my baptism was not valid; I needed to be rebaptized in the name of Jesus only. I reminded him, "I am an old man, uneducated, and I love Jesus. I do not understand what you are trying to tell me, but all I know is that Jesus is my Saviour—that is enough for me."

Wow! That is what we need. Not to be swayed by every wind of doctrine which people bring to us, but only to know Jesus and His teachings and example. When John baptized Jesus, God the Father said He was well pleased, and the Holy Spirit came down on him.

Two Dying Twins Bring Their Mother Life

'Cruelty' and 'sadness' seemed to have been born with Ploscovia. Abandoned as a tiny baby by her own mother she was cared for by another lady. When 11 months old her mother returned, saw her daughter, and decided she should 'repossess' her. As soon would be become apparent, it had nothing to do with love. When she was old enough to do the tiniest thing (African girls begin taking responsibilities at a very early age) she was used as a literal slave by her mother. School attendance never entered any equation for Ploscovia's future–indeed her anticipated future only revolved around her mother's need for a worker.

Kept in the bondage of fear and constantly reminded she was useless, stupid and incapable of functioning on her own, she came to believe it and resigned to her life of servitude to her mother's street lifestyle. In exchange for a small room and meager food, she peddled her mother's home brew. Naturally that brought her into constant contact with the less desirable elements of society. At barely 19 years of age she became pregnant with twins. A habitual thief, the father spent most of his time in prison so Ploscovia continued on in the quagmire of her mother's dominance. The twin girls posed an unwanted interruption to her work schedule and as a result immediately elicited feelings of hatred by their grandmother. Straight out murder, although not such an uncommon occurrence, could have some legal consequences. Starvation and neglect would be a better covering. Thus, often when the twins were given porridge the grandmother would swat it out of their hands and have it spill on the ground. In their malnourished state the twins were especially susceptible to multitude tropical diseases. One day both the twins became very ill and Ploscovia knew she had to take them to a local medical clinic or she would soon be burying them on her mother's small house lot. Penniless, desperation drove her to carry the limp little bodies to the clinic. Each day Ploscovia's plight is multiplied hundreds of times at the clinic. The, "No money–No service" rule must ruthlessly be enforced if the clinic operations are to survive. There are too many penniless mothers at their door each day. Dejected, broken hearted and without energy to return home to watch her babies die, Ploscovia sat in the dirt by the road side, covered her head and babies with her shawl and cried. A passing mother's instincts told her that this young mother was in desperate need. Hearing her plight she told Ploscovia that there is a missionary lady who helps babies like hers.

As was often the case, LouDell returned home late from Starehe Children's Home, exhausted and still facing her household chores. Parking the vehicle and coming around to the front door she found mother and babies literally lying on the doorstep overcome by exhaustion and malnutrition. Emotionally and physically drained herself, LouDell was in no condition to re-start her day. Justifiably, her first reaction was to tell this lady to come back in the morning. Here the declaration of Jesus that we will always have the poor among us, stepped off the pages of the Bible right to LouDell's front door.

However, ministry is a calling, not a job. LouDell, strongly prompted by the Holy Spirit, sat down beside her and listened to her story. Listened with her ears and heart. In this case it seemed that money would solve the immediate problem. She was told not to return for help again since we have so many people who need help.

A week later she returned anyway. This lady was desperate. LouDell gave her money so the twins could continue to receive medical treatment. Their names were sent to the Pentecostal Assemblies of Canada's child sponsorship program (Child Care Plus). News arrived that they had been sponsored. Ploscovia was asked if she wanted to get out from under her mother's domination and abuse. Very much against local culture she agreed. Secretly one day she was moved to another area of the city where one of our local pastor's congregations agreed to nurture her.

Daily selling small amounts of tomatoes, charcoal and peanuts on the roadside, helped her establish independence and a realization of self worth.

Today Ploscovia and her twins are a happy thriving family. Each time we visited their church the twins would rush to LouDell and give her a big hug. I love the motto, "It's The Caring That Counts".

Tithe Giver (almost)

I was preaching on the blessings of tithing in the small rural town of Mugumu. It is an area settled by the Wakurya tribe, who are known for their fierceness and expertise in making and using poisoned arrows. It seems they are always prepared for a fight. Cattle rustling is a favorite pastime of theirs, so they often get to enjoy their fighting pastime. In a moment of inspiration (and after a sermon on tithing), the pastor publicly announced that he was giving one of his cows as a tithe to the Lord. Could it be one of some he had stolen; I don't know. A few months later, I met him and asked about the cow. "Oh," he said. "I have decided to keep her for the Lord and allow her to have a calf, and then both the cow and the calf can be given."

The next time I met him, he informed me that both the cow and the calf had died. I am a firm believer that God blesses good intentions, but in this case I am not sure if the pastor ever had good intentions. For that tribe to give up a cow is like pulling out your own teeth.

Klabu Elimu (Education Club)

Paul, who was in Tanzania running a very successful rural adult literacy program, began a ministry showing videos in the city of Mwanza. The upstairs rented room was smack in the middle of the business section of the city. The room was rented from an elderly Seventh Day Adventist man. Later his son, Mr. Kazi (literally translated "Mr. Work"), took over the building and later still became the government member of parliament for the Mwanza Region. Tragically, he was killed in a road accident a few years later. Unfortunately, these accidents happen too frequently for several reasons: government people regularly travel by road after dark (they always employ a driver); vehicles are not always maintained the best; and roads are treacherous with loose sand, ruts, washouts, and a lack of signage indicating sharp curves. Also adding to the potpourri of the chess game to arrive safely are: people dressed in dark clothes walking on the road; many bicycles with no lights or reflectors; cows, goats, stalled buses and trucks; vehicles with no tail lights and mere candles for headlights.

Other more practical reasons for after-dark travel is that most villages in Tanzania are best reached by road. Night driving is a must since all day is spent visiting villages, plus darkness falls at close to 7 p.m. all year round. Therefore, to "redeem the time," many people travel after dark. It is also well documented that most Africans love the dark, so being out in the wee hours of the morning is not unusual for them.

In those days, video (pronounced "vi-**day**-yo") was a brand new thing. Only a rare few in the city owned a TV. Certainly none existed in the rural area; lack of electricity ensured that. Those who had a TV had to have it registered with the government, and the license had to be renewed each year. A video player also had to be registered and have a license.

As most of the shoppers were people from the surrounding villages, this video centre called *Klabu Elimu* (Education Club) was a huge hit. Videos were shown right at midday when most stores closed for a rest period. There was little to do while waiting for the shops to open again. The upstairs room held about 100 people and was always packed. Videos of all kinds were shown: health related, Canadian Grains and documentaries among them. During the last 15 minutes, a gospel presentation was given. At first Pastor Peter directed it, and later two local pastors from the area were in charge of the showings.

I took over the ministry when missionary Paul left, and we were averaging 80 conversions a month. All of these decisions were registered with names and addresses. These were later passed on to churches in the area from which the people came. Paul continued to raise funds in Canada to pay for the expenses, a huge part of which was the room rental. Paul's funding for the project largely came from friends who believed in him and his integrity in using donated funds. Slowly people stopped donating because Paul was no longer personally on site. Reluctantly, Paul informed me that the donors had dried up and he no longer had any funds to pass along.

For nearly a year, I kept meeting with Mr. Kazi, telling him that we needed to vacate his premises because we had no money to pay him rent. He kept reassuring me that the work we were doing was helping a lot of youth who would otherwise be on the street, so we should just keep on running the video centre and pay if we could. Finally, after we were a year behind in the rent, we thanked Mr. Kazi for his patience, which was most honorable, but told him that we could not foresee any possibility of paying the rent. Sadly, he accepted our termination but never once hinted at asking us to pay arrears.

In fact, the income tax department demanded that he pay taxes on the rent money which he had never received. In spite of receiving letters from us verifying that we had paid no rent for over a year, they still demanded payment. They were sure we were paying rent to him by depositing money into a bank outside of Tanzania.

Africa is known for such cross-denominational unity of purpose in seeing people come to Christ. I believe it is part of the reason we see such a great move of God on that continent. Theological nitpicking is left to the offices of the hierarchy while "the little man" keeps on preaching the gospel and bringing sinners into God's kingdom.

God's Protection

In a number of dangerous situations, LouDell and I have experienced God's protection. Sometimes it has been evident; at other times, no doubt, angels have cleared a path for us without our knowledge. In our travels we have suddenly come upon hordes of people parading in a circumcision rite of passage. Most are high on emotion, local beer, and potent drugs. As we slowly drove through the crowds, there was much banging on our car windows. Some danced in front of us with menacing mock spearing.

Even more dangerous were times when we came upon political demonstrations taking up the whole road. During election run-ups, occasionally innocent people were dragged from their vehicles and badly injured or killed. The fact that you were not involved in politics didn't matter; you were just there — an outlet for their emotions.

On a number of occasions, fully loaded buses careening at full speed ran us off the narrow, sandy roads. In most cases the ditches were not deep, so we were able to take evasive action.

Something to constantly be vigilant about was wild animals darting across the road. Smaller animals can be taken in stride, but things like zebras can do major damage. Even the unsuspecting can suddenly get motivated. Donkeys are the favorite beast of burden in many areas. They wander along many roadsides eating grass. Seldom do they ever even blink when a vehicle passes by within a few feet. However, one day as I was nearing one, he suddenly raised his head and charged across the road. For years that little incident caused many tense moments as I approached a somber donkey. On those occasions, I always let out a little sigh of relief that the donkey didn't move before I passed.

Wartime in Tanzania was very tense. Everyone, especially young soldiers with newly issued weapons, were nervous and unpredictable. Something to always keep in mind was the importance of not making any sudden moves, which could be translated as aggression. Sometimes we assumed that firearms were just being carried as show with no firepower. However, that proved to be a misconception on our part. Every government building, bank, hospital and bridge had a guard with a loaded firearm.

Arusha Technical College is the place where I took my vehicle for maintenance. I got to know the "arms bearing" lady at the gate on a first-name basis. I jokingly said to her that she was probably just carrying a gun with no bullets. "Oh yeah?" she said and pulled back the lever of the gun, ejecting a live round of ammunition. Wow! I had to reprogram my assumption rather quickly.

Socialism, together with an extreme mistrust of everyone, was at its height in the 1980s. Soldiers with loaded guns stood guard everywhere. A guard at the NBC (National Bank of Commerce) Clock Tower Branch in Arusha dropped his gun, which discharged and blasted a big hole through the 50-foot high dome-shaped ceiling. I was not in the bank at the time, but a friend said that everyone was on the floor. Each time I went into the bank after that incident, I would stop and stare up at the hole. One time when was I again staring at the hole, I looked down, and the guard was also staring up. Within a few weeks they fixed the hole.

Guarding is a rather strange phenomenon for a mostly friendly and unassuming culture. Walking up to a bank guard, I engaged him in conversation. He was not happy. I complimented him on the fine condition of his weapon and shared that I hunted and appreciated guns. He began to relax. He was carrying a shotgun, and I talked to him about my guns and hunting ducks in Canada. I asked him how many shells his gun held. Pardon the pun: he stuck to his guns and wouldn't answer. But I knew his weak spot.

"What tribe are you from?" I asked. Having lived in Africa a number of years by now, I already had a pretty good idea. Facial features, as well as certain markings such as teeth missing or shape of face or even skin colour, are pretty clear indications of tribe. I then greeted him in his tribal language.

"What village are you from?" I continued.

"I know that village; I have been there many times. Do you know so and so?" Of course, everyone knows people in the village. Before long we were buddies since we had a number of things in common. Not long after that, I asked if I could hold his gun to get a sense of how it balanced in my hand. Without any hesitation, he handed me the gun. I had easily penetrated his human side, which was what his culture was really all about. The moral of the story is that we always need to be on our guard, no matter how friendly a proposition may be to let it down.

Aside from physical danger, we also encountered spiritual danger. It is to be expected that when your work involves spiritual things, the other side—spiritual darkness—also exists. These forces, seen and unseen, are not happy to surrender any ground. On a number of occasions, we found things placed in our home that were intended to bring the presence of a curse upon us. That is the reason we so appreciated the prayers of people. Besides this, the Bible tells us that when we do not believe a curse, then no harm comes upon us (Proverbs 26:2). There were times when inexplicable things like sickness or depression did hit us, but knowing the possible root of it, we were able to hold steady in the battle.

Being "white," Western and rich meant that we were targets of petty theft. Any momentary slip, and the thief had your things. Just going to the back of the vehicle, all doors had to be locked. Air conditioning was a huge benefit as all windows could be kept closed. Once we were slowly driving through a street showing visitors around, and a thief ran alongside our vehicle and yanked open the door. Fortunately, there was nothing close at hand that he could easily grab.

I'll Shoot You All

A sense of humor is something greatly appreciated in Africa. We enjoyed as good a sense of humor as the people did—sometimes without their knowing it.

Stray dogs were a constant nuisance in the city of Arusha. At nighttime they would be in their glory trying to out howl the pack down the road. Sometimes this made people afraid to walk alone at night. With the number of dogs on the loose, it was inevitable that disease was also rampant. Rabies especially became a concern for officials.

The American government kindly issued ammunition to the city police to greatly reduce the roaming dog population. Many of the dogs belonged to people, but yard fences were not maintained enough to keep the animals from exploring the neighborhood at night. They were probably also out looking for scraps of food. The *pop pop pop* of small calibre gunfire became a normal early morning wake-up call.

Our yard was large by American standards. It had been a former British area, so yards were never less than half an acre in size. An open-backed jeep drove by and then backed up to our gate. It was unlocked, so I invited the policeman to come in. I met him partway up the driveway. He began to explain what I already knew: they were shooting stray dogs. However, it seemed that they were running out of targets and were now going to shoot dogs within yards.

He lectured me that I needed to make sure that our dogs—five German Shepherds—had up-to-date rabies vaccination certificates. If police came into the yard and did not find these, they would shoot the dogs.

I responded that no such thing would happen. Taken aback, he wanted to know what I meant. I said that I had a gun and would shoot the first policeman who came into our yard. He motioned toward the open Land Rover filled with policeman and said, "But there are many of us!"

I responded, "I'll shoot you all!"

He turned and hollered at the policemen in the Land Rover, "Don't come into this yard. This man has a gun and will shoot us all!"

Imagine doing that at home!

I explained my reasoning to him. The first thing that would happen is that some thief would get hold of a police uniform and, on the pretense of checking for rabies vaccinations, would be there to steal. He was satisfied with my reasoning, and I never had any police at our gate again.

Sorry—No Loans!

LouDell and I know what the biblical king experienced. We have lived that scenario dozens of times, "Can you help me? My child is sick; my wife is sick; my brother had a bad accident; my child is being expelled from school tonight if school fees aren't paid, and they also need money for a school uniform. As soon as the rice is harvested, I will repay the money." Usually that did not happen because, in between, too many other emergencies had raised their heads.

Finally coming to our senses, we saved ourselves a lot of anxiety and bookwork. When people came for help, we forgave the loan before they came to ask for forgiveness for not being able to repay. We learned that the word *loan* has one meaning in our Western world and another meaning in Africa. Their defaulting on the loan was surrounded with circumstances we could not even comprehend. It was with a thankful heart that we gladly turned the loan into a gift.

Bring on the Chicken Blood

Elections and sports are two highly emotional triggers. Beatings, gang fights, bludgeoning and deaths sometimes precede such events. Chicken blood is poured in front of the opponent's goal-post (I guess it doesn't work on their own team when they switch ends). Before an election, local newspapers deride those running for re-election by showing their white Mercedes parked in some muddy back street near the most famous witch doctor in the city.

To the delight of the electorate, a little "vote for me money" is passed out. Who can't use extra money for school fees or some much-needed medicine? Usually the country is awash in smaller denomination bills for months after an election. The higher a politician climbs, the more danger awaits—if disagreement with official party policy is expressed. It may be a head-on collision or a run off the road or poisoned food or drink. None of these can be proven, but it is strange how these things seem to increase during an election.

Outstaring a Stare

Local people can outstare Westerners very easily. They lock their eyes on you, and no matter how long you try to stare back, they outlast you every time. Being that it is a rude thing to do in our Western culture adds to our soft resolve not to continue the match. By trial and much error, we discovered their weak spot. The way to outstare a national is to simply keep looking at their feet. Within a few minutes, they begin to shuffle around and then leave. Apparently, they don't think their feet exactly add a lot of dignity to their profile.

Revenge

No matter where one goes or what one does, there are those who feel they have been wronged in some way. They will go to extremes to exact revenge. This can make life one long "looking over your shoulder" existence. Within the culture, signs can be read, so certain "red flags" begin to wave. If you are not of that culture, you had better have a trusted friend who can read the red flags. In my own case, I had a few such friends who, in actuality, were my protectors. No big fanfare was made of it, but my friend said, "When you go there, be careful what you eat." Even if asked, he would not elaborate; he expected you to read the language. He was warning me that the friendliness of a certain person was a cover-up for possibly putting poison into my food. Other times the Lord gave a warning within my spirit that danger was lurking.

Pastor Patroba somehow got onto another person's hit list, and his revenger had chosen his place for the best angle. Unfortunately, Patroba's daughter stepped out of their mud hut first and was killed with a poison arrow meant for her father. Most tribes have developed their own poison concoctions. One kind is used for hunting. The poisoned portion can be cut out, and the remainder of the meat remains untainted. Other poisons are meant to kill by affecting the nervous system. Interestingly, every tribe has an antidote for their poison.

As missionaries, we were not spared the wrath of people. We had 57 staff to operate Starehe Children's Home. Many of the orphans were babies, and they take a lot of care, day and night. A worker was disgruntled about being found out that she was bringing things related to witchcraft onto the premises. She was dismissed, but she came back the next morning very upset. As she was speaking with LouDell, she grabbed her arm and bit her, drawing blood. Whether she was carrying the HIV virus we did not know. We thank the Lord that LouDell did not suffer any medical consequences, although she was a bit traumatized by the event. Getting bitten by an employee is not an everyday occurrence.

No Ordinary Offering Time

Joyce was the pastor's eldest daughter. The church was unable to pay her dad a living wage, often resulting in her and her siblings going to bed hungry. They largely survived on the food people brought to their door. Growing up, she never knew a time where food, clothes or school fees were sufficient for the day. When she was old enough, more out of sympathy than need, we hired her to work on our yard. In this way she could have some money to share with the family. In the process she blessed us with her quick wit, cheerful disposition, and faithful work.

Because they do most of the manual labor, women are stronger and better workers than men. Seeing Joyce react to bugs or snakes or anything resembling either, you would not guess that she was native to Africa. Anything out of the ordinary would send her on a straight path of placing as much distance as possible, in the least amount of time, between her and the dangerous "whatever." On one occasion she was hauling water on her head, which is the least strenuous way for African women, and was passing close to a room where I was unpacking books. Near to her normal path, but just out of view, I laid some black plastic banding. As she came along, I gently moved it, making it appear like a snake. She let out a wild scream, let her pail of water find its own way, and bolted for places unknown. After about 50 yards, she looked around and saw me doubled over in a fit of laughter. She was a good sport.

The day came when Joyce informed us of her engagement to a young Rwandan man. We were both delighted and skeptical. Knowing the African culture, where men customarily return home to their countries of birth, we asked to meet with them both. During the course of the conversation, my wife and I raised the subject of the possibility of the young man getting married and then wanting to return home to Rwanda. There, Joyce would be a stranger. He promised not to take her into that country where memories of atrocities were still vivid. Joyce only spoke Kiswahili, not Kirwandese.

Sadly, a short time after their first child was born, his family traditions prevailed and they moved to Rwanda.

We feared the worst for her and later realized that our fears were well founded. We had a number of acquaintances in Rwanda and inquired as to Joyce's well-being. The latest we heard was that Joyce had four children. Over the years, reports of her abuse—even by her husband—filtered down to us.

One day we heard the African version of our doorbell: *Hodi, hodi!* ("I am here. Is anyone home?"). It was a shaky, unrecognizable, weak voice. We could not believe our eyes when we answered the door. There was Joyce, a mere skeleton; the life in her eyes had vanished. Her forced smile could not hide her condition. She was totally destitute, emaciated, weak, and near collapse from hunger. She was as happy to see us as we were her, but the bitterness of life had sucked any semblance of joy out of her. At her side stood a young, fearful, trembling girl of five clutching at her mother's beanpole legs. In her mother's arms was a two-year-old. It was the two-year-old who now vainly sucked at her milkless breast. What a heartbreaking sight!

As she sat on our couch drinking a soda (pop), she relayed the sad, sad story. We were like parents to her, and she poured out her story, so numbed by the experience that she hardly had the strength to show emotion. It was a long, sordid tale. Full of hope and excitement, she had been married (we knew that as we were at the wedding, but that is the way a story is told in Africa—from the beginning). Life had been wonderful with her husband selling used clothing. Finally, she did not have to go to bed hungry. She could even help her parents and younger sister. Then the unexpected move to Rwanda came. There they lived with her husband's mother on their small piece of land in a rural area. Due to its high population density, few people in Rwanda own much land.

It wasn't long after her arrival that her mother-in-law began showing disrespect for Joyce. Many times, tribes and countries exhibit extreme loyalty; strangers are not readily welcomed. First of all, Joyce could not speak Rwandese, and she did not know how to dig in the family fields. She had grown up in a city and knew nothing of digging by hand, planting by hand, and harvesting by hand—nor did she know how to milk a cow. Joyce stoically bore being called a foreigner, even by her in-laws, and being degraded because she was not a village girl. In essence, she was shunned and became a slave in the household. Following a general custom, her husband began to beat and mistreat her. Even the bearing of four grandchildren did not soften the mother-in-law's heart towards her. Joyce was alone in her suffering and rejection.

Next to not being able to bear children, she endured the greatest shame an African woman can: she came back to her parents in Tanzania. After months of efforts at reconciliation, her husband arrived and they made a new commitment to their marriage. Joyce returned to Rwanda. A short time later, to add to her already great sorrow, her husband began to lose his mind and roamed naked and unkempt around the villages, scavenging food from garbage piles. Joyce was left to fend for herself and their four children.

Finally in, desperation, Joyce made the decision to return home to Tanzania. That is not easily done when you are a slave and entrapped without any resources. After a number of months of pleading with her mother-in-law, it was agreed that she could return to Tanzania if she could find money for bus fare. Another stipulation promised to break Joyce's heart further. Only her two youngest daughters would be allowed to go with her. The oldest two must remain in Rwanda, separated from their mother. She hesitated and determined to continue on in Rwanda rather than be separated from her daughters.

The rains came and her mud hut collapsed. She was in a state of numbness by life's buffeting. She had to leave. Negotiating with her mother-in-law, she reluctantly consented to her taking the two youngest children back to Tanzania; the two oldest would stay in Rwanda. Joyce begged her pastor to help her. He agreed to give Joyce bus fare and a room to sleep in the day before she left for Tanzania. That night thieves broke into the pastor's hut and, right out of the room she was sleeping in, stole literally every little thing she still possessed. In her mentally anguished state, Joyce never even realized there were thieves in her room. The next morning she cried out to God, asking why He had left her and how she could possibly look after her children.

Joyce arrived in Tanzania literally with borrowed clothes for herself and her two small, frail girls. Home held no better news. Her parents were destitute themselves. For a few days she moved in with her sister, who sadly prostitutes her body for money.

That was Joyce's story: an incapacitated husband, a mother-in-law with no compassion, two dear children separated from her, bad fortune at every turn, finally at home and not even a candle-light's glimmer of hope that life would be different. Still starving, still homeless, still penniless, still life was spiraling downward when she thought she had hit bottom long ago. Being able to share her troubles with us was an oasis for her shattered and tattered heart. After mutual tears and sorrow, we prayed with Joyce and gave her some clothing, blankets, money to buy food, and a pot to cook it in.

On Sunday we attended the church service and, as usual, enjoyed the spirited worship of the congregation as well as the colorful dancing choirs. Choirs in Africa seldom sing "canned" music; they write their own songs addressing the things relevant to them. As per Western culture (husband and wife sitting together), LouDell and I were in the ladies section. It was tithe and offering time. We were sitting next to the middle aisle patiently waiting our turn to merge into the flow.

Offering time is actually an enjoyable, if not ritualistic, spectacle. Customarily, two baskets, usually made of woven reeds, are placed on tables in front at the altar. A person stands behind each basket as overseer (I have never figured out what their real purpose was) as people forge into the aisles to go and place their offering into the basket and return to their seats. A lot of congestion occurs, as is normal in many gatherings. It is chaotic, yet surprisingly purposeful and meaningful. Most were giving out of their meager wherewithal.

I might add that having sat through and participated in literally thousands of offering times in Africa, I was in an "automatic pilot" mode on that particular Sunday morning. We were sitting about two-thirds of the way back when suddenly I had one of those involuntary electrical shocks that our bodies sometimes produce — except mine was a spiritual electrical jolt. Joyce had just walked by our chairs. I literally shot up in my spirit, maybe even in my body—I don't know. Joyce was going to place her tithe into the basket. Certainly the help we had given her a few days prior was a drop in the bucket when compared with her need, but here she was tithing! She was "presenting" her tithes to the Lord. A thousand sub-conscious thoughts and emotions battled for lodging room in my mind. Was this possible? Was it any wonder my mind and emotions reeled when I saw her faithfulness to God? Bitterness I could have understood—but this faithfulness?

Here she was, destitute, penniless, with two small children to feed and no place to call home, possessing only the clothes on her back, *and she was giving her tithes to the Lord* from the money LouDell had given her. What lessons the Lord can teach us through His lowly followers! Later, when we again needed a helper in our yard, we hired Joyce. LouDell arranged for sponsors for her two young children. A happy ending to the story is that within approximately three years, Joyce received permission to bring her other two daughters to Tanzania.

CHAPTER 6

A Lawyer, a Church Superintendent, and Smelly Fish

For many positions in our Western world, dignity is directly proportional to the position held. We do not want to be seen doing something beneath our dignity. Walking down a crowded street carrying a piece of lumber or a rolled-up sheet of iron roofing would be unthinkable for us. In Africa, convenience holds the highest priority.

Our national church was in a court case at the regional high court, which happened to be 120 miles from the city in which the lawyer lived and about 250 kilometers from where the church's superintendent lived. The scheduling of hearing cases happens much like a lightning storm—very unexpectedly and quickly.

Toward noon, the lawyer received word that a hearing in the case would take place the next day. Fortunately, the church's super-intendent happened to be in town, so it was decided that he and the lawyer would travel together to attend the hearing. However, by noon most all transport has departed for destinations up country. It was impossible for them to find a bus. Not to be defeated, they resorted to the next option: to find a truck travelling that way and to hitch a ride. Most trucks, for reasons known only to African drivers, like to travel at night. Their headlights seldom ever shine the same way and flicker like fishermen's lanterns because they are loose in their sockets. It is the time when drunks (local brew) are weaving over the road, pushcarts pulled by muscular men vie for the lane, herdsmen chase their cows, goats and sheep along the road, and women wearing dark clothing and carrying baskets on their heads blend into the night until you are upon them. Other vehicles and tractors frequently drive without any lights or reflectors.

The lawyer and the superintendent finally found a truck going to the town they wanted to go to, but there was a slight downside. There was no room in the cab, so they would have to travel in the back of the truck. That presented the next downside: the truck was hauling loose dried fish! Can anyone imagine a Western lawyer and a church superintendent travelling that way? Their goal was to get there; how that happened was not an important part of the equation.

The third downside was that it began to rain as they were on their way. This was at least a six-hour trip on Tanzanian roads. Water on dried fish is akin to the lady in the Bible breaking the alabaster jar to anoint Jesus, except this aroma was at the other end of the odor scale. To complicate things, people tend not to travel with many changes of clothing. The superintendent later said that he and the lawyer were not the most rested or best smelling two people in the courtroom the next day.

The Cry of a Mother's Heart

Letisia had borne a baby girl when she was little more than that herself. With no employment and her family having disowned her, Letisia knew that taking care of the baby was out of the question.

Pain and guilt were acute as she took the only option available to her: abandoning the baby and praying that somebody would find her and lovingly care for her.

For 26 years she prayed for her baby. Now with a caring husband and a family she could provide for, she began to pray for the opportunity to see her baby once more. Out of desperation, she had physically abandoned her baby but had held her child in her heart every day.

In a little village mud church, we cried tears of joy with her as she introduced her "baby" daughter. God had miraculously brought her to her mother's door the day before. The mother related, "I was digging on my small piece of land when I saw a stranger come down the village path and approach my hut. The Lord whispered into my heart that it was my daughter!"

The old familiar hymn "Does Jesus Care?" comes to mind.

This Was God, Not Chance

Twenty-two years prior to this African safari, a young man had lost his older brother of 16 to a freak car accident. In his memory a church was built in Africa. All of these years, a hidden resentment toward God and estrangement from Him had grown. Now his parents were doing short-term volunteering in Tanzania, and it seemed suitable for him to visit the continent. The remembrance of his brother's death stood in the form of a church building. His apprehension built as the day grew closer for leaving his home in America.

Travel arrangements had been made, his tickets were in hand, his passport had been secured, and his visas were obtained. Tomorrow was flying day. A quick little stop at the drugstore, and everything was set. Returning to his vehicle in the parking lot, he discovered that the car window had been broken and *all* of his documents were stolen. The trip he had been looking forward to—and not looking forward to—now seemed to be all but cancelled.

Dejectedly, he headed to the interstate for his drive home, anticipating at least being able to share the bad news with his wife. Along the interstate highway, he noticed some papers scattered in the ditch. On a sudden impulse he stopped, and there were all of his documents. That is not chance—that is God.

The trip from Mwanza city to the region of the Memorial Church took three days. The first day we drove through some of God's amazing creation—Serengeti Park. The road is long, hot, bumpy and dusty. What a relief to drive through a small rainforest, up over a knoll, and be faced with the modern, if modest, Crater Lodge Hotel. The hotel sits perched on the rim of Ngorongoro Crater. The view down into the crater is awe-inspiring.

The next day was another hot African day, and the roads were even worse than the day before. As the day wore on, our friend became more agitated and quietly kept suggesting to his parents that we turn around and go back to the nice hotel. He had had enough of this. Bone-jarring roads and an experience that he had long attempted to forget combined to throw up a roadblock and vigorously argue for the logic of turning back. Bad experiences do that to us. Even if not buried, they are content to be covered over and left alone, undisturbed. Had it not been for the fact that the young man was riding in the back seat with his parents and I was driving, the tough road—both physically and mentally—would have prevailed and beckoned a welcome turnaround.

Unpleasant experiences tend to come in multiples, and more were to come. We were in a rural area with no accommodations to offer rest to weary Western travelers. Typically, *hotelis* in these small towns offered bars that served up loud music, prostitutes, liberal drinking, carousing, fighting and communal bathrooms. Our choice was to ask for a few rooms in a local Bible school with which I was familiar. The word "sparse" does not adequately describe them. They were dark, and new blood made for happy mosquitoes. We were under the impression that some food might be available, but that hope turned out to be miles distant from reality. Around the few dim, naked light bulbs, we scrounged through our lunch boxes to find at least a bit to eat. The next morning we would no doubt locate a local establishment that served hot tea and bread. Apparently this town majored on nighttime activities only. A "mama" cooking up some rice cakes over a small charcoal stove behind a garage was our salvation.

Breakfast over, we headed to the Memorial Church, another three hours away over roads that only seemed to get worse. Eleven years prior, I had had the honor of dedicating this particular church. Having served in Africa for 22 years at that point, my wife and I had some idea of how soon the memory of bad roads, little food, dust and heat could be refreshed. Driving off the main road and up through a small village path, we saw the church a few hundred yards away. It was completely surrounded by hundreds of exuberant people dancing, shouting, waving, yodeling, singing and smiling brightly. What a sight! Our visitors could not even have imagined such a welcome.

The congregation, and the hundreds who attended that morning, listened intently as I emotionally described the circumstances that had brought about the building of their church. Our friend's son had died in an accident. Now we were coming back with flesh and blood of the very people who had been affected.

Person after person rose in the service to express their thanks for the church building. Death is no stranger in Africa. They expressed that although this day was no doubt a painful reminder to our friends and their only son, they were blessed and filled with gratitude.

That service of gratitude began a healing process in the son's life. At the end of the day, the day he had dreaded turned into one of restoration and healing of a spirit that had been bruised and broken for so many years. Only later did we learn that the pastor had left his adopted son's wedding early the day before so he could be at the service. What an example of sacrifice to show honor to us as visitors. The young man is now happily, fully committed to serving God.

As an aside I want to tell you about the official opening of this church. On the day of the opening, we also dedicated a newly constructed office. Literally thousands of people were present. As the time came to dedicate the Memorial Church, people were instructed that only invited guests would be allowed inside the church building. In order to avoid serious overcrowding, men standing on chairs on each side guarded the entrance door. As people surged forward, the "door guards" wielded heavy sticks and literally had to beat people back. I had never witnessed such a thing in my life—beating people to keep them out of church.

To Kill Her Was the Gruesome Intent[1]

For most Westerners, lakes hold a special draw. Recreation, relaxation, watermelon, wiener roasts, family gatherings, and long summer nights—not to mention mosquitoes—imbed themselves into the "pleasant memories" compartment of our brain. At the next opportunity we want to do it all over again. Lakes seem to have something for everyone: water skiing, swimming, places for children to splash around, lazing on a lawn chair chatting with friends, fishing and sunburn.

Tanzania's Lake Victoria, the second largest fresh water lake in the world, conjures up different recollections. Fun is not among them; they are associated with work and livelihood. As dawn breaks, fishermen have set their sails and are off for a day's work of catching fish. Once the gentle breeze dies down, it will mean a hard row to return to land. Ladies are on the shore filling buckets to carry water back to their huts for cooking or washing clothes. Other women are washing the family clothes right there on the shore. A nice large rock nearby serves as an ideal scrubbing board.

Not far away, naked boys, their black skin shining in the sun, splash playfully in the water. They are unaware of the tiny deadly parasite (schistosome, locally referred to as bilharzia) that is also swimming in the lake looking for a host. Entering through the pores of the boys' skin, they will lodge in their kidneys and, over the years, will slowly suck the strength out of them. Left untreated, early death is the result. Mostly the boys are also unaware of the occasional crocodile looking for an easy meal.

Strolling along the beach for relaxation is not common. However, this early morning found a mother slowly walking on the shore. She had no basket of laundry or bucket for fetching water. She carried a small bundle in a colorful wraparound cloth. Her gaze darted back and forth as though she were looking for a lost article in the sand. She was intently searching for something. Turned out it was a suitable place to bury something.

In desperation, or despair, she had brought her baby girl to the lakeside for a disguised stroll. In the absence of any witnesses, she quickly dug a hole in the sand and buried the little girl face down. To make sure she stayed there, a big rock was placed on her back. This was not meant to be a quick death. Without looking back, the mother ran up through the banana grove and disappeared.

Is there any doubt God saw that struggling little body and sent a Saviour to rescue her? When she was brought to Starehe Children's Home, my wife gave her the name Neema, which means *grace*. Surely the Lord has a special plan for this little girl. Is it any wonder we see Starehe Children's Home as a place God prepared for babies just like Neema?

So You Want Gold And Lots Of It?

Msuya was still reeling a week after he spoke with the witch doctor in the heart of the country's second largest city, Mwanza. We were sitting in the city council chambers waiting for a meeting, dealing with orphans and vulnerable children, to begin. He shared the story in a hushed voice.

It all began innocently enough. Msuya, while waiting for a bus, noticed the claims of a witch doctor. They were hand-painted on a cloth flapping in the wind. The local medicine man was able to give you medicine to find gold. He knew the claim was bogus, but out of curiosity walked down the path to the medicine man's small room. After the usual greetings, including getting the news about oneself, and one's family and livestock, the conversation begins. It went like this:

Medicine Man: "So you want gold?"

Msuya: "Yes."

Medicine Man: "You want lots of gold?"

Msuya: "Yes, lots of gold."

Medicine Man: "Where do you want to dig?"

Msuya: "Arusha."

Medicine Man: "What part of Arusha?"

Msuya: "The Hanang area."

Then comes the inevitable question of fees for the medicine man. A brown chicken or a black-and-white goat is within the usual bartering range; if huge, it may be a cow. The medicine man has seen these in the neighborhood, so he knows what he wants.

However, it was the medicine man's totally unexpected request that left Msuya near speechless and dumbfounded, even these seven days later. The request of the medicine man still haunted him. "Bring me a small child." The price for gold had barely had time to register, but already Msuya's heart raced as his mind asked, *How many have already paid that price?*

Can anyone still have doubts about the need to tell people of a Saviour who satisfies the soul?

As for Me, I Have a Safari

The shrill early morning sound of a referee whistle indicated that an announcement by the hired "crier" was to follow. For greater effect, the whistle is accompanied by much banging on hollow pots. He was announcing the death and burial details of a prominent witch doctor in the area. Newspapers are much too slow for such announcements since burial is a soon-after-death experience. The crier's announcement went like this:

"Hear ye, hear ye! Last night the well-known witch doctor by the name of _____ died. For all of you who were taking his medicine, my condolences [delivered in a sarcastic tone]. You women who were his patients, go to his graveside and wail. I know there were a lot of you, so go! You men, go to help dig his grave. The burial takes place at 2 p.m. As for me, I have a safari."

Witch doctors are not put into crude wooden caskets, as is customary, but carried on a hastily built ladder type of bier. When buried, the witch doctor is accompanied by all of his paraphernalia, and a short sword of authority is held upright in his hand. I don't know how many attended his burial, but the road procession was not too large. Many probably snuck around the bushes and hid in the tall grass near the grave, joining the ceremony so they would not be seen. What a contrast to the death of God's saints.

Bury Him With the Pagans

Pastor Julius, our pastor and caregiver for the older children at Starehe Children's Home, was on his day off when he met a funeral procession. Funeral processions are walking processions with local neighbors and passers-by joining in along the way (something like our fire truck chasers in Canada). People take turns helping to carry the coffin as a sign of respect.

Although he had no clue who the deceased person was, Pastor Julius also joined in with the crowd out of respect. As they wound their way up the path and through tall grass to the freshly dug gravesite, it became evident that this was not a Christian burial; they were in the "pagan section" of the graveyard.

Before the brief burial began, the rough-hewn wooden coffin that had been hastily put together was opened. A small-hinged section opened revealing only the face. Pastor Julius looked in, and to his amazement he recognized the face of a young boy who only last week had attended the Starehe Children's Home church for the first time.

Pastor Julius immediately took charge of the burial, telling all in attendance that this was not a pagan boy. He deserved to be buried in the Christian section! The father proceeded to explain that his church had decreed to bury him with the pagans because he had not fulfilled some church ritual.

Pastor Julius preached a sermon on Jesus' words to suffer the little children to come to Him. It made no difference to Jesus which section of the graveyard this little boy's body was buried in. Jesus loved him. The next Sunday, six of the man's other children were in Starehe Church because he knew that the people there cared for children spiritually as well as physically. The following Sunday, they were there again looking every bit like spiritual sponges— drinking in every word about Jesus.

Dedication: A Trite Custom?

Speaking at a church service, I mentioned that we were preparing to have a service of dedication on our Starehe Children's Home farmland. It is not something normally done in Tanzania. Later, the pastor related the following story out of his own life experience.

While he was a boy, his mother had a garden plot of bananas, *muhogo* (cassava), sweet potatoes and corn. From the sale of these she paid for the children's school fees. She noticed that some of her produce was being stolen, and this cut into her meager profits. An older son offered to watch the garden and catch the thief. "However," he explained to his mother, "just the few minutes when he went to the bathroom or nodded off, someone came and stole the produce."

The mother announced to her family that she was dedicating the garden to the Lord. Whoever the thief was, he would be found stuck in the garden. The father, being a staunch Muslim, derided her by saying that no one has ever heard of dedicating dirt to God. It is only soil and garden, after all, and no God would be concerned with that. He allowed her to pray, but the children were forbidden to join in this silly act.

Some days later the older son had not returned from guarding the garden. So at 3 p.m. the mother, together with all the children, took him some food. When they arrived at the garden, they found the older son with tears streaming from his eyes and his nose also running. He was standing with his hands up, supposedly supporting a large stalk of bananas. The truth was that he was the thief. As he cut the stalk of bananas and put them on his head, they stuck there. He could not remove them!

The mother simply said, "In the name of Jesus, bananas, fall to the ground!" which they did. God still takes dedications seriously.

Police Rough Up the Pastor

We were invited to attend the church celebration of Pastor William's Bible college graduation. What a celebration it was! The church was jammed to the doors and beyond; a festive carnival mood resonated in the building. This, mingled with the very real presence of God, made it an unforgettable meeting.

Four hours of "service" melted away without our realizing how long it had gone on. Brightly dressed choirs of every age group in the church sang their privately written songs. All carried a message of reality in Pastor William's life.

The drama group brought fresh recollection and sober reminders of days when things were not so easy. Seventeen years ago, and very young in the ministry, William began pastoring Mabatini (an area where people have tin roofs, compared to most in those days who only had grass-roofed houses). The expectations that the church would grow were not high; William would, no doubt, only be another in a growing list of "has-been" pastors.

Populated by the working class, the area enjoyed a higher standard of living than most. Along with that came bars and lives addicted to all the vices of the world. Also very present were the power of darkness and witchcraft. This was the factor that had defeated most former pastors. Spiritual warfare is relentless, exhausting and spiritually draining.

From the outset, Pastor William recognized this force and determined to be victorious in that area. The drama group ably depicted his farewell to his wife as he was leaving to minister in another village. He had barely left when the witch doctors broke into their home, trying to put curses on his wife and small child. She cried out to God, and they fled.

Two very husky policewomen, looking very fierce with batons beating the air, were on duty the day William and the first few members of his congregation went to a nearby hill to pray at night.

There was absolutely no mistaking it: these were bad people up to no good, and a thorough beating by the police would disperse them. William's best efforts did not convince them that he was a pastor and simply praying; the policewomen beat them all until they had to flee. Some time later, these two policewomen got converted. What a celebration that was!

Many spiritual struggles also brought many victories. That fact was evidenced by the well over 100 people who crammed into the tiny schoolroom. The tropical sun beat down mercilessly on the tin roof only a few feet above everyone's heads, but that did not deter people. Children sitting on the floor, inches from the speaker, were probably in the coolest place.

Pastor William announced to the congregation that he was believing God for a car. The special speaker that morning (who happened to be me) had the last chance to make an announcement and publicly counseled Pastor William that he needed a church building first. The consent Pastor William gave was genuine and born out of a humble heart. Thus began an exciting adventure with God and church growth that continues to this day.

A lot was purchased, but again with counsel abandoned because it was inaccessible to vehicles. Next the vision was born to begin buying up houses and lots situated around the pastor's house. (A former missionary had purchased the pastor's house about 20 years previously. Missionaries need to be visionaries). That was no small feat considering the church members' meager resources, plus most of the neighbors were antagonistic toward evangelical Christians, not to mention those loud Pentecostals who conducted all-night prayer meetings once a week.

Over the course of four years, miracles were in evidence as the relatively small congregation bought four lots and houses. The houses were demolished to provide space to erect concrete block walls. A donor helped them put on the roof. In all of the work, that was the only donation they ever received from an outsider.

Not long after that, Pastor William felt the Lord leading him to open a branch church. That led to opening another branch church, and then another, and then another. These churches were not opened and left to flounder on their own. Pastor William's church supplied church planters, competent pastors, and finances to help the churches get established. Three of the pastors were supported in their studies through to graduation at the Bible college. Pastor William's church also sponsored him in his studies at the college.

The very latest is that Pastor William's church—now numbering over 300 people—has begun construction of the largest Pentecostal church in Mwanza city. Where is it being built? Not at his church! It is at a branch church pastored by a man whom he supported to attend Bible college. What an unselfish heart this man has. Recently I joked with him that he was getting rich now with so many well-to-do people in his congregation. His reply? "On the contrary! I just sent four of my richest families to a branch church so they could help to establish it."

Oh, yes! Pastor William is still riding a bicycle; some day he will drive that car he thought about getting 10 years ago. *Update: William's congregation purchased a car for him in 2009.*

Robbed of a Leg[2]

Frank Joseph is eight years old. His life in rural Tanzania is filled with all kinds of things that "asphalt kids" can only read and dream about. Together with the responsibilities of looking after the family's cows and goats, there are the wide-open blue skies under which to roam. Hundreds of interesting bugs are waiting to be examined, plus limitless hills and rocks to explore. Wild animals may be lurking in a bush, but that is all part of the adventure. More common are the vast variety of snakes that blend into the colors of the grass. One must always be on the alert for them. The puff adder is sluggish and needs sun, so lying on a narrow path is the best place to soak up the rays. A clear rule of thumb is not to take your eyes off the path.

Soccer games, played in the pasture with goalposts made from tree branches that are bent out of shape, are a pastime for the village boys every afternoon. Playing barefoot with plastic rolled up and tied somewhat into the shape of a ball, they kick up dust as they seriously fight for control of the ball. After the game, a naked dip into the nearby muddy watering hole (also used by cattle) will serve as a bath. The mud may just replace the dust, but that is OK.

This particular afternoon seemed ideal for another soccer game. Young Frank was riding his Chinese bicycle (these are tougher than any bicycle produced anywhere in the world) to the regular meeting place. In his eagerness to meet the other boys, he was not as observant as he should have been. His bike hit a rock, and he tumbled into the grass. Only his leg and ego were bruised, but not enough to stop him from joining in the soccer game with his friends. Sometime during the game, his leg gave out and he couldn't even stand. On the back steel carrier of his bike, he was carried over the jarring, rough, winding footpaths to the nearest town hospital. A crude cast was put on, and he was sent home. But soon the family noticed a distinct putrid smell; Frank's leg was also turning a deathly black colour.

Miles away, there was a mission hospital with a European doctor. Although the journey there would be horrendous, it seemed like the only choice. Back onto the steel carrier of his bicycle he went. His pastor pedaled for dear life under the hot, broiling African sun. Immediately upon Frank's arrival, the doctor took an X-ray of the leg and sadly declared that it needed to be amputated. Gangrene had set in, posing a danger to his life. The pastor had seen all kinds of people with one leg, so he knew what lay ahead for Frank's future. It wasn't encouraging. Africa reserves no special treatment for invalids.

"Please don't cut off the leg just yet" was the pastor's plea. "Give us time to pray." Against all medical training, the logic of experience and reason, the doctor agreed. All night the pastor and Christians interceded for the boy's leg. The next day dawned bright and clear—perfect for soccer—but Frank had probably played his last game. After another X-ray to reconfirm the need for amputation, the doctor came out amazed. The X-ray revealed that the gangrene had reversed and new skin tissue was already beginning to replace it. In medical disbelief, he cut away the old decaying flesh, applied some disinfectant, and wrapped the leg. Within two days Frank was walking.

The town dump was robbed of a leg, and a small boy is happily playing soccer with his friends again because Jesus is still the Healer.

Smoker All Fired Up

Donna, a single mom of three children, desperately wanted to volunteer at Starehe Children's Home. Her past life was not one of total commitment to the Lord. From youth, smoking had been a habit that stayed with her, even after she became a follower of Jesus in mid-life.

"No smoking" is a criterion for volunteers coming to help at Starehe. Now Donna was in a dilemma! The desire to volunteer was so strong that she decided to quit smoking before coming to Starehe Children's Home. Lots of activities, excitement, new culture, and fulfillment in working with the children pushed her cigarette craving into the background.

Coming home to the familiar old routines, she found that smoking had been a big part of it. With renewed determination she continued to remain smokeless; however, she always had a pack ready in case she was totally overwhelmed by the craving.

Driving down the freeway, she could not resist lighting up. No sooner had she taken a few puffs when the song "Breathe"[3] ("This is the air I breathe ...") came over the radio. "That's it, Lord! I hear You speaking," she exclaimed. Opening the car window, she flicked the cigarette out—or so she thought. Suddenly she realized the wind had blown the cigarette back into the car and into the hood of her jacket. She was on fire!

Hastily she pulled over, jumped out of her vehicle and rolled into the ditch. A good Samaritan passerby stopped and smothered the fire. That put a real damper on her cravings, and Donna continues to live without smoking.

Starehe Children's Home has greatly impacted the many lives of our volunteers, some in dramatic fashion!

Last Rites Scare

LouDell had been admitted to Nairobi Hospital for gall bladder surgery. Together with another lady, she was in a semi-private ward. She had been "prepped" for surgery that was scheduled for the following morning. Her doctor was a Seventh Day Adventist. As he made his evening rounds, he inquired how she was doing and explained the upcoming procedure. He then asked if she minded if he prayed for the operation to go well the next day.

After he left, LouDell's neighbor patient, who was also scheduled for an operation the next morning, sat wide-eyed on her bedside. She obviously had some Catholic idea of prayer, and her wide eyes were a result of what she had overheard. With obvious incredulous concern on her face but not really wanting to hear the answer, she asked, "Was he giving you the last rites?"

The connection had suddenly struck this patient like a bolt of lightning. *This lady is having a small operation, and the doctor is giving her the last rites; I am having a similar operation, and my doctor said there was nothing to worry about!* LouDell assured her it was not a last rites ritual. Followers of Jesus receive comfort and assurance in prayer.

Could You Bring Me a Steak Knife?

During the years 1973 to 1993, Tanzania, under President Julius Nyerere, was experimenting with socialism. Most Tanzanians loved the concept as it promoted less involvement in physical labor, which most men, at least in the rural areas, weren't doing much of in the first place. To his credit, *Mwalimu* (meaning "teacher," a highly respected title) Nyerere gave the government 10 years to make it work. Things kept getting worse, but President Nyerere held to his socialism course. As a show of solidarity, he walked from Musoma, his hometown, to Mwanza. He warned the people that they would have to tighten their belts. Some retorted, "We don't have belts; we have already eaten them." Dissenters were allowed freedom of expression as long as it did not get out of hand. In spite of the general dislike of the policy, Mwalimu Nyerere was well liked by the people. He was a common man and did not see it as beneath his dignity to dig in his garden plot along with other villagers. To the best of my knowledge, he never needed bodyguards. On a number of occasions, I saw him up close, and he was always at ease in public.

The idea of everything belonging to the state had its limits, as the majority are dyed-in-the-wool capitalists from birth. It is a crucial ingredient to have if you want to survive in a tough world. When an opportunity presents itself and offers a quick buck, they cannot resist it. It may come dangerously close to bordering on the illegal (their borders can be a bit broad when it comes to this), but a person has to be flexible in these things. At other times an opportunity arises that actually involves taking advantage of another person's misfortune. Business is business, and everyone has to live. Brother stealing from brother or son stealing from father, if looked at closely, was simply a transfer of funds within the family.

The socialism experiment led the country to a place of great lack. Generally, people had little incentive to work. True to his word,

President Nyerere introduced reforms after the 10-year experiment came to an end. During those days of experimenting with socialism, some missions working in Arusha collectively ordered 250 barrels of petrol from Kenya. It arrived on a flatbed semitrailer. This was not good since everyone in the town could see the "hot" item. I received 50 45-gallon barrels and stored them in an iron sheet shed that I had built. To this day, I am amazed that nothing was ever stolen. It was worth a lot of money on the black market.

During this time, I was also building a number of churches in the rural areas and needed steel re-enforcing bars. Along with everything else, they were very hard to find. A friend of a friend knew a friend who worked in the steel rolling mill in Tanga. Through them I managed to order 40 tons. Buying directly from the mill also gave me a great price. No such thing as payment on delivery existed. I had to pay up front. Months and months went by, and I never saw any steel. After a year and a half, I had given up and chalked up my prepayment as yet another bad business experience. Two years later a truck pulled up to our gate unannounced with 40 tons of rebar. Although prices had skyrocketed since my order due to the shortage, it was delivered at the price I had paid. Needless to say, I had to keep it locked up so it would not miraculously grow feet and walk away some dark night while the watchdog was devouring a nice chunk of meat thrown over the fence. Over the course of time, I sold the excess, 40 feet at a time, and made enough money to greatly reduce the cost of the church building. Again, none of that steel was ever stolen.

Arusha is known for its international conferences and its proximity to world-famous animal parks. I was in the dining room of a hotel that had once been "upper class." Get the picture right: it was a hotel dining room, but very little food was available and what there was appeared suspect, as were the waiters.

As I was enjoying my omelette, a very well dressed African man came in. Now there were two of us in the large dining room. By his speech, dress and demeanor it was obvious he was from another African country and well cultured. He ordered steak. When it finally arrived with the usual chips and "vegetable in season" (eggplant was always in season), he requested a steak knife. This was after he had tried the regular table knife and soon realized that this beef was from another era. No trouble chewing your food 30 times here, especially steak. Immediately I saw the two waiters exchange glances as if to ask, "What is a steak knife?" As a side note, waiters in Tanzania were aptly called "waiters" as they were constantly standing close to your table watching you eat. This was especially true when they brought you the bill and you pulled out your wallet; they would come even closer to see what you had in it.

Back to the steak knife ... the more time that passed, the more I anticipated the *maarifa* (ingenuity) these waiters were going to come up with. Telling the customer that they did not possess a steak knife would not even be considered an option. One never admits to defeat, even if defeat is obvious.

As the customer's steak continued to grow cold, the boys finally emerged from the kitchen. With some flair, they presented the patron with a large butcher knife. You could see it in their eyes— "Aha! You thought we didn't have a steak knife."

I marveled at the diner's self-control. His first sight of the butcher knife set him back as it was beyond his wildest imagination of what a steak knife in Tanzania might look like. Then he just gave the waiters a look as if to say, "I don't believe this" and took the "steak" knife. If the truth be known, the man was so utterly taken aback by the chasm between what he knew a steak knife looked like and this offering that his vocal cords were rendered utterly paralyzed. The two waiters retreated with that triumphant gait of having been able to completely satisfy a discerning customer—again.

CHAPTER 7

There Are No Fish In That Vat

The tropics are full of wonderful things to do, and I was always on the lookout to add one more to my experience. Our yard was carved out of a steep hill that sported many granite rocks. Some were rocks, and lots were boulders hidden beneath the soil.

In front of our house, there was a nice rock garden hidden beneath majestic, huge spruce trees. What happened to them is a story for another time. I began to imagine a lovely fishpond right in the midst of the rocks. Leaving the natural rocks protruding out of the ground and cementing all around them would create an ideal environment for the fish. It would be a nice addition to my cage full of lovebirds and twelve turtles. How I came to be in possession of so many turtles is another illustration of African entrepreneurship. Young boys, by sheer coincidence, approached a Swiss lady with a turtle they had found. Turtles in the wild are plentiful, so paying for one would be absurd. The Swiss lady had a special soft spot for animals. Not wanting to see this turtle come to some untimely end at the hands of the boys, she paid them for it. What a find! They kept bringing her turtles. She had no place to keep them, so she brought them to me – thus my huge collection.

Africa is awash with craftsmen of every imagination. I employed a friend to build a rock wall having an approximate diameter of 15 feet. After that he laid wire mesh and plastered the bottom with waterproof cement. The pond looked great! After filling it with water, it was allowed to sit for about a month so that any harsh chemicals from the cement would have leached out by that time. The pond had an overflow drain installed, so it was easy to keep the water circulating.

Another friend was commissioned to bring some fish, which he swore he was easily able to do. I wanted the colorful cichlids, of which there are hundreds of varieties in Lake Victoria. The lake was visible from our house, so safely transporting them would not be a problem. It took this friend about a week to make three deliveries. You see, he was working for another white couple just down the road from us, so he had to time his trips when the couple would be away from home for a while. In our world we would say he was moonlighting. In Africa, he was just being innovative to earn a little extra money.

As he emptied the fish into the pond from the very dirty water in his pail, he reassured me that they were cichlids and they would soon be reproducing quickly. After some months I realized that instead of the fish producing, they were decreasing in number. Among all of the fish were some varieties that liked their fish dinners too. Eventually I drained the pond and got ready to start over.

For this fresh start I was going to be very professional. No more of this *kinyeji* ("local cleverness") business. A meeting was arranged with a member from the fisheries department to get some tilapia fish. The man was very officious and had to come to inspect my pond to make sure it met all government requirements for raising fish. The inspection having passed upon payment of a few shillings, we were in business.

A date was set where I would pick him up (few lower government workers actually had access to transport) to go to the fisheries compound. As breeding stock, I was promised five tilapia fish. These would be caught from one of the fisheries' many holding tanks. The tanks are three feet high and 20 feet in diameter. Scientific studies are carried out on the fish. One thing has to be said for the African people: they have a lot of patience and are very dedicated to the task at hand, particularly if a little "appreciation" money is involved.

Tilapia lie flat and hide in the bottom mud when they feel they are in danger. To make the task of capture easier, the drain spout on the tank was opened. The reduced water level would give us an advantage over the fish. Time is no great issue in Africa, so the people who manufactured the tank installed a drainpipe. Who is going to be in a hurry to drain the tank, right? There may be some mathematicians out there who can calculate the gallons of water in a tank the size mentioned above. Then they could calculate how long it would take to drain all of that water through a ¾ inch pipe by simple gravity flow. All I know is that it took a *long* time.

But who cares when you are leaning over the vat and are with African people? There are never shortages of subjects about which to chat. And who could ask for a better environment as you lean over the tank and watch the water lazily drain out? Naturally, the lower the water level became, the more slowly the water flowed out due to the decrease in pressure. Again, the mathematicians can figure out the flow rate as the water lowers and the pressure decreases. A few hours into the process, another African gentleman came over and joined us as we watched the waters recede. It rather reminded me of Noah and the floodwaters receding. After some time he casually commented, "There are no fish in this tank." What does he know, right? The fisheries man is there, and he knows there *are* fish in that tank. His office records show they have been stocked. Records are also being kept about food intake, growth rate and size.

Three hours later, only a few inches of water remain in the tank, and there is a surprising lack of flapping fish thrashing about. Eventually we managed to catch three fish that were desperately trying to hide in the muddy bottom. Not taking into account some worms and tadpoles, it seemed that was the total of the tank's inhabitants. The fisheries man was perplexed until we asked our visitor friend what his job was around the place. He was a watchman—that cleared up everything. It also explained his prophetic gifting. He was fishing out of those tanks at night, so he knew that our particular tank was a "fished out bad hole." One should always listen to the one who knows!

Itinerating in Brandon, Manitoba

Two Pentecostal churches in Brandon combined forces and did a joint missions weekend beginning on a Thursday evening. A total of six missionary couples participated. Unbelievably, the churches had managed to schedule us from morning to evening every day. During the day we visited seniors homes and schools, participated in youth functions, and did local TV interviews. On alternating evenings, services were conducted at the churches. On the final Sunday morning, each church gathered separately and had a missionary speaker. Then both churches came together for a joint dinner meal. It was a great combined effort of unity and purpose.

Missionaries were spread around so that as many tables as possible had a missionary. Talk around my table was animated, with people enthused about what they had been hearing from missionaries over the weekend. One man was especially excited about the morning service he had attended and how he was so blessed by the message from the missionary. I asked him which church he had attended. Then I asked which missionary had spoken. He couldn't remember the name exactly. Then I asked which country in Africa he was working in. The man replied, "Uganda, I think."

Right there missionary deputation had done its job. The message held the impact, not the messenger. I had been the morning speaker at that church, I did not work in Uganda, and I was not recognized even while sitting face-to-face with the church member. I never let on anything. Why spoil the moment? Besides, it had always been my hope in itinerating that people would be impacted by the message, not the messenger. If they remembered me and were impacted by me personally, that would fade away soon enough. However, if the mission message impacted them, that would be life altering for them.

Drive This Truck to the Police Station

Volunteer Simon was minding his own business on the road when a truck ran a stop sign and hit his vehicle. The on-the-spot police investigation showed that the rather large truck had no brakes. After the preliminary write-up, the traffic police, who seldom have a vehicle at their disposal, got into the truck and directed Simon to follow them to the police station across town. The fact that the truck didn't have working brakes was a small, insignificant matter. The police needed a ride back to the police station.

Police can be very officious, but they can also demonstrate some common sense. Being new in the city, I was not aware of some streets being one-way. Most of these streets are only wide enough for single-lane traffic, so my going against the flow could cause a problem. I realized my mistake too late. Sure enough, just ahead stood a traffic policeman already waving his hand for me to stop. I wondered who laundered their strikingly white uniforms. Very rarely did I see one that was dirty. Politely coming to a stop, I rolled down my window. I knew what my infraction was, but I needed to add some innocence to it. The police officer intoned, "This is a one-way street!" I looked him in the eye and said, "Sir, I am only driving one way. Do you see me driving two ways?" This Western logic was too absurd for him, and he waved me on.

Police check stops on the road were a regular occurrence. With so many public vehicles missing some important piece of equipment like turn signals, working lights, or tires with actual treads, it was constant business for them. Other times the load capacity was 25 passengers, but the minibus was actually carrying 45. The excess passengers had to disembark (no refund, of course) and try to catch another minibus able to accommodate them. While on a bus trip, I was counseled to line up directly behind the passenger standing in front of me so the police would see a single person instead of two.

Our orphanage was 10 kilometers from the city, so I knew most of the policemen—and they certainly knew our vehicles. On some occasions I would feign good neighborliness and simply wave at the police as they waved for me to stop. On the way back into town, they would wave me to a stop again, which I did. Then they would chastise me for not stopping the first time. Good-naturedly I would say, "But I thought you were just waving good morning!" They would chuckle but tell me to stop next time. "OK," I told them. "I will."

Santa Claus Was in Bethlehem

We attended the Christmas pageant at the Kirumba Valley Assemblies of God church in Mwanza. All of this Christmas drama is foreign to Tanzanians. The church is progressive and aware of what is going on in America, so they wanted to make Christmas more realistic to the congregation. Everything went well, with many African improvisations to the story and characters. As the scene with Joseph, Mary and Baby Jesus unfolded, Santa Claus, dressed in his red suit, was right there peering into the crib, beads of sweat dripping from his brow. Tropical heat and Santa Claus suits are not compatible. At least most of the Christmas account was accurate. Sunday school teachers could straighten out the meshing of the biblical account with the traditional trappings.

Theft in Front of New Mwanza Hotel

LouDell was parked in front of New Mwanza Hotel ("New" referred to 25 years ago) and, through the car window, was buying some fresh produce from street boys. Another young boy came to the window and grabbed her wallet. These boys are as quick as lightning. LouDell had a visitor with her in the car. She said to her, "You stay here. I am going after that boy!" With that she jumped out of the car and began chasing the boy, calling, "Thief! Thief!" Other street boys joined the chase and caught the boy. LouDell had her wallet back with nothing missing. A very unusual occurrence in Africa—that is for sure.

LouDell's visitor had only been with us a few days. She didn't have to be encouraged to stay in the car; she sure wasn't about to go anywhere!

What Do You Want to Shoot?

Arusha was a hunter's paradise. Animals were everywhere and only people in the area hunted. During a time of extreme short-ages in the country (around the early '80s), I was in the city trying to obtain permission from the regional commissioner (the highest government official in the city) to get a permit to buy lumber for building churches. I arrived home around 11:30 a.m. and found Ron, an Assemblies of God missionary stationed in Arusha, with a visitor from the USA. He said they wanted to go hunting. They needed a guide, and I knew the areas around Arusha pretty well. As I was trying to find supplies for church building construction, I was really busy and didn't have the time; besides, going hunting at noon is a waste of time. It was nearing lunch, so we invited them in for food. The whole time Ron was badgering me to go hunting. Finally I asked, "What animal does your friend want to hunt especially?"

"Cape buffalo," he said.

Cape buffalo is a full day's hunt, and morning is the best time— not the hot afternoon. Eventually I agreed to go but emphasized that our chances for success were nil to minus for a Cape buffalo. We drove into an area 35 miles from Arusha that was made up of a lot of scrub brush and smaller trees. We saw very few animals and certainly no sign of buffalo. Around 6:30 p.m. (in Tanzania it gets dark at 7 p.m., and there is no twilight period—boom, it is dark!), we decided the day was over and started heading home. At 6:45 p.m. we came around a corner and, sure enough, there were two Cape buffalo. They ran off into the bush a little ways, but one was still visible. Cape buffalo are not to be messed with, and the gun of choice is a 458 Magnum. Ron had a 375 calibre—still acceptable but not ideal.

We got out of the vehicle, and Ron's friend started to aim. Apparently he had not hunted much but was enamored with the romance of big-game hunting in Africa. He was waving the gun around and trying to get the buffalo in the scope sights. Meanwhile, it was literally getting darker by the minute. Finally, he managed to rip off a shot and actually hit the buffalo. However, with his inexperience, he was holding his eye far too close to the scope; when the gun went off, it kicked back and cut his head open just above the eye. Now we had two bleeding things to deal with. The buffalo went a bit deeper into the bush, but we could still pick him up using our vehicle headlights by this time. As we maneuvered the vehicle around, he kept hiding in the shadow of the tree. Eventually, I had to risk my life and went in after it on foot to make the final shot.

The fun was over. Now we had to start gutting and skinning. This buffalo was in the neighborhood of 1,500 lb. Buffalo skin is very thick, and you have to pull with all your might to get an opportunity to keep on separating the hide from the flesh. In the end, your hand becomes numb from the constant exertion. No matter how far you go into the bush in Africa, you will find people. Naturally, people living many kilometers away heard the shots and began trekking toward us. Lights are also highly visible in the rural areas, where no lights beyond kerosene lamps are ever seen. As we were skinning, people began to arrive. They were salivating for the meat—well, to be perfectly honest, their preferences start with the intestines and then move to the lungs, the heart, the stomach, the liver, the kidneys, and last of all, the meat—the *steaki*, as they would say.

We loaded everything into the vehicle, plus a passenger to show us his village about five kilometers away. There we dropped off most of the meat and *all* of their goodies. We got home well after midnight! About a month later, I was in the area again visiting a church. On the way back, a fellow waved me down on the road. He asked if I had gone hunting again because they had just finished the buffalo meat we gave them! My mind began computing ... hot weather, no freezers or refrigerators, one month. How could that meat not have begun to walk on its own after that period of time? However, from experience, I know that if meat is hung in a large tree, it lasts for a good period of time. The local people tolerate well-aged meat much better than we do.

On another occasion, Ron again brought a visitor who wanted to go hunting. It seems that every visitor wants to hunt in Africa. It was the dry season, and animals were scarce. It was also close to the city; therefore, the animals that were still around had been targeted by the military, so they were off into the hills pretty fast. All that was on the plains that day were Thomson's gazelle—referred to as "tommies."

Animals know the difference between the Maasai walking across the plains and a hunter. Maasai warriors are everywhere, it seems. If you stop your vehicle anywhere, within a half-hour a

Maasai warrior is sure to appear. Most of the time they are alone, sauntering across the open plains.

As we were trying to get close to some spooky tommies, sure enough I saw a Maasai coming our way. By that time we had been hunting for a few hours, and things did not look good. Every once in a while, I had a mischievous streak, so I suggested to the visitor that if he was going to get a Tommy, there was only one way to do it. I would call the Maasai over and convince him to change clothes with the visitor. The Maasai would put on the lovely jeans, and the visitor would don the red Maasai wool wraparound—which probably had not seen water in a good long time. Maasai are not known for their need to take regular baths. The deal was done. Behind the vehicle they went and exchanged clothes. I was nearly buckling over with laughter because I knew exactly what would happen within a short period of time. As anticipated, the happy hunter started his trek toward some animals in the distance, all the time trying to walk nonchalantly like a Maasai. He had not gone more than 20 yards when he started to itch and scratch like a baboon. I knew that the combination of wool (having not seen water for months), the odd flea, and certainly the 85-degree temperature on the open African plains would be a recipe for some hilarity at the expense of a US hunter desperate for a hunting story to tell when he got home.

After 20 minutes of sweating and scratching, he decided his preference for jeans outweighed a Tommy story at home. Convincing the Maasai to part with the jeans was a tougher sell.

This Time It's Your Turn

I purchased a type II short wheelbase Land Rover from a Catholic mission station just out of Arusha. Although it had a few problems, it was in generally good working order. Being excellent on fuel consumption was a plus in a country where fuel cost more than steak. Late one afternoon I took our yard worker and went to hunt some meat for our guard dogs. About 10 km out into the bush, the vehicle just died. I tried everything, but no dice—it wouldn't start. I figured it was out of gas. The only thing left to do was to leave the worker with the vehicle while I hiked the seven kilometers back into town. He wasn't too crazy about that because he was very afraid of the Maasai, many of whom were in that area. Coming back with LouDell and some gas, we poured it in and the Land Rover started right up. Problem solved! "OK. See you later."

We could still see her dust when the Land Rover stopped again! Then I knew it was not a fuel problem. My feet were still sore from the last hike, and now another loomed. I checked all the wiring, but to no avail. It was trip time again! This time the house worker had to make the trek. Now he was more scared than ever since he would be out in the open in Maasai warrior country. Remembering the incident still brings a smile to my face.

For some unknown reason, our worker had brought along an ankle-length coat. Darkness was not far away, and there was no way he was going to get caught on the open plains with Maasai all around. He was about 20 yards down the road when I looked back at him walking as fast as he could. His coat was blowing back. He looked exactly like Jesus' disciples walking in old Bible movies.

Their coats were always blowing behind them indicating urgency and speed.

What Did He Use?

Saddam, a young Muslim teenager, frequently came to play soccer with the children at Starehe Children's Home. Once he came to the Sunday meeting but stayed only until the singing was over.

I was unaware of his presence on another Sunday when Pastor Julius asked me to pray for all of our students who would be writing exams the following week. I was also unaware of the fact that Saddam was expected to be in the Muslim mosque where the local Muslim mullah had scheduled a prayer time for the very same purpose.

Three weeks later I heard the full story—well, almost the full story. Saddam's mother was furious with him for not having shown up at the mosque. The Muslim mullah was also not happy, especially when he found out where Saddam had been.

A week later, there was a fiery sequel to the story. All of the children who had been prayed for in our service at Starehe passed their exams—including Saddam. *Not one* of the students who had been at the mosque prayers passed. The mullah was now furious but also intrigued, so he called Saddam and wanted to know what the "pray-er" had used. Obviously, there must have been some witchcraft-like object. Saddam said that only once had he cautiously opened his eyes to look at the person praying, but he did not see anything. "All the 'pray-er' used was the name of Jesus." The Muslim mullah kept on interrogating him, but Saddam stuck to his story: the "pray-er" used only the name of Jesus. He had peeked once, but the man still had his eyes closed and was talking.

Jesus, the Master Planter, has so many ways to reveal Himself to those who are closed to the gospel. Fortunately, the Holy Spirit does not know that terminology. He approaches every heart with an opportunity to become a follower of Jesus.

God Calls Third World People?

I thought that we were the missionaries, and that God called us to share His gospel with the people of Africa. Local people were to be the "receivers."

The Third World is synonymous with poverty, huge debts, and receiving, receiving and receiving without end. A friend of 35 years said that from childhood he had heard and experienced nothing but "giving" from the developed world. As a child in the fruit and vegetable market he was offered tea, cigarettes, coffee, sugar cane—all for free. "Just try it!" In addition, missionaries offered clothes, Bibles, school buildings, church buildings, chairs and free education, both secular and religious. They came giving.

Living in such a country, one becomes stoic about our being on the giving end and their being on the receiving end. Let me set the stage; let your imagination go.

The vast majority of people in Tanzania live hand to mouth. Living conditions in the towns and cities consist of the following: you board (get pushed in by the conductor) a crowded minivan with a carrying capacity of 16—already 30 people have thumbed their noses at the manufacturer's specifications; after a half-hour ride and stops too numerous to count, the bus turns up a narrow, dusty side road. Westerners call it a "side" road, but residents see it as a main road; after all, a vehicle can negotiate it.

It is narrow and made even more so by immobile taxi cars and loitering taxi drivers. Goats at will, are eating anything they fancy, like paper, corn, dried sardines, cloth, or drying rice. Chickens, too, are residents everywhere. Vendors are selling handfuls of charcoal (few can afford to buy bigger quantities), roasted peanuts and corn, dried sardines, live chickens, bracelets and bangles — a virtual Wal-Mart, but all lined up along the dusty road, leaving barely a single lane for anything with wheels. Then there are the masses of people making the road surface invisible because of their sheer numbers, giving the effect of a slow-moving lava stream.

From that parade of every imaginable object, the person turns and jumps over garbage-filled ditches and heads toward his/her dwelling place. That path leads between low mud structures built barely three feet apart which always have an aged rusted iron sheet roof, a curse the Westerners brought. The house becomes a virtual sauna under the hot, relentless African sun. The pathway is never straight or level, nor is it devoid of garbage and people.

Finally reaching your dwelling, you duck through a corrugated iron sheet gate and into a dirt courtyard. It has up to 10 small connected mud dwellings facing the courtyard. Built side by side, your wall also acts as the other side of the neighbor's wall. Your conversation penetrates even into the third dwelling.

The toilet and shower are communal. Cooking areas are communal as well.

Inside your "home" you have a six-foot-wide living/dining room extending 10 feet. The bedroom is six feet by eight feet. It has a small iron sheet window. This houses a family of six and even a few visitors. Oh, did I mention that there is no electricity or running water, and no paint on the walls?

Here is Helena's remarkable story. Her husband owns a very successful corn grinding mill business. He has a house built out of cement blocks sitting on its own lot (only a low percentage of Tanzanians fall into this category). It has seven rooms, each the size of a whole dwelling for an average person. There is electricity and running water. Helena does not have to work by selling roasted corn or charcoal or roasted peanuts by the roadside. Helena is blessed to have a loving and industrious husband. As a very young girl, God called her to work with orphans. Wherever she could, she would help with orphans. Then she heard about Starehe Children's Home, and the Lord just seemed to confirm in her spirit that that was where she should be.

That sounds wonderful, but Starehe was hundreds of kilometers away and at least a four-day drive. Helena has a husband and children. How could she fulfill this "call" of God? She shared her burden with her husband, and he willingly agreed that she should go. He would look after the children and the business.

She arrived at the gate of Starehe Children's Home with nothing but the call of God. No letter of introduction from her pastor; no written story of how she happened to be there. She simply mingled with the dozens of other ladies, the majority of them single moms, who daily stand at the gate desperately hoping for a job to alleviate their miserable financial troubles at home.

We had received new babies, and LouDell was conducting interviews for an extra worker. Out of the dozens of ladies waiting and hoping, she hired Helena. We never heard her story until a year had gone by. She was just a common African mother whom God had called, sent, and placed at Starehe without anyone but Him and Helena knowing the details—serving Him unannounced as it were. What a confirmation that the Lord builds His church in every land with indigenous people!

Feelings of superiority can easily become part of life if you allow the notion that you are the "called one," sent to help, to take root in your mind. Few CEOs are willing to take advice from the fellow on the assembly line. This notion of superiority was forcibly brought home to me when I mentioned to a fellow worker that I had learned a few things from the national people. His haughty retort was "What can these people teach you?" In actual fact, over the years, I learned many things from them.

Musoma airport – the plane leaves – our Africa experience begins

Enjoying the beauty of the tropics – Ronald, LouDell, Rhonda & Stephen Posein

Six months of total immersion to learn Kiswahili – author in foreground

My favorite setting for a church service

Water baptism taking place in a water hole

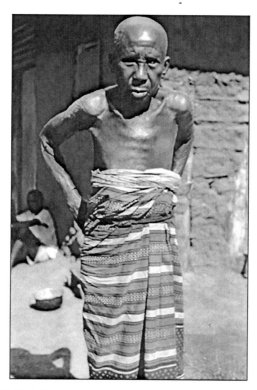

Witch doctor who surrendered her life to Christ

Typical hut of the Iraqw tribe

Carrying water is part of the daily routine

Cooking tea and rice

Everyone in the family has a duty

143

Many varieties of bananas are grown

The termite hill goes deep down into the ground as well

144

The bridge I crossed with the Toyota Land Cruiser

The road sign – turn left down this road

Fishing is a livelihood of many living near Lake Victoria

Bath time

Ploscovia's Twins (front) Margaret & Grace

Plead Guilty? No Way!

Justice in Third World countries is sad as the rich can win any case, be it embezzlement of large amounts of funds, driving an unsafe vehicle, or even murder. It can have a comical side to it at times.

Our school bus driver, Robert, was very conscientious about his responsibilities. He treated the bus like his own and went out of his way to accommodate the children. Every morning he checked the oil, inspected the tires, and kept the insurance and road licenses up to date. He was never late at any school to pick up the children, and he never left until every student was on the bus.

One day as he was coming up the hill to the orphanage, he was following a pickup truck loaded with lumber. It was signalling to turn left (in Tanzania, people drive on the left side of the road). No problem: Robert pulled over to the right to pass him. Just then the pickup driver noticed a road on the right which he decided suited him, and veered to the right. Albeit not a serious accident, Robert collided with him. He called me, so I picked up a police officer from the police station. (Junior police officers don't have access to vehicles.) We drove to the scene of the accident so the police could file a road accident report. By that time, the pickup driver had fellows there who had unloaded the lumber; he didn't have a red flag on the over-length load.

Without question, if you rear-end another vehicle, you are at fault, no matter what the circumstances are. Robert was charged with dangerous driving. There would have to be a court case sometime in the future for the judge to determine what he would have to pay. From that time until the court case, Robert was bombarded with phone calls encouraging him to pay "a little something" so he would not have to appear in court. On a Sunday evening, Robert got a call that he was to appear in court the next day at 8 a.m. Our lawyer and I had been instructing Robert to plead guilty so the fine could be paid and the matter finished.

I went to the farm to do some plowing and returned to the orphanage around 4 p.m. I didn't see Robert, so I asked if anyone had seen him. The workers said he hadn't returned yet. From my experience, this did not sound good. It didn't take long to get to the courthouse to find out where Robert was. The clerk said his case hadn't come up yet, but as it was getting close to quitting time, he no doubt would be coming in soon. Finding out in which court chamber he was to appear, I sat on the bench in the four-foot wide corridor, waiting. Suddenly I heard shuffling and loud talking. To my complete Western amazement, I saw Robert coming down the corridor in prison garb and shackled together with four other men. Not the least bit concerned (remember, he was to plead guilty, right?), I waited outside the open door until all came out again, still shackled, and went off to the holding cell. This was for a traffic violation!

I gave it a few minutes and confidently went to the court clerk, asking how much we owed in fines. She said, "Oh, there is no fine. He did not enter a plea, so he is going to jail until the case is heard again in three weeks' time!"

I said, "*What?* No, he can't go to jail. He is our school bus driver; we need him."

I called the lawyer, who couldn't believe his ears either. Now it was time for some serious negotiations. The problem was, it was 5 p.m., and everyone was going home. We managed to locate the judge who had heard the case and met with him. Negotiations resulted in his declaration that the case had never been heard. The original transcript would go through the shredder. Robert would be brought in again, and everything would go smoothly. The lawyer and I spoke with Robert and again reminded him to plead guilty. When the judge asked him how he pleaded, he wouldn't say a word. After a few tries we convinced him to plead guilty, which he did very hesitantly and in a very low voice.

Later I got the full story. It turned out that the fellow to whom Robert was shackled originally pleaded guilty to not having his vehicle keys in a safe place. His son had taken the vehicle and been in an accident where a boy was killed. *Ten years in jail* was the judgment. After hearing that, there was no way Robert was going to plead guilty to anything. The case was finally over, but then it took us until 10:30 at night to get his clothes and belongings back. Asking for "a little something" continued the whole time.

CHAPTER 8

Where Did Those Parts Come From?

While working in the Arusha Region, we imported a Volkswagen Westfalia Camper directly from Germany. Things in a Third World country are negotiable at times. Going to register the vehicle, I noticed the numbers were getting near to 500, so I thought that would be an easy number to remember. We travelled a lot, and over and over again needed to record the vehicle registration number on various forms.

A vehicle receives an initial registration plate that stays with the vehicle until its dying day. So the fellow said that if I wanted that number, he would call me when the ones before it had been taken. After about two months, that is what happened, and I got plate ARH 500. "AR" represented Arusha and the "H" represented moving through the alphabet as numbers reached 999. Since each city registration carried the first letters of the place, when seeing a vehicle license plate, you could always tell what part of the country it was from. Dar es Salaam, being the capital, was an exception as their plates were designated as "TZ."

East Africa is hilly in places, so I ordered the vehicle with a 2,000-litre engine. Normally they came with a 1,600-litre engine, which was painfully sluggish, especially on hills and when trying to pass another vehicle. On the roads in Africa you need all the power you can get. The Westfalia was a great vehicle as we regularly spent a total of six months each year away from home visiting churches in the rural areas. With the sleeping capabilities, we could visit churches in an area and park in a pastor's yard at night and be self-contained as far as sleeping was concerned. Custom dictated that we ate with the national pastor. We enjoyed the local food and always appreciated the time of the fellowship. However, they never eat until 9 or 10 at night.

Supper consists of a meal of *ugali* (pronounced uuh-gaa-ly), which is *very* heavy on your stomach. Some boiled meat and spinach or beans round out the meal. Ugali is made from finely ground corn flour that is boiled in water until it is stiff. From the pot it is molded into a bread-shaped loaf. It is eaten with the right hand by digging the fingers into the loaf and twisting off a piece. The secret, for those of us who do not have "blacksmith fingers" capable of taking the heat, is to just peel off the outer layer of the loaf since it cools the quickest. The portion taken needs to be just large enough so you can roll it around in the palm of your hand and fashion it into a small ball. Into that, an indentation with the thumb is made. That spoon- shaped piece is then dipped into gravy and hopefully reaches your mouth without dripping or your fingers touching your mouth. This is very important hygienically as up to six people may be eating from the same main dishes. In our culture we say, "No double dipping." Rice and meat are meals prepared for special occasions and visitors.

The roads we travelled were extremely dusty. In fact, in some places the dust lay in potholes. When you drove into them, the dust rolled in front of the tires like very fine flour. Located at the back of the vehicle, this was a bad recipe for the Volkswagen engine. That is where the most dust boiled up from the road. The air filter was outmatched by the fine dust which managed to sift through. In essence, as the fine dust mixed with the engine oil, it became a wonderful grinding paste.

My first engine gave up at exactly 40,000 kilometers. Arusha at that time boasted a wonderful technical college, courtesy of the German government. Having such a facility overseen by expatriate Germans was an unbelievable blessing. Being a teaching college, the automotive part held every tool made by Bosch.

Billowing smoke like an old-time washing machine motor, I knew the end had come. Taking the Volkswagen to the college garage, they stripped the motor down and found that the rings were mere slivers of steel. They were so thin, in fact, that the head teacher sent them to Germany to inquire whether they were using different size rings now. No parts were available, but they listed everything needed and said my best bet was to try the Volkswagen dealer in Nairobi—Cooper Motors. I also had a Toyota Land Cruiser, so thankfully I was not stuck for a vehicle.

On our next trip to Nairobi, I gave the parts department a long list of the parts needed. Telling the parts man that it was a 2,000-litre engine, he simply shook his head and said, "There are no 2,000-litre Volkswagen engines in East Africa—only 1,600s." Once a parts man has told you what he knows for sure, it is very difficult to convince them to take a tiny step toward actually checking it out. Uncharacteristically, he agreed to actually check their stock at the back.

After waiting for a very long time, I knew he had simply humored me by taking the list and disappearing into the cavernous rows and rows of cubbyholes containing parts—from the floor to 15 feet high. The time taken very clearly indicated that he had simply stuffed the paper in his pocket, ducked among the shelving, and disappeared out the back door to the ever-present back alley cooks. Instinctively I knew what was happening: at this very minute he was enjoying a cup of tea, having stirred in five tablespoons of sugar. He didn't have to add milk as the tealeaves were already boiled in it. Drinking a cup or two and chatting with friends would take up just the right amount of time to legitimize his absence. Having by this time spent eight years in Africa, I knew the drill well: listen to the customer, disappear, go drink tea, and come back. With a "straight face" say, "As I thought, we do not have any of those parts."

To say I was surprised does not do justice when expressing my next reaction. I heard a rather loud thump on the four-foot-high rubber-covered parts counter as the parts man plopped down a cardboard box. Dust that had settled on the parts it contained lazily rose out of the box. The first surprise was that all this time he had actually been looking for the parts. Going to the counter, I asked what he had found. All he said was "I don't know where these came from. You see, they are all covered with dust. I have been working here a long time, and I have never seen them. We only stock parts for the 1,600-litre engine. I don't know who ordered them or when, or why, but here they are!" *Every* part listed on the paper was there. It would be interesting to hear the rest of the story about how they actually got there. This is another most improbable and unexplainable miracle that we experienced in Africa.

I Knew It Was A Waste Of Time And Energy

Part of making the gospel message relevant is to present it in a way that draws and excites people. I was visiting churches near the Uganda border in the Kagera Region. It got its name from the fact that the Kagera River runs as its border with Uganda, Rwanda and Burundi. This was also the river that carried the slaughtered of Rwanda like so many logs floating through its rapids at the convergence of the four countries.

During the '70s the Bukoba region was known for its coffee and banana plantations, 87 inches of yearly rain, and its bad roads. Constant rains led to road erosions and washouts that were impossible to repair due to their severity and number, not to mention very few road repair machines. Road challenges meant that it took a long time to get to where you were going and could necessitate staying in the rural area for a week or more, sleeping wherever it was convenient for the local pastors and me. Most times, for me, that was in my vehicle. This presented a lot of inconveniences since I would travel with bags of used clothing, spare tires, a loud-speaker, and extra fuel. Another part of the luggage was cinema equipment. This consisted of a 16mm projector, reels of film, a projector stand, a 1,000-watt generator, electrical cords, fuel for the generator, and a white sheet used as a film screen. Strung up between two poles, this was rather ingenious since the film could be viewed from both sides.

On these excursions my typical day consisted of waking up with many small faces peering into the car window. An adult was always vigilant nearby, and as soon as they noticed movement in the vehicle, they would shoo the children away so I could maneuver around to get dressed. Standing outside by the vehicle door was a small pail filled with hot water for washing that was always welcome. Waiting around at a discreet distance were dozens and dozens of children (and adults a little farther away). It was so interesting to them to see how the white man washes his face. The act of brushing teeth never failed to solicit many curious eyes and snickers.

It was a five-star stay if the area had an outside toilet, known in those parts as the "British long drop." They did not look like anything that we from the West would imagine. More about that later. Most of the time, a bathroom was consistent with one of my first experiences regarding this amenity. Asking a fellow pastor where the toilet was, he looked at me a bit quizzically and eventually waved his arm in an arc, indicating it was wherever I wanted it to be. After that, we did not just casually drive up to a place or park our vehicle but conducted an immediate search for a hideout—a small bush or a dip in the landscape or, if we were really lucky, a stand of banana trees. Now I have really digressed from my story.

After the preliminaries of waking, a hot cup of tea was served with a boiled egg that could have been snuck out from under a brooding hen. Then it was time to load the vehicle, cramming in as many people as space permitted, and then off to hold a service. My schedule consisted of visiting a church in the morning, having a meal, visiting another church in the afternoon, having a meal again, then showing a film at night at a third location.

A film show was a real production that entailed a whole string of procedures. First, a loudspeaker was hooked up to my vehicle's electrical system. I am not—no pun intended—"wired" for understanding the connections between fooling around with the electrical system on the vehicle and what can go wrong. A number of blown fuses and incapacitated gauges attested to that. Eventually, I got pretty good at knowing where the fuse box was and which fuse to check.

Late one afternoon I connected the loudspeaker to the vehicle's electrical system. Excitement was all around as pastors eagerly waited for me to choose four of them to accompany me for the tour through the countryside. Three sat in the back and one in front. To be in the front was the crown jewel of positions, as that pastor would be holding the loudspeaker and doing the narration, inviting people to the cinema show that night. Sometimes they sat on the roof rack, which allowed the pastor to point the speaker in different directions. It was always an enthusiastic event where the pastor was tempted to take the opportunity to hear himself preach. Sometimes I had to remind them that announcing the evening service was the point of our driving around. African evenings are typically very still, and sound travels a long distance. If we found a hill, we knew that hundreds and hundreds of hut-dwelling people heard the announcement.

Being just a few degrees shy of the equator meant that at 7 p.m. it was dark. Before the darkness arrived, everything was set up and ready to go. Within a half-hour there would be hundreds of people sitting on the ground. Loudspeakers had to be in place before anyone arrived. An improvised screen consisted of tying a white bed sheet to flimsy sisal poles. Finding some of those and securing them was a challenge. From there, the distance could be calculated to set up the 16mm projector. The picture needed to cover the full screen. Donated by the Canadian High Commission, my 1,000-watt generator, besides generating power, also generated a fair bit of noise. The point of the evening was that my voice should be heard.

Trying to keep the sound of the generator at a minimum, it would be placed behind a bush on the other side of the vehicle or in a slight depression in the ground. Being that far away necessitated running a long electrical cord. That added another challenge as many people would be walking over them. Finally, everything was prepared, people had arrived, and it was time to get the generator fired up. A local pastor greeted everyone and informed the people what was going to be happening throughout the evening. Now was the time where I started to hold my breath. Let me explain.

My 16mm projector drew just under 1,000 watts of power. The generator produced 1,000 watts if it was firing on all cylinders. That meant it was always under full load and the slightest surge of power usage by the projector choked it. That created a correlating problem. The generator was so hot from running at the constant load that it needed a half-hour to cool off before it would start again. That part of the generator was totally aggravating. Clouds of red dust would be generated as people began milling around. Blips in the film were a common occurrence as bugs were drawn by the light and got into the projection lamp.

Finally, things were back to normal with the film progressing. Then another little interruption: a "homebrew" drunk would wander around and invariably stumble over the cord running from the generator to the projector, and off went the projector. Everything would go black, and people would begin physically wailing on the 'stumbler', together with prolonged verbal abuse.

The films were an hour long and in English, so the audio was translated into the local language through a loudspeaker. Early on

I remember at one showing, I was calling the donkey a zebra. The two words are very similar in Kiswahili. *Punda* is "donkey" and *punda milia* (literally "crying donkey") is "zebra." I don't think anyone minded, although they were probably wondering if there was something wrong with my eyesight.

By the end of the evening, I was wasted. At this particular place, I was going to sleep in the mud church. Sleeping in churches in this area was not too uncomfortable because a very fine grass was used to cover the dirt floor. They didn't have doors or windows, but that was no big deal. Past experience had taught me to first ask an important question: "When was the last time the grass was changed?" Together with all the things that little children do naturally, tiny grass fleas loved the cozy grass if it had been stationary for a few weeks. They loved your tender skin even more than the cozy grass.

I had packed everything up and, with some pastors, was sitting on the grass around the old faithful kerosene lantern (moths also faithfully zooming around the light). I was waiting for the ladies with children to vacate the church so I could roll out my sleeping bag on the grass floor and go to sleep. Slowly the pastors slipped into the dark African night to return to their huts, some of them many miles away. Distance never seemed to be an issue for them. I was left alone.

I began to evaluate what all of my labors could have accomplished. Most probably nothing compared to the amount of energy I had expended. Especially these films. What could people, living way out in this rural area with no phones, no electricity, no vehicles, the nearest town many kilometers away, ever get out of this? I was sure no one was getting anything out of these films except the experience of seeing men walking on a bed sheet. It was a waste of my time and energy, not to mention a nuisance to the drunks who were being beaten upon without their having the faintest idea as to why. My mind was pretty well made up that this would be the last film showing.

I could see that the kerosene lanterns had all disappeared from inside the church. Just as I was getting ready to move, a lady appeared out of the shadows. As she came into the circle of the lantern light, I saw that she was carrying a radio. That is not unusual for Africans. They love listening to a radio.

In a gentle voice she apologetically asked to speak with me. This was probably 10:30 in the dark African night! "You see this radio? I stole it and tonight I was convicted and went home to get it. I want to return it." Who knows how far her hut was from the church? Why couldn't she just have waited until morning? Probably because the Lord had a message for me right that evening. It couldn't wait.

After the service she had followed the path weaving through tall snake-infested grass and banana groves to her house, a number of kilometers away. She had returned this same night, and here was the radio she had stolen a few months ago. This is a very good example and lesson: to act immediately on what the Lord is speaking to us about. Waiting gives time for our corrupted intellect to begin negotiations: should I or shouldn't I? Why can't it wait until tomorrow? The problem arises because we are the questioners as well as the answerers. Most times it is not beneficial for us to begin allowing our "reasoning" process to kick in. It can mask, and in some cases nullify, the voice of God.

A life had been turned around. When I made an evaluation, my time and energy was a waste; but when God evaluated it, it became valuable in some person's life. Very seldom after that did I ever try to evaluate the impact my life or ministry may have had on anyone. I left that to the Lord! Never again did I show a film thinking it might be a waste of time.

I Don't Know How You Made it Here

When ordering our Volkswagen Westfalia Camper, we decided to have it shipped to the port in Tanga. An advantage was that missionaries lived there, so all of the necessary paperwork for clearing could get done as information became available.

Waiting for the arrival of a new vehicle was nearly as exciting as waiting for the birth of a child. Tanzania, during its period of experimenting with socialism, did not allow anyone to possess foreign currency or to order anything from outside the country that required payment in foreign currency. Vehicles required being paid in foreign currency, thereby reducing their meager reserves. The Bank of Tanzania spent foreign currency only on the most pressing items. Literally, the only vehicles being imported were the ones that mission organizations ordered directly from overseas. It also meant that one never saw even one new vehicle in a car show room anywhere in the country. It was like being on sugar withdrawal. Foreign currency reserves were low due to the decline in world prices for coffee, tea and sisal, which were export items for Tanzania. Later, revenues from mining greatly helped that situation.

Finally, we received word that the Polish ship *Wladyslaw Jagiello* was scheduled to dock. On October 27 our Volkswagen

Westfalia Camper arrived. What anticipation as we saw the brand-new tan colored Camper sitting on the dock! Clearance would take a day or so, and we would head back to Arusha, five hours away.

Once the papers were all in order and with gate pass in hand, we got into the Camper, inhaling the wonderful aroma of a new vehicle. Now, to turn on the switch and experience the exhilaration of actually driving it! The engine cranked over with sharp energy, but it would not fire. In the end we had to bring in another vehicle and tow it out of the port. Fortunately, Tanga had a Volkswagen dealership, and we towed it there. The next day we found out that all of the electrical wires from the engine were missing. They were obviously stolen on the ship. Since the dealership did not have any experience with a 2,000-litre engine, they were unable to figure out where to connect wires. To further complicate matters, the 2,000-litre engine had dual carburetors that needed to work in tandem. At the time East Africa brought in only 1,600-litre engines, but they were not very powerful on the road. That is why I ordered a vehicle with a 2,000-litre engine.

We experimented by putting wiring here and there. Finally it fired but was not running well; running less than smoothly, however, it could move under its own steam. We headed for Arusha, jumping and jerking and not being able to get much speed. When the German technician at the Arusha Technical College heard my story and looked at the engine's wiring, he could not believe that the car even ran, let alone that it had made the five-hour trip from Tanga. Wires had not been hooked up properly, and some were totally missing. The mechanics in Tanga couldn't be blamed; they had never seen such an engine.

Then another problem showed up. The gas gauge was not working. Possibly during manufacture, a wire had not been attached. The German technician looked in his Volkswagen manual, did some measuring, and drilled a hole through the body directly above the fuel tank. Seeing a hole being drilled in my new vehicle simply by taking a few measurements was not very reassuring. Sure enough, the wire to the gauge was lying unattached.

Knock, Knock, das Auto

The Volkswagen Westfalia was a tough vehicle considering the challenges of African roads. It was also a comfortable vehicle on trips where roads were tarmac (paved). Sitting up in the higher seats gave you a better picture of what might come running out of the tall grass along the road—zebra, ostrich, a cow herder with his cows, snakes or donkeys. Due to our workload we seldom had the time to do much leisure travel. There weren't many places you could go that were less than a day's travel.

On one occasion we accepted an invitation to visit the missionaries in Eldoret, Kenya. We were looking forward to it as we were going to spend a few days at some very bad road spots watching the East African Safari Rally. Seeing these professional drivers sliding through corners at high speeds and then plowing through water-filled mud holes was exciting. At night their spotlights lit up the road as though it was noon. The roaring of the engines added the element of the feeling of sheer power.

As we were enjoying the trip and looking forward to our visit, we suddenly heard a strange knocking sound coming from the engine area. Stopping to do an inspection, everything from the outside looked OK. The engine had enough oil, and all the wires were attached. Starting the vehicle again, the knocking got louder. Pulling into the nearest service station, we called the missionary, who came and towed us to their place. We had been in Africa long enough not to let this little thing be a setback to our few days of holidays. The car sat idle for the two days as we watched the Safari Rally. Our hope was that possibly things had righted themselves. Hesitantly I started the car. No change—the knocking was still there.

I called my friend, Granger, in Arusha to come and tow us back to Tanzania. It was only about an eight-hour drive, which is not much in Africa. As I was driving the Toyota Land Cruiser and pulling the Volkswagen, I could occasionally feel Granger putting the car into gear instead of allowing it to freewheel. Suddenly he started to blink the lights. Stopping and going back to the car, I asked him what was wrong. He said the motor had quit knocking! Sure enough! We both drove home under our own steam, very happy that another motor job was not needed. Later, a mechanic explained that occasionally a little piece of carbon breaks loose and lodges in the piston, not allowing it to close completely, thus causing the loud knock.

Silent Sentinel

Fellow missionaries lost their baby girl at a few months of age. She was buried in a cemetery in the city we later lived in. It offered me the opportunity to lovingly care for the gravesite for 18 years. Never, without deep reflection, did I go to the gravesite to clean up and plant flowers. These missionaries had sacrificed beyond words.

Inevitably, in the future, they would return to Canada, but that little grave would forever remain in Africa. We received many visitors who were interested in seeing our orphanage. Without telling them where we were going, I would take them to the gravesite and let them feel first-hand what commitment is. Commitment can exact a heavy toll, but it always ends with contentment and few regrets.

Paradox or Miracle?

Wherever we visited churches, people were asking for Bibles. It was far out in the rural area of Arusha Region that Maria asked for a Bible. It would have to be on the next trip, probably a year away. For me, it has been a life discipline to always honor a promise or commitment.

Indelibly etched in my mind is our next trip to her church. We were visiting churches about 150 miles south of Arusha in an area that basically had a main dirt road with many small trails leading off to villages. Word had been sent ahead that we would be visiting Maria's church. In the stillness of rural Africa, a vehicle can be heard a few kilometers away. As we followed the track and came around a bend, we saw hundreds of people off to our right. No sight can compare to seeing a large gathering of African ladies dressed in their colorful "wraparounds," called *kitenge*, singing and dancing. Men usually wear more subdued colors.

The church, being 100 meters off the track, had no roadway entrance for a vehicle. People either walked or a few had bicycles. However, the congregation had slashed a 10-foot-wide roadway through the four-foot tall elephant grass right up to the church. As we slowly drove along the prepared way, people were singing and dancing right alongside our vehicle – and stirring up a lot of dust. We had barely stopped the vehicle and rolled down the windows to shake the many hands extended to us through the open windows when I spotted Maria. Over the greetings and singing, she mouthed whether I had brought "it." I knew exactly what she meant and said *ndio* ("yes"). I can see her now, dancing, laughing and excitedly shouting to her friends, "He brought it! He brought it!" She never learned to read—which is very common for ladies in the rural areas—*but* the Lord had opened her mind and she could read the Bible. We witnessed this on a number of occasions.

Trim the Dead Branches

"You can't teach an old dog new tricks" is a very old saying, but many times it described our experience. As Westerners, we have ingrained habits and ways of doing things that can override anything learned about another culture. I am not referring to having learned cultural characteristics in a classroom from a teacher. I am speaking about having experienced culture first-hand. Many times we sat back and laughed at ourselves for not having learned well. Actually, we had learned, but on the spur of the moment had forgotten to put it into practice. Usually it happened when we were in a hurry and gave instructions on the fly.

As a result of hurried instructions, our yard was basically cleared of old, majestic pine trees. Trying to help a student with his school fees, I called him to delimb three spindly, dying pine trees. The four healthy ones would remain. Later, the delimbed trees would be cut down and sawn into lumber. Absolutely sure that the difference between the healthy and the nearly dead trees could be distinguished, I pointed (which a person should never do when giving instructions) to the diseased trees and instructed him to cut off all of the dead branches. I didn't have time to go and actually hug the trees in question. Coming home two hours later, I could not believe my eyes. All of the healthy trees had been delimbed, with a few branches at the top left as crowns. The dead and diseased trees stood as wonderful sentinels guarding the fish pond beneath them. Thus ended our lovely grove of trees.

CHAPTER 9

Ingenuity

In the West, Third World ingenuity at best would be

labeled "deception." With no new spare parts being allowed into the country, keeping a vehicle maintained was challenging. I say challenging because there were a number that belonged to that family of challenges. First, as I said, there was the lack of spare parts. Secondly, mechanics—and I use the term loosely—were adept at removing from your vehicle a spare part that another person needed and for which they was willing to pay a nice price. A lady I knew imported thousands of British pounds' worth of spare parts to upgrade her Land Rover. "Mechanics" diligently installed them all and—smile, smile—kept her old parts. Sometime later she took her vehicle into this "trustworthy" garage for service, which took a little longer than usual for some reason.

The reason, of course, was that the "mechanics" removed all of the brand-new parts installed a few months earlier and replaced them with her old parts; they got more money for the new than for the old. Another favorite trick was to give you an oil change including filter. Invariably your old filter was removed, nicely washed in gas, and reinstalled. Lesson: mark your old filter. Good profit margin, that. Expatriates know that they regularly pay for their white skin—and they pay a lot.

"Chalk it up to another bad experience" was a favorite saying. Come to terms with it, foreigner—it is part of being a visitor.

Currency exchange is something we got used to very quickly. Every country has its own currency, although it may possibly bear the same name as currency used elsewhere. Tanzania has the shilling; so does Kenya, and so does Uganda. All carry a huge difference in what they are worth, however, compared to the U.S. dollar. Their value is also varied when it comes to trading within their zones. Moneychangers were always at the borders between the countries constantly badgering you to change your U.S. dollars or Tanzanian shillings into Kenyan shillings. Short of the story, never do it. These fellows are masters at sleight of hand. Change with them and you will find a stack of paper in the middle of the bills. All of this is possible because of the fistful of local bills that a U.S. dollar buys.

Thank You for Your Concern

Murphy's law is like the international Doctors Without Borders organization. It is all over the world and shows up with a smile and a hello! Tanzania was in the midst of preparing for a war with Uganda's Idi Amin as well as implementing the experiment with socialism. Both issues birthed a heavy emphasis on security and suspicion.

We were on our way out of Arusha for our regular weekend church visits in the rural area. Suddenly, a banging and clattering came from the engine area of the Toyota Land Cruiser. Making the best of the few kilometers of paved road, we were clipping along at 90 kilometers an hour. In many areas along the roads, the ditches are gently sloped and grassy.

Without hesitation I slammed on the brakes and drove into the ditch to avoid a potential accident on a narrow road. The vehicle having barely come to a halt, I popped the hood latch, bailed out and opened the hood. Within minutes heavy army vehicles and burly, frowning soldiers surrounded us, firing questions staccato style. Seeing them come barreling down the road, I realized that we had just passed the main gate to an army base and were now parked just outside the base fence. No doubt the field glasses had been on us as we came down the road, and our sudden actions spelled some sort of sabotage. Having seen them coming and immediately knowing the reason gave me time to gather my wits and not show any signs of excitement. As they fired questions, asking our names, address, business, and why we had stopped, I calmly answered them. This may have heightened their suspicions. Finally, convinced that we presented no threat, they prepared to leave. As they were getting into their vehicles, I still played the unassuming, naive, regular guy and thanked them for coming to check on us. I was met with very stern words: "What do you mean by that?"

I answered, "You know people these days. They can see that you have vehicle trouble on the road, but they just drive past, not caring at all. You men showed real caring in coming to see if we were all right."

All I heard was a gruff "Uh!" Only Murphy's law could have arranged for our fan belt to explode right at the army base where suspicion was highest. After all, there were 150 other kilometers where it could have happened. The Lord helped us by sending a vehicle that towed us into Arusha. We got a new fan belt and started our safari again.

On another occasion, we were 75 kilometers out of Arusha, heading to visit churches. Our Volkswagen was loaded to the roof with used clothing, books, Bibles, and other literature. The engine began to sputter and miss. It lost all power. Thankfully, there was a wide area where we managed to coax the vehicle. I opened the engine compartment to see if anything obvious, like a loose wire, was visible. It sounded like an electrical issue. I made sure the spark plug wires were firmly connected. Nothing. Once again the Lord helped us. Within a few minutes, a man came along who was a mechanic. He opened the distributor cap and saw that a contact on the points had fallen off. What are the chances of that happening? It never happened before, nor after, in all my years of driving.

You Owe Us for Fuel

Obstructions, deviations (detours), and holdups, along with things being on the road that should not be there, can make for a "hiccup" trip. When a 10-hour safari lies a head of you and the tropical sun is beating through the window, you want to get out and get on with it. Two of the constant road annoyances were the "slow bumps" or "speed bumps" on a nice tarmac road and "slow to 50" signs through dozens and dozens of villages built alongside the road. These two things alone could add a couple of hours to your already torturous trip. Some wise fellow from South Africa made a handsome profit by selling radar guns to the Tanzania Police Force. Then some sharp-eyed individual realized that a Western hair dryer looked exactly like a radar gun. Unless you asked to

actually see the speed the gun had registered, you were most probably making a handsome profit for the owner of the hair dryer. The
dangling electrical cord should have been a tipoff. Most of the time,
police could be detected by the oncoming flashing lights of other
motorists. Other times, if you were going fast enough, you could
pass a police checkpoint before they could mobilize themselves to
scramble out of their chair, set under a shady tree on the side of the
road. Or you could just wave at the officer, thinking he was giving
you a friendly wave. Normally, police on traffic duty didn't have
access to a vehicle, so a chase was improbable.

It was the last gesture that got me into trouble. Two kilometers
past my last friendly wave and to my great astonishment, in my
rearview mirror I saw two Land Rover vehicles quickly gaining on
us, with little blue lights flashing. Pulling us over, they hemmed us
in as though we were huge criminals. OK, time for negotiations.
I had been in the country long enough to know that the first scare
tactic used is that you would have to return with them to the nearest
police station and probably face a court case. Acting as though that
part of the conversation has not been heard, you begin the negotiations (explanations), all else failing, you eventually asking them
for possible solutions to the impasse. This can take some time.
Eventually I paid a fine—cash on the spot—for exceeding the 50
km/h speed limit through the village. Now what? They were following us and stopping us again! "Oh, by the way, you still owe
us money for the fuel we used to chase you." Talk about entrepreneurs! Who thought they could make socialists out of these people?
They were capitalists through and through.

On another long trip, for some distance I had been following
a truck that was slowly lumbering along. I still had a long way

ahead of me. Passing was not safe on the windy road. Eventually, on a straight downhill stretch, there was an opportunity to safely overtake the truck. For reasons known only to someone else, there was a solid line. Sure enough, at the bottom of the hill sat a traffic policeman. OK, time for negotiations again, and hopefully common sense would prevail. This man was the toughest policeman I had met in a long time. He simply would not budge from his position. Trying every angle I could imagine, he still was as cold as steel. Early in the conversation, I had told him that I was a pastor. Many times an apology and a show of contriteness were sufficient to avoid a long delay. Not this time. At the end I said to the officer, "Now, sir, if I was your pastor, would you be trying to detain me so long?"

"OK. Go!" was the curt, unhappy reply.

Who Has the Power?

It happened in the Nshamba area of Bukoba Region where I was visiting churches. Due to the number of churches, seldom could I visit each one more than once every year and half. This meant that the place was always packed with a variety of people. Some were genuinely interested in hearing God's Word, some were curious to see a white man, and the Holy Spirit strictly drew others. It was an afternoon meeting: hot outside and, combined with the smells of Africa, hotter inside. Some people sat on backless benches made of tree slabs set on rocks. Many simply sat on the grass-covered dirt floor. Children, even less than adults, exhibit no sign of shyness. Seldom did I preach when I was not surrounded by children sitting very close to my feet.

In this particular service, I noticed a girl of about 20 years of age enter the grass-roofed church shortly after I had started to speak. She sat at the very back. It did not take long to recognize that she was demon possessed. Facial expressions of hate, plus eyes that are cold, penetrating and evil, are pretty familiar signs. I had seen it all many times, and it did not intimidate me as I preached. At the close of the service, the children were shooed back and people who wanted prayer were invited to come to the front. It was with more than a small bit of astonishment that I saw the girl from the back come forward for prayer. She was the last to come and wound up at the end of the line.

The line of people wanting prayer stretched the width of the 20-foot-wide church. Their needs are varied. Some are sick, some live with a sense of darkness hanging over them, some have family issues; for others, their fields need rain. Women who are barren carry a heavy burden, and God hears their prayer. It is with much jubilation that they show you their baby the next time you visit. Seeing their joy, and knowing what barrenness means in their culture, brought joyful tears to my eyes each time.

As we reached the girl and began to pray, demons manifested themselves and the Lord graciously delivered her. The change in her countenance was remarkable. Her face was radiant and whereas she originally slunk into the church, she left with a spring in her step. The power of God drawing her with His love was stronger than all of the evil spirits that controlled her. They could not keep her from responding to that love.

On another occasion, as we came to pray for an obviously demon-possessed girl, she let out a loud shriek, saying, "Don't touch me!" Jesus touched many people and they were healed. This girl, too, was set free as we laid hands on her and prayed. God is so faithful to His needy people.

Decapitated

Weather in the tropics seems not to have much of a happy medium. It is either dry or wet. Normally, Tanzania experiences short rains in September, October and November and long rains from February to May. Sometimes the short rains are just as heavy and long as the long rains. If we depended on the dry season to travel into the rural areas, there would be six months at best available for work. Rains never deterred me much, but they did result in some interesting times on the rural muddy roads. Drainage is a constant problem, with the roads actually becoming rivers at times. Although the long rains that year were very heavy, I took my trusty little short wheelbase four-wheel-drive Land Rover up into the steep hills of Mount Meru to visit a church. Arusha lies at the base of Mount Meru, which eons ago blew its top. The north side sports a huge crater. Looking at the trajectory of the mountain still intact, it is actually thought to have been higher than Mount Kilimanjaro 35 miles to the North.

Most of the roads leading up into the rainforest and lush fields are nearly impassable during the rainy seasons. Slipping and sliding is the order of day, and with steep drop-offs travelling is not for the faint-hearted. That is why my wife, LouDell, did not accompany me on that particular Sunday. She had been on these roads with me before. Regularly, I assured her that things look worse from the passenger's side. Seldom did it convince her. Home was the place of comfort.

As the church service progressed, so did the rain. Big raindrops hitting the *mabati* (galvanized corrugated tin sheets) seemed to gain strength in sound as it vibrated into the church. The tin roof sheeting is only about two feet above your head. Together with the sound, there are the inevitable drips that find their way through previous nail holes or rusted out areas. For the most part, there were no raindrops; it was a steady downpour. Having a strong voice, I managed to make myself heard as I preached. Toward the end of the service I noticed a man, drenched to the bone, come in and sit at the very back. The timing seemed strange, but many very normal things to them seem strange to us.

Hospitality is a huge part of the African culture, and no amount of rain is going to dampen that party. Knowing the steep and muddy road I had traversed did cause me some stress as the rains continued to pour down. However, I had been up there before and I knew another road that I thought might be more passable. The men chatted while the ladies were cooking the plantains and beans. Eventually, the "latecomer" came in and, after the usual greetings, asked pardon for interrupting. This nationally accepted greeting from a younger person to someone who was older, from one in a lower position to someone in a higher position, or from an ordinary citizen to a more important person stems back to the slave days when the slave was forced to respectfully greet his slave master (usually a rich Arab) with the word, *Shikamoo* "I fall (grab) your feet." The response by the slave master was *Marahaba*—"do it many more times." These days, people continue to use the greeting without any master-slave connotation. It remains a greeting of respect. In fact, the government wanted to ban the use of the greeting. There was such a big uproar in the country, however, that the proposed legislation had to be withdrawn.

The latecomer had a request: would I come by his place and transport a friend to the hospital in Arusha? Now I know that 99.99 per cent of the people do not live near the road. Going off the road to a mud hut meant either going up a hill, which you later had to come down, or go down a hill, which you later had to come up. The thundering downpour hitting the roof was a constant reminder of that. Even after a number of rejections, the man persisted in begging. Finally I said that I would agree, but the sick person had to be brought up to the road. I was leaving in an hour and told him that if no one was on the road when I passed, I would continue on home.

True to my word, within the hour I set off, slipping and sliding with the four-wheel drive fully engaged. Rounding a bend, I saw a group of men carrying the obviously sick man. A short wheelbase Land Rover only has two front seats and a small back cargo area. Since the sick man was in no shape to sit up, they unceremoniously maneuvered him into the back of the Land Rover. When I saw him, I immediately knew the sickness from which he suffered. No matter where you go, local homebrew is readily available. It has a very high and potent alcohol content. An incident was reported in the local newspaper of two women sitting in a room where a 45-gallon barrel of beer was brewing. It blew up under the pressure, and both women were killed. To get the alcohol content consistent to what people are producing in the villages, breweries test the local beer and then produce theirs to stay in that range.

Back to my sick patient ... No doubt he had been drinking and things got out of hand, resulting in a fight. His cheek was slashed open by a *panga* (machete) and his teeth were clearly visible. His head sported another nasty gash. In the normal scheme of things, this would not have necessitated a visit to the hospital; but in his case, the infection in the wound was clear to see. This fight had taken place a number of days earlier! Back on the main road a police report was necessary as I was transporting him over regional borders. That took some time.

As it was Sunday, I also knew that getting any attention at the local hospital was going to be a challenge. There are ways of getting some adrenaline flowing in the veins of the nurses, and I resorted to allowing my imagination to dictate what that might be in this situation. The hospital looked totally deserted. Erase from your mind the picture of a Western hospital. This is a low-slung building, much like we would see saloons portrayed in the Wild West. Once in the spacious entrance, you notice doorways leading off in each direction. Each one has the customary sheet of cloth hanging on it, acting as a door. Obviously nurses had heard me drive up and they hid or—probably more likely—had not awakened, not having heard the vehicle arrive.

I called, *Hodi, hodi!* (the African term meaning "I am here"). After an embarrassing amount of time and short of checking the halls myself, I noticed a head slowly peer around a bed sheet serving as a door, then another. Seeing a white person tweaked their curiosity, and they came out. "I have a decapitated corpse in the back of my Land Rover, and I need your help." My goodness, the place came alive! Nurses emerged from everywhere. Orders were given to get a stretcher, and off to the vehicle we marched. Everyone was in high spirits by now. Reaching the Land Rover, I opened the back tailgate to reveal the live corpse, its head still intact.

"Aw, he's still alive," the deeply disappointed head nurse drawled.

"Oh, yes. Seems I made a mistake, but since you are all here, could you just admit him?"

I have never seen a more disappointed group of nurses. They probably never trusted a white man's word again.

Witch Doctor on the Plains

Africa was much as I had dreamed it would be. Mud huts, wildlife, open plains, and time rolled back many years. Having national pastors who quickly sensed my interest in everything African gave me opportunities to learn and experience. We were heading to a church and driving across the lovely open, dry plains in the Mara Region. The fact that all land belongs to the government has its advantages. No fences exist, and you can drive wherever you like. As we drove close by a yard, Pastor Stephen commented that a witch doctor was at the hut tending to a sick person. How did he know that? "See that pot on the grass roof? See the way the cooking utensils are arranged in the yard? See the person lying on a mat in the shade of the hut?" See that rock? See the small piece of burnt paper? Residents reading signs is different from the way we read signs. They read things instead of words: a piece of paper stuck on a thorn bush, a bent branch, trampled grass, or a faint a bicycle tire track.

As we passed by the hut where the witch doctor was preparing his potion, Pastor Stephen asked if I wanted to speak with the witch doctor. I said that of course I would, and it was agreed that on the way home we would stop by. Within the cultural practice, that is perfectly acceptable. Three hours later, we did just that. It turned out that this particular man had been overseas and studied medicine.

He combined his medical knowledge with local herbal knowledge and added a spiritual component. The treatment price was paid in chickens, a goat, or even a cow if it was a "tough" case. He had no objection when we asked if we could pray for the sick man.

Pastor Stephen was only a child when the first Pentecostal missionaries came to Tanzania. In his early teen years, he committed his life to Christ and immediately began preaching. His sermons were animated, full of stories and practical lessons—oh yes, and long! He was a quick learner and put biblical principles, along with some missionary cooking recipes, into practice. Visiting his family hut for the first time, we were not expecting to be served an omelette. We were also not expecting it to be from an ostrich egg. Ostrich eggs have volume equivalent to 24 chicken eggs, so only one has to be cracked open for an omelette that will serve a number of people. They are made for a threshing crew—that is for sure.

To prevent disease, he practiced the hygiene he had been taught. He boiled the family drinking water. No one in the village would ever think of doing that. For hundreds, or thousands, of years, people had never boiled their water, and some of them lived to be 80 years old—but some only to eight years. Water, especially during the dry season, is scarce, and one cannot be too particular about where it comes from. If a small muddy pond exists, that is a good source. It doesn't matter that the cows, sheep and goats use it to quench their thirst. As those familiar with animals know, they are not limited to drinking from the perimeter. They like to wade in and, if having the urge to relieve themselves, they do not think of getting out of the water to look for the nearest bush. Being that the muddy pond is within walking distance, the household also use it for drinking and bathing. It is called a multi-use facility.

Pastor Stephen was very unique in that he dared to break tribal custom in honoring his wife to follow the biblical pattern. It was something that dramatically stood out in his village. It meant totally breaking a cultural custom that had existed back to a day where no one could remember it. He, a man, took turns in working the family fields. One day he worked while his wife tended their children. The next day she went, and Pastor Stephen looked after the children. The elder village wise men, along with others, ridiculed him, but that made no difference. He saw his wife as equal with him in God's sight, and he was not ashamed to practice this outside the mud walls of the church.

Western values, however, do have some strange boundaries. Something Pastor Stephen did not learn was family planning, Western style. He had 19 children. In the African setting, that was good family planning, as he would have many children to look after him and his wife in their old age. It was their retirement pension.

Traditional responsibilities dictated that the woman fetched the water, carrying it on her head while a baby was on her back. A ten-kilometer walk is not considered far. Other responsibilities included gathering firewood, pounding corn in a pestle, digging the farm fields—again with a baby on her back while cooking food. Men carried the responsibility of protecting the family. Jointly with men from the village, Pastor Stephen was obliged to grab bow, poison arrows, and spear to follow cow thieves. Sometimes this dragged into days and ended in a fight with the raiders. An ideal time for thieving cattle out of their thorn-bush fence close to the hut is during a very heavy downpour when the sound was muffled and tracks were obliterated. The ever-present high-pitched yapping of inbred dogs, curled up somewhere out of the rain, also contributed to the noise.

Another major responsibility of the man is to keep the grass roof in good shape. Typically, a grass roof lasts five years, and then it has to be totally replaced. By that time snakes have found a good home in it and it is saturated with smoke from the cooking fire. Grass that is around three to four feet tall is gathered and tied into bundles. These are brought to the hut site and allowed to sit for a few weeks to completely dry out. Then, beginning at the base of the roof, layers of up to eight inches are laid out. Small tufts are tied to the existing twig latticework, each layer overlapping the other. It is a huge job requiring a fair bit of skill to ensure that the roof will withstand strong winds and heavy rains. Each tribe has a unique way of building a hut and shaping the roof. With the introduction of galvanized corrugated iron sheets, the men were able to shed this regular task. The sheets are expensive, but purchased over a period of time, it is possible to have a good roof. "Good" is a subjective term because it is a good labour-saving item for the men; however, under the hot tropical sun, the low-roofed hut becomes a literal sauna. When the men were not carrying out their responsibilities, their time was spent sitting and chatting in the shade of the largest tree in the area.

Who Is This Terrible Man Who Needs Prayer?

Dr. Ivan was a Christian doctor who worked at the Aga Khan Hospital in Mwanza. Third World hospitals have their own character. They are low, sprawling one-story structures. Walking into the spacious foyer, you are nearly overcome by the overpowering smell of disinfectant with which the floors are washed. Dr. Ivan was very up front about his faith in God. Each night as he finished his shift, he went to every bed and prayed with each patient – what their religion was didn't matter. During lunch break he also held a Bible study, to which he invited me. On a number of occasions I did the prayer rounds with him.

One afternoon he called and asked me to meet him at the home of a mutual Hindu friend and businessman. Apparently, his wife suffered from spells of amnesia, and just the day before had been found wandering the streets. East Indians typically live on the second floor above their business. Going in from the alley, we climbed the steep and winding stairs to their apartment. The first things we noticed were the smell of curry, the small windows, and the dark blue 14-foot walls – they love the dark blue colour. Before prayer we talked at length about who Jesus was, and that the reason we pray to Him is because He is alive. We explained that Jesus loves us and hears our prayers. Then we prayed for his wife.

Having finished praying, the husband began to describe a very bad man who needed prayer as well. This man had fits of rage, was often angry, was divisive, swore a lot, was lazy, and stole frequently. In order to pray more specifically, we asked, "Who is this man?"

I came as near as I ever did to falling off my chair when he pointed to his friend, who was calmly sitting beside me. "That's him right there!" He was as taken aback as I was because he never protested, and we prayed for him. The Lord graciously touched the wife and she was healed of her amnesia. They later moved to Toronto and I lost track of them. I don't know how the other man fared.

Don't Greet Me

Greetings are a high priority in the Tanzanian culture. *Jambo* ("What issue do you have?") is the most common. A friend of mine who had just arrived in the country thought people were talking about a jumbo jet. The plural, *mambo* ("Do you have any issues?") is also used. The reply is *sijambo* ("I don't have any issue") or the plural form, *hatujambo* ("We don't have any issues"). *Habari*, short for *habari gani* ("What is the news?") is also common. The reply to that is *nzuri* ("It is well"). Walking down the street, people greet each other in a friendly manner.

Greetings are not to be rushed either. If people have not seen each other for a time, they are full of questions: "What is the news… of your family; husband/wife; children; chickens; goats; cows; parents; other family members?" An employer may be very upset with a worker, just waiting for him to show up for work the next morning to give him a lecture. No sooner has the worker stepped in the door, and the employer begins to rail on him. On and on he goes. At the end the employee says, *Habari za asubuhi* ("Good morning"). He hasn't absorbed a word that was said because he has not been greeted. By that time, the employer is frustrated and in no mood to repeat himself, so he just says, "Get to work!"

Looking to build a church in Arusha city, we found an empty plot that we felt was in a suitable location. "Plot" is the common term used for a building lot. As usual, it requires a lot of negotiation to determine the price. Once a price has finally been agreed upon, you breathe a sigh of relief, only to be informed that the price now has to be presented to the rest of the extended family for their agreement. "The rest of the family" may be spread all over the country.

As we were in the midst of our negotiations— meeting each day on site for two weeks—I became aware that the man I was dealing with was a witch doctor. They are also referred to as "medicine men." Each morning he had enthusiastically greeted me with the words, *Bwana Asifiwe* ("Praise the Lord"). This is a very common greeting among Christians and is used to such an extent that it has really lost its impact. The day after I discovered his profession, I gave him a strict lecture on why he had no business using the Christian greeting. "Don't greet me with that greeting again" was my instruction to him. From then on he used the generic "Good morning."

It is impolite not to say, "I am fine" when being asked, "How are you?"—just like in our Western culture. However, within the same breath, people follow that up with "But …" Then they go on to tell you what is really going on in their lives—which is not fine.

Most Probably Electrocuted

In our Western world we have grown accustomed to calling a craftsman to look after things that have broken down. Many of those things we can actually tackle ourselves. During the 1980s, there were literally no people to call. Consequently, the broken down washing machine became a "learning on the job" assignment. It was made easier by your wife's threatening to send you to the river to scrub the clothes on rocks. Even dangerous things like electrical problems were at times not out of bounds. Did I mention that all Third World countries use 220-volt power? The cardinal rule is this: if, after repairs, you turned the power on and started seeing sparks or smelling smoke, quickly turn the power off again. Shortage of electrical wire resulted in wires being run, but they were all the same colour. The initial house wiring may have been less confusing because the installer was working from scratch and had a diagram in his head as he moved along.

However, years later, it became a challenge to figure out which wire was hot and which was neutral and which was ground. An electrician was working on a plug in our house, when suddenly his big toe nail began to smoke. He said that is the reason electricians wear flip-flops. Another thing that makes our hair stand on end is that electricians seldom turn the power off while working with live wires.

Wanting to give a Bible college student some help with his school fees, I had him come to Starehe Children's Home to exchange a smaller electrical panel with a much larger one. It was his trade, and he said he could do the job. He was to call me to shut off the power when he was ready to disconnect the wires. Having busied myself with other tasks, I nearly froze with fright when I noticed an hour had passed and he had not called. I was certainly expecting to see a shrivelled corpse as I hurriedly ran to the room. Unbelievably, Stanley had removed the small panel and was already installing the larger one. Like porcupine quills, bare, live wires were poking out everywhere. Rather annoyed that he had taken a chance which could have been deadly, I remarked, "You didn't call to cut the power!"

"I think I am OK," he said! Fear and 220-volt live power don't seem to go together for them.

On another occasion a welder was repairing a steel security door and couldn't seem to get any power to the welder—it wasn't humming. I don't suppose it had anything to do with 100 feet of electrical cord that was kinked and had more bare spots than not. Added to that was his tinkering around at the meter trying to insert bare wires into a connection ahead of the meter. In all of my 35 years in Africa, I do not recall ever seeing an electrician's cords having ends on them. It makes sense as you can use whatever end is available to thread into a socket. Getting frustrated, the man who had carried the heavy, open welder on his shoulders was told to grab the wires and call when he felt a tickle. *Are you kidding me?* Feel a tickle with 220 volts?! Sure enough, as the electrician pushed and jiggled wires at the meter, suddenly the man called that there was power at the other end. Going over to the welding machine, I noticed his bare feet were calloused and had cracks in them like a split watermelon. Maybe they served as insulators—I don't know. Shoes are worn when you go to meet a special person, not necessarily for working in, so he wasn't wearing any. By the looks of his feet, I doubt any shoe would have fit him.

CHAPTER 10

Don't You Remember Me?

Together with all of the other do-it-yourself projects

was one that involved cutting a rather large limb (10 inches in diameter) off a tree. Under normal circumstances that would not cause too great a challenge, except that this one protruded at least 10 feet over the house roof, which was made of asbestos sheeting. Send that crashing onto the roof, and you have a whole lot more DIY (do it yourself) projects. With the long rains on, it would not be a nice scene in the living room.

Kick in Mr. Maarifa (make it work somehow). I reasoned that if I sent Simon up onto the branch—I have never met an African who is not an excellent tree climber—then I could hook it up to my four-wheel- drive Toyota Land Cruiser. The "failsafe" plan was this: as Simon sawed the branch, I would see it begin to break, gun the vehicle, and pull it past the house as it fell. I couldn't see any holes in my reasoning, so we went for it. Simon climbed out on the limb with a saw and fastened a chain around the limb. I hooked the other end onto the front bumper. This way I could observe and would be able to time the breaking of the branch with pulling it hard enough to miss the house. Keeping the tension on the rope was crucial. Tree branches break suddenly and without warning. That did not happen this time.

In my mind's eye seeing the heavy branch crashing through the roof, I began to put pressure on it a bit too early. As I saw it breaking, I really began the big pull. By this time, the branch was arched in the shape of a bow and arrow. As it broke off with some force, the other end naturally sought its original position. Unfortunately, Simon was not anticipating this. The branch acted like a slingshot, shooting him some distance from the tree. There was no swearing, no trembling, and no accusations. I believe he was just happy to still be alive. He went straight to the workshop, gathered his clothes, and without a word walked out of the gate. He never even came back later for his owed wages.

Four years later, coming back from leave in Canada, we stopped in to see some missionary friends. The yard worker was very friendly, but neither LouDell nor I could put him into the slot of people we knew. Finally he said, "Don't you remember me? I am Simon. I used to work for you!" Then he mentioned "the tree." Ah, yes—it all came back, and we had a good laugh remembering the episode of "the white' man's plan" gone wrong.

You Are Such a Kind-hearted Employer

We had a house worker in Bukoba who had previously worked for the Canadian High Commission in Dar es Salaam and who was recommended to us. When we went on safari, he would hot-wire LouDell's Volkswagen Beetle and drive around town. A friend complimented us on our generosity in letting our house worker drive LouDell's vehicle while we were away. That was the end of that!

He was still with us when we were getting ready to go on furlough and packing our things into barrels for safe storage. We knew we were not coming back to work in the Bukoba Region. As we packed, he stole things from just under the top items in the barrels. We had listed everything we packed but did not discover that things were missing until we unpacked after coming from furlough. Those were some of the joys of having workers. There were many more which I won't go into right now.

Only a Taxi, It Seemed

For two weeks Pastor Stephen and I visited all of the churches in the North Mara District. After a few days I was "fed up"! Everywhere we went we got a huge breakfast, food after the service, food at the next place we went, and again after the service. We visited two churches a day. At times we were asked to stop at the pastor's house to pray a blessing over him, his family and home. Having someone bless their home is very important to Africans, probably something we don't take seriously enough in our culture. The practice became very meaningful to me.

A lot of millet is grown in that area, so every place we went, we ate stiffly cooked millet. It is a deep chocolaty colour and is very sticky. Tanzanians call it "medicine to stop you up." They know what they are speaking about—that is for sure. Heading for home, the roof rack of my vehicle was full of 100-kilogram sacks of corn and millet that the people had given to Pastor Stephen for his family. The African people are very generous. At times that generosity made us feel guilty. Out of their meager possessions they insisted on giving us gifts. Live chickens were a favorite gift. A chicken meant a lot to them, so in giving it they were showing much respect. To reject a gift, even in our culture, is very demeaning. Eventually, we found a way around our guilt complex. We graciously accepted the chicken with sincere gratitude and appreciation. Upon coming home, the chicken was kept until the next pastor came to visit. Then we would give him the chicken. We honored the givers and also honored God in blessing others.

When we got home from the church visitations, I spoke with Pastor Stephen and let him know that I would never go with him on such a trip again. Usually it is better to be honest in a kind way rather than to carry a grudge. Everywhere we went, he spoke only his tribal language of Kijaluo. That tribe couldn't care less about others who didn't understand their language. They say that their language is the most important thing, and it doesn't matter if others do not understand it. A very rude practice in our culture for sure, but not in theirs. There are some things in a foreign culture which one never can reconcile.

Only a 35-Kilometre Trip by Foot

I made an emergency trip to the Karagwe area near the Uganda border to deal with a pressing church matter. The round trip would take 12 to14 hours, so I did not want to encounter any pastors on the road waving me down to speak with me about their church issues. Their accounts are always long and detailed, demanding a lot of time. Culturally, coming to the point too quickly is not well accepted. A good discussion can last through half a dozen cups of tea. On this trip I had neither the time nor the patience needed for added in-depth problem solving. My experience that news in Africa travels at the speed of light was no idle, trite saying; it was true. No doubt pastors along the road would get word that the missionary had passed through, and they would be waiting for me at the side of the road on my return trip.

Taking another road home would certainly outsmart them. It was very unusual in Africa to have an option of two roads going to or coming from a place, so this was a splendid option. Alas, my ploy did not work altogether. Suddenly, ahead on the road, I saw a pastor waving madly as I came along the dirt road. My first question was, "How did you know I was coming down this road?"

"I just thought you might return to Bukoba down this road."

Between the road I now was on, and the one I had used that morning, was a distance of 35 kilometers! Distance is of little consequence for people who are used to walking. I had to give this man full credit for having tapped into the Westerner's logic.

Otherwise in Good Condition

CASPAIR, a charter company, was the only way to fly in or out of Bukoba. Everything else was grounded due to the breakup of the East African Community. As a result of mistrust and hot political rhetoric, all the armies in Tanzania, Uganda and Kenya were on full alert. Flying a route between Kenya and Tanzania, while all borders were closed, offered a lucrative business venture. Flights were always fully booked. Once, unfortunately, a plane came too close to an island in Lake Victoria which had a Tanzanian Army garrison. It was shot down. The Tanzania *Daily News* reported that they found the body of the pilot with no arms, no eyes, and one leg missing, but otherwise in good condition. Needless to say, the company did not challenge the Tanzanian skies again.

Learn to Type!

Ephraim served as the general secretary for the Pentecostal Assemblies of God. General Executive meetings were long and covered many issues. Minutes were recorded by hand and had to be typed so they could be run through the Gestetner machine to make copies for all the pastors. In our society as well, people can be elected to a position more because of popularity than for their ability to do the job. Ephraim's job required typing skills which he did not possess. He had always convinced a missionary to do the typing for him. Then along came a young missionary (me) to replace the old one. Ephraim had not informed anyone of the arrangement and the 'old' missionary had not informed the new missionary.

After the next meeting, Ephraim happily came to the office to drop off the minutes. I began to ask questions as to the meaning of it. His common sense dictated that there was no other way since he could not type, and the minutes had to be typed. I was in no position to add another job to my portfolio. Laying a bit of a trap, I asked him who the PAG general secretary was. It was him. "Then you do the work of the secretary" was my logic. If I were going to do his job, I would also hold his title. Since that was not going to happen, he needed to learn how to type. In the end we compromised. I agreed to type these minutes, but the next were up to him. To help him, I gave him a book on learning how to type. He happily agreed to the arrangement and left, thinking that I would buckle under the pressure the next time around again. Three months later, he brought the minutes to be typed again.

An agreement is an agreement, and in spite of his getting very upset, I did not give in. I knew that learning to type would be an asset to him, so I was willing to take the bad press to see this accomplished. Ephraim was very upset and left my office complaining loudly about my Christian character as a brother. For months he did not come back to my office. When he did, he had learned how to type. A number of years later, he acknowledged that my strictness had helped him a lot. He thanked me. Teaching self-reliance is not always a piece of cake, but it is rewarding in the end.

No Passports

Living in Bukoba on the west side of Lake Victoria felt like we were cut off from the rest of the country and the world. And we were. Our telephone did not work, mail took a month or more to arrive, and newspapers took at least a week to 10 days to get there. Airplane flights were erratic, so they could not be counted on. If heading out of the country, Kenya was our usual place to go. We could get there via Uganda; however, the borders were closed due to Tanzania and Uganda being at war. That meant travelling to Mwanza, then over Serengeti Park and into Kenya. Mwanza was a 16-hour hard drive over very rough roads. From there, Nairobi was another 10 hours.

Our children, Rhonda and Stephen, were attending boarding school at Rift Valley Academy, an hour out of Nairobi. Due to the distance, it was impossible for us to see them during midterm break. They spent the break with schoolmates or missionary families. This was a family hardship for sure. The time came for end of semester break, and we were on our way from Bukoba to Nairobi. After a long, tiring day's drive, we stayed overnight in Mwanza at the African Inland Church guesthouse. In my exhaustion, I forgot to lock the vehicle after we arrived. After supper I checked on the car and discovered that a thief had stolen my briefcase, which contained our passports, personal documents, and legal land documents for the mission.

All our documents still showed the Musoma address as we had just moved to Bukoba. Our children were expecting us, so it was impossible to return to Bukoba or take the time to wait for new documents from the Canadian Embassy. Times like these require some grit and taking one step at a time. We drove to Musoma, which was only slightly out of our way. During our time in Musoma, I had built up a trustworthy relationship with the local authorities. Without hesitation, they agreed to help us. All I needed was to have a passport picture taken and present it to them. That done, they issued us a temporary pass into Kenya. We got new passports in Kenya at the Canadian High Commission. All of it was a big hassle but was another example of God helping us in a difficult situation.

As a side note, may I suggest that the next time you are speaking with missionaries, ask them about some difficulties they face or have faced. They may be reluctant to tell you, but you will gain a new appreciation for them. Let me share an example. Friends of ours had completed their home assignment in America and were on their way to Tanzania. All of their things had been packed and shipped. On their way out, they visited a church in another locale. One of their sons stayed with friends. Arrangements were that they would meet a few days later at an airport. Two days later, they received the devastating news that their son had accidentally been electrocuted. They held the funeral and, with hearts breaking, proceeded to the mission field. Can you imagine the agony they endured as their belongings from America arrived? Unpacking their son's personal items was extremely heartbreaking. There are many accounts of tough times for missionaries. Keep them in your prayers.

Confuse a Girl for Me

Lucas was a bright young man who pastored our Bukoba city church. Evangel Publishing House in Nairobi needed a person who was qualified to work in the press. They informed the Tanzania church leadership of the position. Certainly Lucas was a keen learner, and I knew that although he did not currently possess the qualifications, he would quickly learn. We would miss him, but the opportunity to move on into an important role was not to be missed. His fluent English and command of the Kiswahili language would be huge advantages in the publishing house. He was accepted and did not disappoint in his ability to learn the work.

We kept in touch, and sometime later he wrote a letter asking that I pray for him. He was not married and was looking for a girl. This is how he expressed it: "Please pray that the Lord would confuse a girl for me at our upcoming youth rally at Goibei."

"It didn't happen and I am disgusted," he later wrote. The words people used as they wrestled with a foreign language were a refreshing look at ways to express things. Our Western expressions can become very dull and unimaginative at times.

I wonder how many times I gave them a chuckle when using their language.

The Meeting Is Cancelled

Coming home after spending four years on the field launched us into visiting churches. A small church in Southern Alberta had a church name, but certainly did not have a reputation in the community as being a church. A longtime deacon was called of the Lord—I am being facetious—to remind the pastor who the real leader in the church was. In reality, he wanted to be the pastor.

The pastor and I were attending a conference out of town. At the close of it on Saturday morning, I was going back with him to speak at his church. Those were the days when people actually didn't mind coming to church, even on a Saturday night, and were eager to hear the speaker again on Sunday morning. We got home to his church around 5:30 p.m. The pastor decided to call a few people to remind them of the special evening service. He quickly found out that the deacon had been calling people and telling them that the service had been cancelled.

In the small church it wasn't hard to reach everyone and get things back on track with short notice. At the close of the service, the pastor and I stood at the back greeting people and shaking hands. The deacon stood right by the exit door so he could get in the last handshake. Sadly, power struggles have lamed many a church.

This deacon often carried on a loud argument outside the church with the pastor or other leaders. You can imagine that the neighbors' view of the church was not positive. It was no great surprise when, four years later, it was necessary to close the church at that location. It was reopened in another part of the city with new management and did well.

On another occasion, I was scheduled for a service at a rural church. I arrived early and visited my uncle and aunt. After a while, I suppose something seemed strange to them, so they asked why I was there. Nothing had been announced in the church that I was coming. They called the deacon, who said that the superintendent had not informed him of my coming. I said that I had a letter from the superintendent saying that he had informed the church. "Wait a minute," he said. "I have the letter here." There was silence for a number of minutes, and then he said, "Well! He put it into the last sentence of his letter, but I didn't read that far!" Again, it was not difficult to get everyone together and the service went on as my schedule showed. I thoroughly enjoyed itinerating and sharing what was happening on the mission field. I also enjoyed getting to know the individual pastors and learning of their struggles. Being in the same profession produces a strong bond.

Arusha City Is a Big Place

We were back in Africa for our second four-year term and anxious to take up our new posting. Missionaries picked us up from the Nairobi Airport, and two days later we drove with them to Arusha. Our children, Rhonda and Stephen, remained in Kenya attending Rift Valley Academy.

Former missionaries had been renting a house from the Lutheran church up the hill on Mount Meru in a village called Ilboru. When we asked about it, we found out that the previous missionaries had informed them that the house was no longer needed. It had been rented to someone else. The fact that there were absolutely no rental properties available in the city made the news extremely disappointing. Later, when they were asked about it, they said, "We had to find our own housing; let everyone else do the same." Even your own people can be cruel at times.

Our search for a home began by asking people at the Mount Meru Hotel about housing. People just laughed at us. Some said they had been staying in a hotel for two years already. Good luck finding a place to rent! We heard of a house for sale by a Swiss couple, but when we went there, they had just sold it. They were so upset because they had not been able to sell it to a European. It was a beautiful place with lovely gardens. A very rich Ugandan man purchased it, and he continued to keep it up very well.

We stayed for two days looking for housing and then drove to Mwanza. Before leaving, we went to a transport company and arranged with them to meet us in Bukoba, three days' drive away, to pick up our household belongings. They asked where the goods should be taken when arriving back in Arusha. Truthfully, we answered, "We do not know. Just take them to Arusha." The driver responded, "But Arusha is a big place!" We knew that—and that was all we knew.

From Mwanza we drove to Bukoba, met the truck, and packed our household things. LouDell had a Volkswagen Beetle that we had purchased privately. On my many trips out of town, she was left without transportation, so this vehicle was a blessing for her especially.

We arrived back in Arusha—about a week before the truck. By following a lot of contacts we were able to temporarily rent a classroom at the Greek school. They had a lovely compound with nice vegetable gardens and lots of fruit trees. We were literally camping, which added to the urgency of getting a permanent place to rent.

Finally we signed a lease for two months with the Baptist Mission College at Ngarenaro. From the house we had a lovely view of Mount Meru. When the truck with our things from Bukoba arrived, LouDell cried, and I had to be very brave and try to remain positive. The truck body was not airtight, so fine dust covered all of our belongings. Being the proverbial entrepreneurs and trying to make a little money on the side, in Bukoba they had piled burlap sacks of tealeaves on the top of LouDell's car. It was a total mess, with the paint rubbed off in many spots. We had the truck back up to the lawn and offloaded everything outside. Then we took the water hose and washed everything down, including sofa and chair and all the furniture, which should not see water. Miracle of miracles, it all dried nicely and didn't seem to be any the worse for wear. To this day we believe it was a miracle.

After a month our hopes were dashed when we were informed that we had to vacate by December 15. House hunting started all over again. Rhonda and Stephen arrived home from boarding school on December 1. Where we would spend Christmas we did not know. Having been away from us for three months, our children needed some stability in their lives. We count it as a miracle that the house that we were initially trying to rent from the Lutheran Mission became available to us. We were able to move into it almost immediately. The house was extremely small, and we had boxes piled to the ceiling along the hallway. Hallways became tunnels between rooms.

Negotiations began as we received word that the Assemblies of God Mission were going to sell a house of theirs. It was a struggle: naturally, they wanted the best return, and we didn't have the funds. At first our Canadian office flatly refused the house purchase.

However, we were desperate and indicated we would finance a portion of it ourselves. The mission was able to finance the house through a special project fundraiser.

Sitting on an acre lot and facing Mount Meru, it was a beautiful location. An Italian man had built the house for his son, who later decided to move to another country. The mission owned the house for many years, and missionaries who followed us were blessed to have a permanent place to live. After 18 years, I as Field Director, sold the house for 3 times of what we had paid.

"Oh, I'll Just Go Home"

We were having a regional committee meeting at the village of Entasak. It was not unusual for such meetings to go near midnight or beyond. The length of the meeting was in direct relationship with when food was served. There is nothing that people enjoy more than to eat a hearty meal of cooked cornmeal, meat and beans, and then engage in leisurely conversation—but it was supposed to be a committee meeting. As the night wore on, some talked, while some nodded off and joined the debate later, oblivious of what had transpired earlier. All of this was happening by the light of an old kerosene lamp in a mud hut, which soaks up light like a dry sponge soaks up water. After the meeting, which ended well past midnight, I spoke to Pastor Marko and asked him in whose hut he was going to sleep. I had our Volkswagen Camper, so overnight accommodation was no problem for me. But I know that beyond housing for the immediate family, the huts have little room for any visitors.

Pastor Marko answered, "Oh, no. I am going home."

"What?! You live 30 kilometers across country from here, and the way home goes through some forest with wild animals!"

"That is no problem. It won't take me long."

I was totally dumbfounded by his fearlessness and disregard for distance in the dead of night.

Follow the Birds

People love wild honey and regularly collect it in the forest. Bees build their hives in hollows or in holes in the trees. The trick is to find the tree. A method Tanzanians have developed is to search for a certain bird that also loves honey. Then they slowly follow it as it flits through the trees until it gets to the tree with the bees. Very ingenious. In order to get the honey, they build a fire with wet grass at the opening. In that way they smoke the bees out and later collect the honey. This method of collection has also led to a proverb about handling other people's money. It says, "One can not collect honey without some of it sticking to your fingers."

When Is the Next Trip?

Our mission had an Isuzu seven-ton truck which only had three very low-geared gears and no split shift. Around the Arusha city area, seldom could I reach enough speed to get it into third gear (empty). If there was the slightest incline in the road, it powered out. It was actually an excellent truck for a national driver because you knew he could not speed, no matter how hard he tried.

It was a challenge to send the truck, loaded with church building materials, near the Ngorongoro Crater in the Karatu region. During these years everything was in short supply—that included lumber. We became accustomed to the challenge of everyday living. Realization of how we had adapted was brought home when Len and Lorie arrived in Tanzania for their first four-year term. They stayed with us for a few days before I took them to Musoma to attend language school. The drive was a 12-hour affair through the very hot, dry and dusty Serengeti Park.

Before leaving, they needed a few supplies, so I took them uptown. A shop I was familiar with tended to have a few boxes of cookies, some cooking oil, and flour. To make the most of sales, the shop also sold fresh cow's milk. We stepped into the small, hot store, and the smell of sour milk hit us in the face. Len and Lorie began to laugh as they thought I was playing a joke on them, saying I was taking them shopping and going to a place like that. The reality was that it was the best place we had. Leaving a few days later, I had to carry a 45-gallon drum of gas because no fuel was available on the whole round trip. With all of the shaking and rubbing, the drum developed a pinhole leak which we sealed using a bar of soap (a helpful trick I had learned from a national). More than once I used soap to plug a hole in my gas tank when I hit a rock.

In the rural areas, these shortages were multiplied. I would buy lumber in Arusha city and take it to our yard (it was all green stock, with water literally dripping out of it at times). Then we would load the truck until the overload springs were about halfway depressed. On the way to the Karatu area and beyond, there were two especially steep hills. The fellows would climb until the truck powered out in low gear. Quickly, a big block of wood was placed behind the wheel, and they off-loaded half the lumber. From there, in low gear, they continued up to the top of the hill. Off-loading the other half, they went down to pick up the first half load. Back they went to the top of the hill to reload the second half, then off they went to the next steep hill, where they repeated the process. Talk about patience! Only they would have the patience to cheerfully do that.

Flat tires on the truck were a regular occurrence. Changing tires was a huge job as everything was done by hand. A large pipe was used as an extension to break the wheel lug nuts. The jack was similarly large to lift the heavy truck. It was most difficult and time-consuming when the inner dual tire went flat. Then both wheels had to come off. No worries! Every time the fellows came back from a trip, they immediately asked when the next one was planned! Can you imagine?

Mixed-up Plans

In the spirit of renewed co-operation, Tanzania, Kenya and Uganda revived the defunct East African Community. It worked very well again until selfish politics got in the way. All countries closed their borders with each other. Kenya was the country we had the most dealings with and, with the border closed, it created a huge hassle. For a letter of permission to cross the border, we had to apply to the Chief of Police in Dar es Salaam. If granted, this permit was sent to the Regional Police Commander in Arusha. Notification required a physical trip to the police station to check if the application had been granted. That was a painful experience. The officer in charge took a stack of permits, probably six inches thick, and began rifling through them to find your name. First of all, Western names are not familiar, so unless pronounced very slowly, it could be misunderstood. Very frustrating were the times when the "rifler" was distracted by another police officer. He would turn his head and speak to him, all the while still flipping pages of permits. You quickly learned to recognize your name while looking at it upside down.

Most of the time, the letter did not come until two or three days before our requested travel date. We found out that if the reason for travel was stated as being medical, then permits were more readily issued.

There were upwards of 35 missionary children from Arusha, Moshi and area attending Rift Valley Academy in Kenya. I was given the job of getting a border-crossing permit for all of them. Each parent let me know if his or her child was coming to Tanzania on the school bus. Invariably, some who had intended to travel to Kenya to pick up their children had a change of plans and wanted to quickly get their child's name included on the permit. Eventually, I put all of the students' names on the request, no matter what the parents said their plans were.

The Rift Valley Academy (RVA) bus could only come up to the Kenya side of the border, and then the kids had to walk through no-man's land to the Tanzania side. That was about 500 yards. There was a mix-up in communications, and Rhonda and Stephen thought we were coming to RVA to pick them up, so they didn't get on the bus. That fact we did not discover until other children gave us the news. We drove back to Arusha to the police station and, after much negotiating, they gave us an emergency permit to cross the border. Phones were a hopeless cause, but we did manage to reach the children and inform them of our plans to pick them up much later that evening.

We arrived back at the border about midnight and discovered that the kids had left their passports at school in the mix-up! Again, with much talking, Immigration allowed them into Tanzania. Going back out to school a month later, they had to do a lot of explaining as to why they were in the country without a passport. The Lord protected us as well, since if any emergency had occurred, it would have been disastrous for the children not to have their passports.

You Stole My Soda!

For a year we looked after the Tanga churches while resident missionaries were on leave in Canada. Eli worked as yard worker for them and remained to look after the yard and the guard dogs. Grass had to be cut and the house needed to look as though it was occupied. Soda drink (pop) was really hard to buy in those days. Everything in the country was in short supply. In fact, it was the manager of the soda plant to whom you went with your request for one or two cases (24 bottles) of soda. That meant first getting permission from his secretary to get through the door to go to the second floor to stand in a long line waiting in the hall to be called in. To this day I wonder how the manager of a large operation could feel fulfilled in signing requisitions all day for one or two cases of pop.

The first time we went to Tanga, we were overjoyed to find a whole case of soda in the house. Good-naturedly I thought, *Wow, our fellow missionary must have gotten saved to leave us a whole case of soda,* considering how difficult it was to get it.

Eli stretched my patience. We could be in Tanga three days or more and then, just as we were in the car, already in gear, getting ready to leave for Arusha, he would come with all of his issues. In spite of repeated instructions to talk about things before I got into the car to leave for home, his practice never changed. Once I got fed up and very upset. His issues would have to wait until my return next month. A big issue, it turned out, was that we had drunk his soda! Here I thought the missionary was the nice guy.

On another trip, I took some wild meat for the dogs. I didn't have room in the car, so I put it in a small roof carrier that the Volkswagen Westfalia had right in front. With the temperature being 100 degrees plus, the meat started to defrost. When I stopped to fuel up, I suddenly heard people saying, "He must have hit a person!" I never corrected them. Blood was running down the side of the vehicle.

CHAPTER 11

Saved From a Maasai Spear

Our children, Rhonda and Stephen, were coming home for summer break. About five miles from the Namanga border, a Maasai herdsman chased his cattle across the road in front of the vehicle. It was on a rather steep hill, so I could not stop the Volkswagen Kombi in time. The cattle had come out of the bush at a dead run, giving me no chance to avoid them. All I heard was the rumbling sound under the vehicle as I plowed through the herd. For just such a situation, we had a sturdy steel grill installed on our vehicles. I proceeded to the border. As I was completing the paper-work, the Maasai herdsman arrived. He was livid! Had no one had been around, without giving it a second thought, he would surely have put a spear through me. His eyes were cold and steely.

We went back to the scene with the police, who wrote up a report. There were eight sheep and goats lying dead. More Maasai herdsmen were sitting on the side of the road. The police assessed the situation and jotted down notes. Back at the border, the police called the Maasai into his office. He dismissed him, then called me in. The Maasai herdsman was warned that it definitely was his fault for chasing cattle across the road at an unmarked crossing.

Therefore, the policeman still had not made up his mind whether the Maasai would be charged as well as having to pay the repair costs on my vehicle. The Maasai was given this message in order to discourage him from trying to contact me later to demand payment for the dead sheep and goats. The police said, "Ah, they are butchering them right now and looking forward to having a feast tonight. Nothing is lost."

Of all of the tribes, Maasai exhibit the least fear. Fatal animal diseases crop up from time to time. During a severe outbreak of anthrax, cattle owners were instructed to immediately burn any animals that were suspected of dying from the disease. The spore can lie dormant in the soil for up to 10 years if the cow is allowed to simply decompose. It goes without saying that the meat was not to be eaten. However, regularly Maasai would be found feasting on an animal that had died. When reminded of the danger, their reply was that one or two might die, but those were the chances. For them, seeing meat go to waste is like our ignoring a $100 bill lying on the street.

London's Big Ben Tolls

We picked up Rhonda and Stephen from Rift Valley Academy just outside of Nairobi and were on our way home to Canada. We stopped in London, England, for a few days. We were all so excited to be heading home after four years of living in austere conditions as far as amenities went. In London we stayed at the 'House Of Rest', a missionary "rest house." These are well-maintained places offering food and shelter at a very reasonable price.

Arriving early in the day, we wanted to make the best use of our time, so we immediately set off to discover London. The manager of the rest house provided us with some information, and off we went on a tour. Supper would be served at 6 p.m. sharp and would not begin until everyone was present. In the city, we had to sample the famous British style fish and chips served on newspaper. It was a scrumptious meal. Our tour took us past Big Ben, the financial district, Madame Tussauds and the queen's palace. Arriving back at the rest house around 5:30 p.m., jet lag was biting at our heels. We were totally wiped from the flight and our wanderings around London. The time change finally caught up to us as well—London is three hours behind Tanzania.

In spite of stern warnings that we all needed to stay awake, the children and LouDell succumbed. I did too for a few minutes. Then suddenly in my subconscious I heard this big BOOM, BOOM! I dreamt it was Big Ben chiming, but it was actually the huge four-foot brass gong at the bottom of the stairs being struck, indicating time for supper. It took a minute or two for me to regain consciousness and realize what was happening. Quickly rousing everyone was a chore. Finally, we stumbled down the stairs and into the dining room, where everyone was waiting for us! Grace for the meal was never said before all guests were present. Good British exactness, but very embarrassing for us! Rhonda never did do well waking up quickly, so she needed to be pushed along in a daze. Yes, we had fish and chips for supper that night! Rhonda was not in a good mood and protested the whole time that she did not want fish and chips again after having had them for lunch.

The next day we toured London again, going to Buckingham Palace, Westminster Abbey, the financial district, and other touristy places. Taking the underground was a challenge. We were not used to the train doors closing so quickly. Stephen was on crutches, having torn some ligaments in his ankle playing rugby shortly before our trip. To protect it, the doctor had put his leg in a cast. The first time we went to the underground, people behind us were very upset because we took so long to get on the escalator. Part of it was Stephen on crutches, but the other part was that the escalators move very quickly — and they are very steep. From the top you cannot see the bottom. We had not been on one for over four years!

The Falklands War was just on, so when people looked at Stephen I would say, "Falklands." They usually saw the humor in it.

The Tax Man Comes Calling

Many customs and dealings in Tanzania reminded me of biblical accounts. I went to call on a Ugandan national whom I normally hired to do electrical work. On the side he ran a fish export business supplying Dar es Salaam, and even Uganda, with fish. Walking into his small one-roomed office, I saw another man sitting across the desk from him. I excused myself for interrupting but was told not to worry and to come on in. He motioned for me to sit in a chair just a few feet away from his other visitor — you can't get too far away from anything in an office that measures 10 feet square.

After some discussion, my friend opened a desk drawer and handed over a large bundle of money to the man, who then proceeded to leave. Seeing the puzzled look on my face, my friend said, "Oh, that was the tax man." I said, "Wow! That was a large sum of money you gave him!"

He replied, "It was a good deal for me. If I had to pay all the taxes I owe, it would have been a lot more money. Now I am happy and he is too." Within a few years the revenue service tightened up their regulations considerably. All money owing had to be deposited straight into the government's bank account. A few workers had to readjust their lifestyle.

Is There A Difference Between Long And Short?

We were getting ready to build at dorm at Starehe Children's Home in Mwanza. Roof rafters were to be hardwood lumber so the termites would not eat the wood. A method used to protect soft-wood lumber was to paint the wood with a mixture of half diesel fuel and half-used engine oil. That meant ugly black rafters.

I knew a man who had a license to cut lumber out toward Bukoba. He supplied the exact amount needed, plus he delivered it right to our yard. I had our two-yard workers help me to pile it up neatly to give it time to dry. We began with the longest boards at the bottom and worked up to successively shorter ones. Local people are generally not adept at doing such things in order. I taught them and explained exactly how I wanted it done.

When I came back from town an hour later, they were so proud because they had finished the job. When I went to see it, the pile of lumber was totally mixed-up: short, long, shorter, longer. I said, "Don't you see how we started—longer and the next shorter?" Yes, they had seen it but were not able to reproduce it once I was no longer there. The correlation between lengths or even sizes is not something they grow up having to deal with. The girl we had helping us in the house could not arrange measuring spoons according to size. These examples again reminded me how our education is learned by the things we used on a daily basis while growing up. These days, students find it difficult to do math in their heads. Why should you have to use your head when you have calculators? The same goes for the local populace. They never had need for doing certain things. From our side, don't ask us to balance a full pail of water on our heads or split a rock along the fault line.

Stop! Shut Off Your Lights!

While Tanzania was at war with Uganda's Idi Amin, security was a high priority. Security people especially were very nervous. Soldiers with guns were stationed literally everywhere in the country: at banks, schools, government offices, and on bridges. Police road checks, at times every 10 kilometers, were common and made travelling very time-consuming.

We lived in Bukoba, which was only 10 kilometers from the Uganda border. Travelling from Mwanza to Bukoba was no picnic as it was the main road connecting Tanzania and Uganda, plus it entailed a 16-hour drive on very rough and dusty roads. There were no hotels, gas stations or rest stops either. In some places the road went through forested areas and was lightly populated. There was always the risk of holdups by thugs. We had heard of many such incidents. Mostly carried out at night, big logs or boulders laid across the road made avoidance impossible. You had to stop, or if you were agile enough, you could make a quick turnaround, but most times it was too late as bandits yielding AK-47s sprang from the tall roadside grass. Trying to catch these bandits was a futile exercise. It was common for local police to rent out their uniforms and guns to them at night. Talk about entrepreneurs at the expense of the general public's safety and well being!

In spite of our knowledge of extreme dangers, we did find ourselves on the road one dark night. It came very close to costing us our lives. We had promised a high school girl a ride to her boarding school, which was about seven hours from Mwanza. Something held her up and we got away late. Then we missed a ferry, which set us back another two hours. It was well past 10 o'clock at night when we turned into the long school lane. By "long" I mean probably a half-kilometer of narrow road lined with huge trees. Rural schools are easily identified by this common layout. Already being late and still having a tiring trip ahead of us, I was wasting no time heading for the school. Suddenly my headlights picked up a barrier across the narrow road. At the same moment a wide-eyed soldier wielding an AK-47 came awake. Bewildered by the sudden headlights, he had no idea who we were or where we had suddenly come from. The fact that you can see headlights at least a kilometer away is a sure bet he had been sleeping on the job. Immediately, I opened the car door so the interior lights would go on and he could see that we were not dangerous. Continuing to wave the gun at us, he shouted for me to turn off the car lights and for everyone to get out. It took some minutes of calmly explaining what we were doing. Try that with a jittery soldier and an AK-47 pointing at your head. It was only the Lord's protection because I have seen people being shot first and the situation being assessed later. Security stress can precipitate that.

In the Name of Jesus

"You don't touch family" is a Western rule of thumb. In today's
society, though, that thinking may be challenged. In African society
it is also taken for granted, but there are exceptions. Pendaeli pastored a small congregation on the hills of Mount Meru near Arusha.
Receiving ample yearly rainfall, it is a lush area with a moderate
climate. The soil makeup and rainfall there are ideal conditions
for growing coffee. Pendaeli had harvested 50 bags which he kept
in his house, waiting for prices to rise. It is a common occurrence
for pastors to be invited to nearby churches for services. Pastor
Pendaeli's brother and an accomplice took advantage of the situation and went to his hut to steal the bags of coffee beans. His wife
was alone at home and, under a barrage of demands to open the
door, she stood her ground. In a final act of brazen thievery, they
threatened to lock the door from the outside and burn down the
grass-roofed hut. Pendaeli's wife shouted at them, "I have a spear
and in the name of Jesus I am coming to kill you both." She burst
out of the door, and the men fled.

For some time, an action of Pastor Pendaeli's had me baffled. It
is not my nature to immediately make a big deal out of something
when I don't understand it. I like to wait and, over time, assess it
to see if it eventually becomes self-explanatory. After a couple of
years, I concluded that this case had to be resolved by asking Pastor
Pendaeli directly.

In the service during prayer time, he turned around and faced
the wall. The area had a long, strong Lutheran influence and I
thought it might have been something he had become accustomed
to seeing in his childhood. One day I said, "Pastor Pendaeli, I have

a question. Every service during prayer time, I see you turn around and face the wall. Is there some spiritual meaning behind this?"

His eyes widened in surprise at my question. "Brother missionary, what other time is there to check whether my zipper is up?" I suggested he try some other time to do the zip check.

What Was in That Letter?

Tanzania has no door-to-door mail delivery service. That is not hard to understand when you realize that there are few streets, as we know them, on which houses are arranged in an orderly fashion. Houses and huts are scattered helter-skelter on hillsides and valleys. Thievery in the post office is a constant temptation, so adding letter carriers would be a disaster. Four-inch square, numbered, green postboxes are part of every post office. Normally, there are walls of them. They are undercover but accessible day and night. It is a place that is always busy with people coming and going.

Once I spotted a man sitting on the step reading a letter. It took him some time to finish reading. It seemed that he slowly read it over two or three times. With a downcast face he folded the letter precisely and proceeded to slowly, methodically tear it into pieces. It seemed like a sacred moment for him, and I did not have the heart to interrupt his anguish over whatever news he had received. To this day, I still wonder about it. It reminded me of Jesus' mother Mary when Scripture says of her, "But Mary treasured up all these things, pondering them in her heart." (Luke 2:19 ESV). There are times when a moment or experience is too sacred to share with anyone else.

Since every "white person" in town rented a postbox, it stood to reason that coming to the post office to collect mail was a regular necessity. That created a perfect place for street boys, and a few street girls, to gather and beg for money. Usually the small boys would use a high-pitched voice and in their best English would say, "Give me my money." They were to be congratulated for their use of English although sometimes they did get it slightly wrong.

It was not a good idea to help them since there were reputable organizations that offered them assistance. Giving them money at the post office undid what the organizations were trying to accomplish in teaching self-reliance. The main thrust was that they should return home to live with their parents. A young lad was using his high-pitched "Please help me" voice in his pitch for money. Jokingly I said that I was sure he was a girl due to his voice. He argued that he was a boy. I insisted he was a girl. Back and forth the argument went. Finally he dropped his pants to show me the proof. He won the argument—pants down.

Maasai Encounters Lion

The king of the beasts in Africa is actually not totally king. Having no natural enemies, the lion roams at will and at ease. One never sees a lion looking over his shoulder or slinking along rivers for cover. Especially at water holes, animals are ultra nervous: when lowering their heads, their vigilant gaze for predators is impeded. But not a lion. He saunters up to a water hole, crouches down, and continues to lap water until his thirst is quenched, never once nervously lifting his head. The nearest tall grass is protection enough from the hot tropical sun as he flops down to sleep. Lazily he rolls on his back and wiggles to rub his back with never a worry in the world. Being in full view is their preferred lodging.

However, let a Maasai warrior come striding along on the African plain and things change. We were watching a placid pride of lions snoozing under the shade of a lone acacia tree. Occasionally twitching to dislodge the ever-present flies, they weren't even bothered by our presence. Suddenly, in unison, they sat up right on their haunches and focused their attention into the distance. A Maasai warrior was following a path that led directly past the tree. The lions were sleeping, but their nostrils are as sensitive as smoke detectors. Subconsciously, their brain jolted them awake. Muscles tense, they were ready for action. Their bodies, super-charged with adrenaline, bolted—not toward the Maasai, but lickety-split across the plains in the opposite direction.

The Maasai's rite of passage, whenever possible, still centers around killing a lion. Although technically unlawful, it still happens with regularity. Ask any nurse working in a dispensary in far-out rural areas. Warriors come in with horrendous festering wounds as a result of maulings by a lion. Depending on the number of young men—Morons—the group going through the rite of passage to adulthood ceremony may range in age from 14 to 18.

Killing a lion is the crown jewel of achievement. When a lion is spotted, the Morons begin to encircle it and slowly tighten the circle until the lion is face-to-face with the spear wielding, shrieking young men. Nothing remains but for the lion to make a dash to escape. Whoever is in its close range pathway does not escape the extended claws of self-preservation. As the lion makes a leap to clear the circle, the Moron directly in its path plants the butt end of his spear into the ground and attempts to impale the ferocious, fang baring, clawing beast that is using him as a springboard to escape. For the rest of his life this successful Moron carries the badge of outstanding bravery within his whole initiation group.

As a result of this regular practice, lions have, over the years, somehow developed a fear mechanism that triggers when they detect the smell of a Maasai. Hunters have become the hunted.

We Want to See Jane

Jane was an orphan who, at the age of three, was brought to our orphanage by social services. This was not a common practice, but her mother had gone to the Bugando Hospital to give birth. The majority of mothers in Tanzania give birth at home. The possibility exists that this mother may have been sick and simply went to the hospital for treatment, not necessarily to give birth. A few days after giving birth, she passed away. This left baby Jane at the hospital with no parent. Sometimes mothers are so frail they simply cannot remember any details about next of kin when registering at admittance. The mother herself may have been a single mom. Giving a contact address is complicated when living in a rural area. At any rate, no relative came to claim Jane. For three years the only home she knew was the bleak, dark halls and rooms of the children's ward at the hospital.

Never having had contact with a family, Jane was fearful and clung to the legs of the lady caregivers at the orphanage. She grew up being a sweet but shy little girl.

During a devotional time with the children, Pastor Julius asked for prayer requests. Hesitantly Jane raised her hand. "I want God to bring a 100-kilogram bag of wheat flour." Wheat flour is not a common staple in the African diet. It is used for making delicious flat bread. The donations that people generally brought were bags of sugar, rice, beans and fruit—never wheat flour. Tempted to inwardly say, "Oh, sure," Julius nonetheless addressed it as a serious matter, and all the children prayed. Within two days a donation of a 100-kilogram bag of wheat flour arrived. A few weeks later, Jane prayed that a case of baby formula would be donated. She specifically prayed for Lactogen 2 (there was also a Lactogen 1). Again, it arrived: Lactogen 2, exactly the formula that Jane had prayed for.

It was my habit to tell people about such miracles demonstrating God's faithfulness to the prayers of children. On this occasion I shared it with two bank tellers at Stanbic Bank. A month later the tellers suggested that the bank make a donation to Starehe Children's Home. Management accepted their suggestion, and the donation of sugar, rice, beans, and cooking oil was brought by 15 of the bank employees. In the course of visiting, the two bank tellers asked me if they could speak privately with Jane.

This was not our accepted practice, but they continued to insist on speaking with her. Sensing there was something more than just an interest in Jane, I asked them why they wanted to see her alone. With some hesitation they said they wanted Jane to lay hands on them and pray for them so they, too, would receive the gift of receiving anything they asked for. A good thing can awaken a wrong motive. No wonder we are cautioned to seek after righteousness. It reminded me of the biblical story of the young girl who was being used for profit and who was set free by Paul's prayer (Acts 16:16-18).

You Mean We Are in Charge?

Our churches multiplied at such a rate that our heads were barely above water when it came to training sufficient pastors and leaders. That resulted in missionaries having to bridge the gap. Until sufficient local pastors were trained, we were the district and regional superintendents, the treasurers, and the chairmen of committee meetings. In one region I was the superintendent and church treasurer of 55 churches and had only one other ordained pastor working alongside me. In every church I visited, I served communion and performed child dedication. Water baptismal services were plentiful. It was also expected that I perform a wedding, if needed. In order to be ready for any situation, I carried my proverbial doctor's medicine bag wherever I went.

This "authority" of the missionary had its shortcomings. National pastors did not become familiar with what happened beyond their sphere of pastoring. They also did not understand the long-range plan, which was that every one of them was to become a future leader.

I was visiting churches within the Rwanda refugee camp near the Rwanda border. The Tutsi tribe had been hunted in their own country by the Hutu tribe. As a consequence, many Tutsi fled to Tanzania. Watutsi, as they are referred to locally, have the distinction of being among the tallest people in the world. They are very quick learners. During the time of the Belgian rule in Rwanda, Watutsi were the favored tribe and held most clerical, administrative, financial and education related positions. Indeed, they are a tribe that is much closer to Western culture than African culture. They readily admit that they prefer to hang out with Westerners rather than other African tribes. Their wit and quick reaction to a Western joke are unusual in Africa.

Not just anyone could get into the refugee camps. I had to have a letter of permission from the Commissioner of Refugees in Dar es Salaam to enter the area. This may be misleading as the camp was actually thousands of acres in size. Rwandans are very industrious and good farmers. They have to be since the country of Rwanda is small, with a high population density. Good rains at the refugee settlement enabled them to grow large fields of corn and beans. It really was an oasis of food. Banana plantations began to spring up around them as well.

Having the letter of permission in hand still did not give me total freedom within the camp. Although trusted to be on good behavior while visiting churches, I did have to return each night and spend it with the government-appointed commandant. These were simple, bush type accommodations. Sleeping on the cement floor at times was uncomfortable, especially when my air mattress sprang a leak. The one naked, dim light bulb in a room devoid of furniture did not add anything to the ambience of the bare cement room.

Getting to the camp was not a picnic either. The roads were bad and purposely kept that way to discourage much travel into the sur-rounding area. After a 14-hour bone-jarring trip, one had no desire to go sightseeing. One place in the road was a particular challenge. For some bureaucratic reason, the road was built through a swamp connecting two lakes. It worked in the dry season, but when the long rains were on, both lakes rose and became one. That meant a 200-yard drive through water. Fortunately, enough trucks had been stuck so, over time, big rocks were dumped to form a solid base. Helping with the navigation was a high spot near the centre that one could aim for. I had some experience in traversing rivers and had learned that just enough speed is needed to keep the vehicle moving. If the speed is excessive, the water splashes up on the motor and gets the spark plugs wet—you are then "high and dry" except that you are wet and in water. In two areas, the vehicle dipped into spots so low that water literally rolled up over the hood. Our vehicle was a four-wheel drive with a high hood.

Having completed one long service, we made our way to another church. Arriving there, we found that food had been pre-pared for us. That was welcome as we had not eaten since early morning. Sitting on low stools outside a hut, the ladies brought us water to wash our hands. They used long, thin gourds measuring about two and half feet. The top is cut off and seeds are removed. In essence they become water pails. As they poured the water over our hands, I detected a distinct smell of home brew. When they took their leave, I motioned to my hands and looked quizzically over at the head pastor of the area. In a low voice he mentioned that this pastor was known for producing the best home brew in the area.

My next question was obvious, to me at least. I asked why the church leadership had not taken any action. He said, "You are the boss, and until you discover something, we have nothing to do with it." The next day a lengthy discussion with pastors in the area took place. It was explained to them that they were citizens of the country and this was their church. They were responsible. My role was a temporary one and one day I would leave. What kind of church would they have if discipline hinged on a mis-sionary who was 14 hours away? From that distance how could he detect a problem and work out a disciplinary plan? If they wanted a bunch of pastors who were brewing beer, among other unsavory things, then so be it—it was their church. When I received a copy of the minutes of their next pastors meeting, I knew they had gotten the message. Seven pastors had been put under discipline pending their repentance and observance that they had mended their ways in accordance with what is expected of a spiritual leader. It was a painful experience, but one that brought positive growth.

"Oh, You Only Want 15 Kilos?"

Rascals exist in every country. Tanzania is no exception. At times they come out of the woodwork in greater numbers when Westerners show up. For some reason the Western "sellers of ice" in the frozen North are lined up, cash in hand, at every igloo— in this case a grass-roofed hut. When a national relates a story, Western common sense is hypnotized, the mind goes numb, the eyes glaze over, and a coma results. Without even a request, the Westerners liberally empty their pockets and, in some cases, their bank accounts. Snake oil seems to come to mind from the deep recesses of my memory. However, once you have been stung 50 times, you begin to acquire the knack of reading unwritten signs.

The 110-acre farm of Starehe Children's Home was near an agricultural research station. All kinds of wonderful new seeds and plant varieties for corn, rice, cassava, mango, papaya, oranges and limes were produced there. To promote better yields, local farmers were given these seeds and plants free of charge. Ground corn is a staple food, and each year we planted acres of it. Early one morning I visited the station. That is what I loved about Tanzanians: they were at work by 7 a.m. or earlier. This applied to govern-ment offices and businesses. There were no corn seeds left, and it was suggested that I return just after harvest. Because we ran an orphanage, we could get all the seed we needed. Returning there a few months later, I got the proverbial runaround. The person in charge was "not in" (literal meaning without lying); he was not in the room, but he was on the premises. "Come back tomorrow" is always the way out of today. Wanting to get the seed, I returned two days later, early in the day. Fortunately, the man in charge was in his office. Now the seed man's long story, which brought a muffled chuckle to my funny bone, began.

There had been very low rainfall in the Bukoba area (a 16-hour drive west), and the government had ordered that all of the corn seed on hand had to be shipped, as quickly as possible, to that area. Planting needed to begin immediately in order for the harvest to be ready within four months and stave off starvation. Therefore, absolutely no corn seed was available. "Not even one kilo?" I inquired. You see, if I was able to get even one kilo, we could plant it and then selectively build up our own seed stock.

A visiting agriculturalist taught me how to select productive seed. The corncob is broken off at the top and bottom, leaving only the middle section. The circumference area that has 13 kernels is kept. As the cob gets narrower, the kernels around it are fewer in number as well. Only seed from the productive area is kept. In that way a good seed is guaranteed. I think the Bible talks about reaping what you sow.

If kernels from the area of the cob that has few kernels in the circumference are planted, a corncob that has few kernels in the circumference is produced. I thought that all of the kernels would inherit the genes of a big cob, but not so; each kernel reproduces its own kind. This is a good example of why children cannot inherit their parents' religion. On the bright side, it is also the reason why children need not inherit the bad traits of their forefathers.

What made me chuckle was the fact that for two years I had lived in the Bukoba area, where up to 58 inches of rain falls each year. "Lush" is the word that best describes the whole area. Second fact: people do not plant corn; they have banana plantations. Thirdly, the Wahaya tribe detests the taste of cooked cornmeal. They exclusively eat cooked bananas and beans—every day! As a side note, plantains are very acidic, resulting in the people's stomachs actually being eaten through like a sieve. A doctor explained this when our house worker passed away with a stomach ailment. He said it was a very common cause of death.

The man at the research station had no idea that I knew he was spinning a tall tale. Just in passing— naturally, not on purpose—I mentioned that I was willing to pay cash for 15 kilos of corn seed. No receipt was needed. Then the man said, "Oh, let's go see the storekeeper. I am sure we should be able to find 15 kilos." Of course they did, and they made a little money on the side. Do you think that any farmer ever got seed or a plant free of charge? These workers had big families to feed, they had medical bills, their relatives had medical emergencies to care for, they needed to buy a plot of land on which to build a house, or the house needed new iron sheets, the second wife (about whom the first wife knows nothing) always needs a new dress, there are school fees—and the list goes on, just as it does here at home. In essence, selling the seed was like working a second job, except you worked it at the same place during the same hours as your first job. The designation "poor, uneducated African" is purely a figment of the Western mind's imagination. It all depends on which standard is used when speaking of education.

Some Died Of Natural Causes, And Lions Got Others

Where the projects began I have no idea; however, spread all over the country, there were dozens of government ranches. These ran into hundreds and hundreds of acres. The intention was that herds of cattle would multiply and bring money into government coffers. As seed stock, each ranch was given an average of 100 head of cattle. Cattle should not be trusted to be homebodies— they forget where they belong. This required hiring farmhands and their families. After 15 years some forward-thinking person in government decided to count their wealth. This required many cattle auditors.

Auditors enjoy a good steak as much as anyone else. The old adage, "You can't take honey out of a bees' nest without some sticking to your fingers," had not just materialized out of thin air. It had some truth connected to it. In this case, *steaki*, as it is locally known, was substituted for the honey. As testimony to the length of time it took to complete the task, it was obvious the auditors loved their job. The final tally showed that most ranches had fewer than 10 animals left. How could that be? Well, some died of natural causes, some were eaten by hyenas, some by lions, some were stolen by cattle rustlers, and some had wandered into tsetse fly country and succumbed to sleeping sickness.

Truth be known, some were also herded off into the auditor's herd to renew the gene pool. Truth two be known, the hired hand, his family and friends regularly enjoyed good meals. I am sure that for a family to eat meat once or twice a month was not part of their regular routine. Truth three be known, some cattle had been hidden in the far reaches of the bush out of the auditor's sight. Ah, the involvement of government in business—always a hugely profitable enterprise for someone.

CHAPTER 12

You Will Not Sleep With Me Tonight!

Stepping out of her mud hut near Mwanza, Pastor

Patroba's daughter took the poison arrow meant for her father. A person can become an enemy for any number of reasons. No matter how trivial the transgression or perceived transgression, the revenge meted out frequently results in death. The lack of sufficient police to carry out a thorough investigation in the rural area contributes to the boldness of taking this route. Then there is always the distinct possibility that witchcraft was involved. Someone has not been feeling well or his crops have failed or cows have died or his wife cannot bear children or the wife attends church against the husband's wishes or .... anything. A person goes to the witch doctor, and he or she is advised, after paying a chicken, a goat or even a cow, that such and such a person is responsible. By whatever means, this person must be taken out of the picture. A poisoned meal or soda pop, a poison arrow, a stabbing, or a spearing will do the job. If police are informed that it was a recommended witch doctor remedy, they back off investigating the case as they do not want to be next on the list.

Pastor Patroba was a successful pastor when I requested the church's national executive to transfer him from Mwanza to Arusha. We were planting a city church, and he had a gift for pioneer work. I do not know any details, but when we returned from furlough, local pastors shared that Pastor Patroba had mentioned that he had some damning information about me. He stated that when he went to the government with it, I would be arrested and deported with "Persona Non Grata" stamped in bold red lettering on my passport. Such a stamp forbids you ever to enter the country again. I do not presume to know why, but a few months later Pastor Patroba was hit head-on by a bus as he rode his motorcycle.

His brother, who was visiting from out of town, had received the news. Not knowing anything about the early morning accident, we went to visit Patroba at his home. Patroba's wife was not home, but we found his brother, who was in shock and mourning. He asked us to share the sad news with her when she returned. What a sad time for this wife and her children.

Increasingly, funerals in Africa are complicated affairs. Before tribes began to intermarry and move out of their customary tribal areas, it was simple. A person died, and before the day was out, he or she was buried. There are no mandatory birth or death registrations. Each tribe has its own custom related to death. In the case of the Wajaluo tribe, of which Patroba was a member, their customs are filled with some gruesome rituals. Years ago, when a husband died, his body was placed in a small grain storage container woven with reeds. This container was placed just outside the hut door. Every morning for a month, the wife had to open the lid and look at his decomposing body. Doing this in the city is illegal.

Wajaluo are adamant that a member of the tribe, no matter where he may have been living, must be buried in their homestead. Some tribes bury somewhere on the property, some bury at the front door, and some inside the hut. Patroba died a 16-hour journey from his paternal home. A private taxi or rented vehicle is the only way to transport a body. These carriers do not possess an ounce of mercy; they have "ice in their veins," as we say. They smell money and charge huge sums for the use of their vehicles. In order to accommodate relatives and friends, a number of vehicles or a bus must make the trip.

Collecting these cash funds to return a body home takes time. Friends, relatives and tribal members are literally forced to come with a donation. It is an obligation they are aware of. Pastor Patroba's body was badly smashed and lay in a cooler at the local morgue. There are morgues the way we know them, and then there are morgues that simply copy a Western term, but nothing else. Washing the body and injecting it with formaldehyde were the responsibility of a friend or relative. Preparing a coffin ahead of time is a bad omen, so upon death a coffin is "made to measure." The body is measured, and a carpenter proceeds to build a coffin to fit.

It took a week to secure all of the funds to transport Patroba's body to the home village. The journey to the place of burial began the next day. I had started the trip a few days earlier in my own vehicle. In preparation for the trip, a friend or moonlighting hospital staff member once more injects the body with formaldehyde. The coffin is brought and the body, wrapped in a linen shroud, is placed in the coffin and the lid is nailed down. Loaded onto the back of a Land Rover truck, it stops at the hospital exit gate. An

open-air service is held, which gives the local Christians an opportunity to bid farewell. From there, as many people as humanly possible cram into the back of the truck with the coffin. A 16-hour trip
over rough roads and oppressive heat awaits them.

Few things happen quickly in Africa. Custom dictates that the
widow spend the night in the same mud hut room as the coffin.
Following a breakfast of cups of hot runny porridge, negotiations
begin. If the widow decides to remain in the area of her husband's
clan, the body is buried in a certain location. If she decides not
to remain there, another part of the family compound becomes
the burial place. A woman does not have much authority, so the
elders have pretty much decided what should happen. Patroba's
wife, having spent some years removed from the strict tribal customs, had a mind of her own. She has decided to return to Arusha
city. Around 10:30 a.m., they begin to dig the grave and gather
sisal poles to erect a shelter for visitors. Mercilessly the tropical
sun beats down on everyone. Sheep and goats await their fate as
food for everyone present is a must. In fact, mourning may last a
month, and everyone needs to be fed for that period of time. How
can people who are relatively poor mooch off their people, who are
also poor? Why is there no understanding of each other's plight?
Chalk that up to another custom Westerners cannot comprehend.

As Tanzania field director, I represented The Pentecostal
Assemblies of Canada mission. I was a guest of honor. But this
came with a price. Member of the tribe or not, I had been requested
to donate a hefty sum toward the transportation of the body. Guests
of honor are seated closest to the coffin. With whiffs of air coming
my way, irrespective of all the formaldehyde which had been used,
I could detect the pungent odor of the decaying body. To put it

mildly, I was shocked when they announced that the coffin would be opened for viewing. This was a week after death. The corpse had been in a cooler and spent 16 hours in a hot Land Rover, then overnight in a mud hut. It was now well into late afternoon of the next hot day. Standing respectfully back in the fifth row of mourners, I thought I would be safe this far away. Immediately the coffin was opened and, even with two people standing at the head of the coffin spraying air freshener, my stomach urgently signaled that I was still way too close. I hurriedly exited out into the fresh air.

Male friends in the Wajaluo tribe have a further custom. Two make a pact that upon the death of one, the other inherits his wife. As a cleansing ritual, he must sleep with her on the night of the funeral. During her speech at the funeral, Patroba's wife went against all the tribal custom. Referring to this pact, she pointedly addressed Patroba's friend and said, "A man here thinks that I am going home with him tonight. Tonight you are going home with a glass of water and nothing else." She returned to their home in Arusha and lived as a widow. Tribal customs are powerful and many times take precedence, but she demonstrated what it meant to be truly committed to Christ.

A Numb Brain

Sometimes it is not until you look back that you realize how ridiculous something was that you did or attempted to do. For a number of days, I had been visiting churches far out in the rural area. The road network was such that I could basically make a loop in the area and not have to backtrack to get home. After the last Sunday service I was ready to get home, have a shower, say goodbye to sleeping in the car, and breathe air that was not filled with dust. Very seldom does one travel alone. This is not necessarily by choice; rather, there are always people needing to go where you are going. About two hours into my homeward safari, we encountered a dry riverbed about 15 feet deep. Not even a four-wheel-drive vehicle had a hope of negotiating the steep sandy banks. Here was a tempting sight: four logs had been laid across the 20-foot span—only the logs (nothing between them). It was decision time. Should I backtrack the eight hours to get home or should I attempt the delicate "wheels aligned on the logs" circus act? From the other side of the riverbed, it would only take four hours to get home. Two strong forces combined to help me make the ridiculous decision: my tiredness and my unwillingness to back away from a challenge. Contrary to African thinking (that everything is possible), even my passengers thought I was crazy.

To demonstrate their unwavering optimism I am reminded of a friend who told me that he had purchased a cow in the rural area. A slight problem was that there were no roads into the area, and a river needed to be crossed. The cow happened to be on the other side. Larger rivers are infested with crocodiles, and a floundering cow would certainly be a welcome attraction for them. I was curious and wondered what ingenuity he would attempt to employ to get that cow over the river. I asked him how he intended to do it. Without missing a beat he said, "Oh, I will bring it across the river on a dugout canoe." Now that is something I would have loved to witness—a typical example of not being afraid to tackle any situation.

Back to my bridge. My plan was that the passengers would go across and take a bead on a log and on my tires. From my higher vantage point, I would have no idea where the tires actually were. My normally brave passengers didn't even dare walk across the span; they hugged the logs and crawled across on all fours. Mr. Murphy and his law must have been away on another assignment because I managed to snail crawl across and was met by wild cheers on the other side. Looking back, my brain must have been numb to tempt fate in that crossing. Had I slipped off a log, it would really have required some ingenuity to get to the other side. My eight-hour detour would most probably have turned into an eight-day vehicle recovery affair. I hadn't allowed my mind to think of failure or any possible consequences that might result. Like many decisions we make in life, we fail to project what the consequences may be.

Don't Ever Think Of Having Children!

Many times I remarked to my wife, LouDell, that we will never be confronted with so much humor at home as we are in Africa. There was constantly something humorous happening. One contributing factor is that Tanzanians have a good sense of humor and readily and good-naturedly laugh at themselves and others. Humor, I found, opened doors to people in that it showed our human side (we were white, after all, and seemingly of a higher class—if they only knew!). As a consequence, they relaxed.

In times of severe gasoline shortages, you bought petrol whenever and wherever it was available. This was especially true if you were travelling beyond your hometown city limits. Government-regulated prices allowed little room for a profit margin. Consequently, the "private pocket" needed it more than any repairs or maintenance the service station may have required.

No smoking at a gas station is pretty well a universal and international common-sense rule. To my horror, I noticed that the attendant was smoking as he began to pump gas. To top that off, a tiny pinhole was merrily spraying gas out of the nozzle which obviously should have been replaced a long time ago. Rather annoyed and definitely fearful for my life and my vehicle, I demanded that he stop smoking. He kept his head turned away and maintained that it was safe, except that when the petrol overflowed and began to spill out of the vehicle, he jerked his head back to see what was happening. By that time, and in serious prayer mode, I had retreated as far away as possible.

Tanzania is approximately the size of the province of British Columbia but has roughly the same population as all of Canada. This translates into people being wherever you are and wherever you look. You can be sure they are watching your every move. What else is there to do? Don't even begin to entertain the thought of "farm freedom," where you can zip around a barn corner and relieve yourself. Employment opportunities are scarce, so idly sitting around is a favorite pastime, if not the only one. Besides, "white" people do funny things, so let's watch them.

As the attendant came to request payment, I remarked that he should never think of having children. This ignited curious questions from the "sitters." Having children in Africa is as normal and expected as our "apple pie without cheese is like a hug without a squeeze" rationale. Even the attendant was overcome with curiosity. By this time everyone's attention was focused on my reasoning.

Sometimes a short lapse in time immeasurably increases the impact of a statement. In a good-humored tone of voice I said, "Because one stupid (in the local language, *stupid* carries a softer meaning than in English) person like you in the world is enough. Don't think of passing on those genes to any children." That resulted in uproarious laughter and opened a barrel of barbs toward the attendant. He took it in good humor and had a few good comebacks himself. There are a few expressions, like *donkey* and *black*, which a national can get away with saying to someone else, but which are totally out of the question for a Westerner to say. Good humor always has to be respectful to the culture of the people, never degrading or humiliating. It also underlines the fact that just because somebody has said something, it does not give you license to say the same thing.

There Is The Commander Of The Army, And Then There Is The Army Commander

A friend held a prominent position in the Tanzania military. He was an accomplished fighter pilot and even deployed to China to help them with their army communication systems. During the war with Uganda's Idi Amin, this man was brought back to command an army unit. His superior instructed him to take his troops over the Kagera Bridge, which joins Tanzania and Uganda. From his previous military experience he knew that this was a bad tactical move. He was so sure, in fact, that he refused to follow orders.

He was summarily dismissed and another commander given the same instructions. As the Tanzanian troops neared the end of the bridge, it was blown up by Ugandan soldiers and everyone died. His failure to act as instructed cost him his job, but in this case saved his life. He said that it was God who had warned him of danger.

God does that in the physical as well as the spiritual realm. It's important to know who our real Commander is.

I Hear "Mekiko, Mekiko"

Sospeter was a lay pastor working with the Swedish Pentecostal Church. When I got to know him, he was helping missionary Aimo to build churches and perform general carpentry. He was a hard worker and a keen student, learning all he could. In time he established his own carpentry company.

When the Swedish Mission released Aimo to be our point man working with the Rwandan refugees, Sospeter became his right-hand man. He helped erect buildings at our outreach station near the refugee camps, put up tents, built school desks, and purchased materials. Sospeter did not know how to drive and learned that as well in his time there. He spent a lot of time basically being in charge of the outreach camp.

The Finnish government became very involved in supporting the work with the refugees, especially in the schooling aspect. When the Rwandan refugees returned to Rwanda four years later, Sospeter transferred over to help the refugees coming out of Congo. Over the years we became good, trusted friends.

Returning from the refugee settlement, he came to visit. Actually, he had a burning question on his mind. While sleeping at the refugee camp, in a dream he had heard a voice saying "Mekiko, Mekiko." *What does that mean?* he wondered. He felt confused. To me it was clear that he had heard the words "Mexico, Mexico."

Our human minds begin to arrange the pieces and put the puzzle together from a picture we can relate to or a scene we have seen. My mind did a quick calculation. This man is married; he has five school- aged children; he has never heard a word of Spanish, let alone knowing how to speak it; his English consists of "Good morning, teacha"; he lives in a poor Third World country—what are the possibilities of his having access to sufficient funds to go anywhere? In addition, he has no contact or connection in Mexico.

I could not have been blamed if, from a purely intellectual and experiential perspective, I had said, "You know what? Forget the voice. I think you may have gone to bed on a full stomach of bad food and had a dream. The possibilities of your going to Mexico are below nil."

However, I have had some previous experiences of seeing God not being bound by less than a perfect resumé or bag full of tools. God is the One who ensures success. He doesn't need all the pieces of the puzzle to be in place before He can move. In fact, He creates pieces to fit His canvas.

I sincerely encouraged Sospeter, gave him some simple ABCs about what to do, and promised to pray that God's speaking would become a clear direction. We prayed together to that end. I was excited to see what would happen. I mean, not a single thing in Sospeter's circumstances pointed to a sliver of possibility that this could ever happen. As we say, the odds were heavily stacked against him. God would have to do some major reshuffling on Sospeter's behalf.

Within two years he was in Mexico, and six months later he moved there with his wife for a two-year assignment. When God speaks, there is no doubt about it.

Let's Have a Coffee

I went moose hunting with a friend in the gas hills of Whitecourt in Northern Alberta, Canada. We left early and could not have wished for a more perfect day. In early morning the snow fell and blanketed everything. This ensured that tracks were fresh! Pulling his eight-wheel Argo machine with his pickup, my friend led the way. I followed in my old pickup with a nice Honda quad loaded on the back.

Roads were a bit of a concern as we got into the steep hills. Among these gas sites the hills are narrow with steep drop-offs and not geared to unsuspecting public travel. The temperature began to change and got to the place where the four inches of snow turned to ice beneath our wheels. With a load pushing you from behind, the secret to keeping your vehicle on the road is to go just fast enough so the load behind doesn't end up in front of you.

On a steep decline, my friend had exactly that happen. At the bottom of the hill, the trailer fishtailed back and forth, and finally gravity took over. His trailer separated from his truck and, together with the eight-wheeled Argo, rolled down a steep embankment. Cautiously I came to a stop behind him, and we got out to survey the damage. It was a *long* way down. We both stared at the Argo.

It seemed like a redundant question, but finally I asked, "What are we going to do?"

He calmly looked at me and replied, "First, let's have a cup of coffee."

Coffee was the furthest thing from my mind right then. However, that decision to let things settle down before acting served me well many times in Africa. Step back; let emotions regain an even

239

keel; take time to survey the situation. During that breathing room, things tend to come into clearer focus. This is a good principle to use in our lives.

Answer the Pay Phone!

A pastor in the U.S. shared an example of God's obvious intervention. He and his family were going on holidays and, for whatever reason, they left home later than anticipated. About 100 kilometers down the interstate, the children spotted a McDonald's sign. You know the rest of that story.

They exited the interstate and parked the car about a block away from the McDonald's. As the family walked past a pay phone, it rang. One has the same reaction to that as a "WET PAINT" sign, right? Hardly hesitating, the pastor entered the booth, picked up the receiver, and said "Hello." The lady on the other end said, "Pastor, I need to talk to you. I am in great distress." That cannot be explained away as a coincidence. It can be explained by reading God's Word, which unmistakably states that God is sovereign and He loves us. He has ways of crossing telephone lines to arrive at His purposes.

Burned Rice Tastes Terrible

Every year at our Bible college graduation, we enjoyed a wonderful menu of African food. It consisted of *pilau*, goat meat fried until it was hard, gravy and salad. African tea accompanies every meal. Every year the rice was burned, and the food was served an hour late.

The first year, by four in the afternoon, I was so hungry that I didn't notice the burned rice. The second year, I actually tasted the burned rice and decided something had to be done.

Before graduation I sat down with the cooks and tried to reprogram their system. In order for the food to be ready on time, they had to figure out priorities and work backwards. That was a novel idea to them. Starting point: what time should the food be ready?

How long does it take to cook three huge pots of rice? "Three hours unless we forget to add firewood every half-hour."

How long does it take to clean 20 lb. of rice (separating all the gravel out of it and start cooking it)? "Two hours." One need not be too meticulous as crunching a few grains of gravel is normal.

How long does it take to deep fry the goat meat? OK, now you know what time to butcher "Billy" and get him cooking in the pot.

How long does it take to make tea for 100 people?

Although it is not in their DNA to be exact, they did get it pretty close to right, and I was forever a happy camper come food time at graduation. In our graduation of life, we need to work backwards as well. Where do we want to end up in life? Having established that, we can work toward it. More important, where do we want to end up after our lives are over?

Rwandan Genocide Prisoners

Large numbers of Rwandan Tutsi tribal people fled to western Tanzania in the late '60s. In the ensuing years, many became Christians, and some of those became pastors. As the genocide of 1996 came to a close, they were offered the opportunity to return to Rwanda. Their former homes and farms were given back to them.

This was a huge undertaking after 40 years, considering who might now be living on those properties and where would those people go.

These were my pastors in the Bukoba region and we had become good friends. I was invited to go to Rwanda for a weekend and minister to their congregations there. As I was driving with a pastor friend of mine, I noticed lots of prisoners working out in the fields. Their garb made them stand out. They were unmistakable in their washed-out pink Kansu (gown) type prison garb. I did not see any gun-toting guards with them. Asking the pastor about that, he explained that these were all Christian prisoners. That really aroused my curiosity. What difference did that make? He explained that they didn't need any guards since the prison guards knew they would all return after their workday. Even in the difficult times, trustworthiness and faithfulness had become a hallmark. Many of these people had surrendered their lives to Christ while in prison. When that happened, they confessed their part in the genocide. They confessed, "I killed 25 people"; "I killed 30"; "I killed 50." When chided that they were simply saying that in the hopes of receiving a lighter sentence, they would take the officials to the spots where they had buried the people they had killed.

Gruesome, but what a transformation!

Freshly Ordained and Humbled

The cheapest mode of travel in Tanzania is by train. It also boasts the least number of breakdowns in the travel industry. It really isn't an industry, though; it is more like buying into a traveler's lottery scheme. You may make it to your destination or you may not. Refunds for not reaching your destination are not in the mix, nor is food supplied in the event of a week's breakdown in the wilderness.

The Pentecostal Assemblies of God Tanzania had just held their biennial conference in Dodoma. Travelling back to the city of Mwanza in their newly pressed suits and ties were five young men who had just been ordained. This means the denomination has verified that they have successfully met all of the requirements to be full-fledged pastors. Now they would be permitted to do water baptisms, engage in offering Communion in their churches, and apply to the government for permission to perform marriages.

This is a life achievement that demands a huge celebration with friends and family. Occasions like this require that it be a memorable celebration, so a cow will be slaughtered in order that all can feast appropriately. After the celebration, good memories remain along with a huge debt. Having achieved this status, however, did not translate into a higher class travelling ticket on the train. They rode third class, which meant jarring, hard wooden seats, dirty windows—if they had glass at all—and space shared with the goats and chickens belonging to other travelers, not to mention hordes of children with differing stages of emotions.

Travelling with the "elite" were three other young men who had not satisfied the ordination committee with their readiness to be ordained. As is normal, the train started into its slow-moving, clattering rhythm and things settled down. The "travelers" then migrated to the second-class section. When the conductor came around to them, they would use their powers of persuasion—sometimes involving a little money—to remain in the second-class area.

"You boys are in the wrong section. Get back to the third-class section!"

"But, sir, we are pastors, and we five have just been ordained by our church; the other three not yet."

Scrutinizing them all, he said, "These guys I believe, but not you. Back to third class where you belong!"

Yes, the five ordained pastors were ordained to make the trip back into third class, and the three who were "not suitable" remained in second class. It became a regular topic of friendly banter between the pastors.

Reminds me of Ecclesiastes 10:7 (KJV) —I have even seen servants riding horseback like princes and princes walking like servants!"

CHAPTER 13

Establishing a Church on Fish Broth[4]

Witchcraft and witch doctors are synonymous with

Africa. In many ways a paradox surrounds them as well. On one hand, they are revered and their instructions are followed to a T, no matter what the cost. On the other hand, in some tribes people are murdered because they are suspected of being a witch doctor.

A fellow national pastor and I went into an area where no gospel message had penetrated and where witchcraft was prominent. Ukerewe Island is a small and extremely fertile island lying within Lake Victoria. Its people grow wonderful huge pineapples, pawpaws, oranges, lemons and mangoes. Every morning large loads of fresh fruit and vegetables are off-loaded in Mwanza and quickly snapped up by eager traders who will haul them off to the main fruit and vegetable market to be sold at a tidy profit. The island was only a short ferry ride away from the regional capital of Mwanza. It was also only a short ferry ride from Musoma, the regional capital of Mara. Church work tended to follow the governmental boundaries.

The island lay approximately in the centre of the regions. Here is where a bit of church politics entered the picture. At the national executive meetings, the matter of establishing a work on the island was discussed. No consensus was reached. Each region claimed the island was in their area of work. The impasse at the national executive level dragged on. There are times when one needs to leave the boardroom chair and get on with the work. Rev. Stephen and I took things into our own hands. Without anyone's blessing, we set a weekend to go over and establish a church.

Thursday morning found us on the island heading through banana plantations and fields of pineapple to a village called Nansio. There we set up camp for the weekend. Pastor Stephen located a family who knew some far-off relatives of his, and they agreed that he could stay with them. Culturally, it is considered very rude not to invite a stranger for lodging and food. I was invited back for food each day.

Only when absolutely necessary did I stay in a hut with a local family. Most family huts are fully utilized for their family members, and if I stayed with them, I knew that the four or five children who normally occupied the tiny room and bed would be spending the night in the entrance, shivering under a thin blanket.

I scouted around the village and made a large, dilapidated schoolroom my Hilton Hotel. Naturally, it sported no lights, no glass in the windows, and no doors. That was already evident by the goat and cattle droppings littering the floor. Funds were not available to build desks for the students, so they sat on the classroom floor. This is why I found an empty classroom. Assessing the layout, I decided to place my canvas cot against the farthest wall, away from the door-less openings. Evenings in the tropics can be chilly when the wind comes up. Chirping crickets, dogs yapping in their high-pitched voices, and people arguing could not overpower my tiredness. A long, hot, dusty trip in the tropics works better than any sleeping pill.

During the night in my subconscious, I felt something was not quite right, but it was not strong enough to wake me up. Early morning as pots and pans began to clatter and children's voices became clear, it also became clear that my "hotel room" had harbored some hidden guests in the night. I suppose I was the intruding guest. My throat was raw and burning. The acrid fumes of caustic bat droppings and urine caused this burning. The attic housed thousands of them! Had I scanned the ceiling the night before, I would have seen the telltale signs of large deep brown circles staining the ceiling tile.

We were not exactly welcome guests on the island and, as mentioned, only the cultural sensitivities of the host family provided us some food over the weekend. However, I had never seen African hospitality dished out as meagerly before or after. They barely stayed within the boundaries of warding off being cursed for a lack of hospitality. Fish boiled in a large pot was the first night's menu. After that, for three nights, we fished from the same pot, but no new fish were ever added—only water. The fishy smell teased our taste buds, that's all. This in a community surrounded by water where fishing is a way of life! With my sore "bat throat" food did not hold much of a draw anyway.

For three days we conducted evangelistic services in the village market, distributed tracts, and spoke with people about the Lord. For three nights I slept on my safari canvas cot in the large schoolroom on the cement floor. There were lots of windows, but fortunately no glass. This allowed the thousands of bats to freely flutter in and out. Morning light found them comfortably roosted in the attic where it was dark and hot.

By Saturday afternoon a number of people had committed their lives to the Lord as Saviour. In light of that, we announced that a water baptism service (full immersion in Lake Victoria, which is randomly visited by crocodiles and presents a strong risk of bilharzia infestation) would take place on Sunday morning.

A very aged female witch doctor had committed her life to the Lord. Forever etched in my mind is a picture of this very old, completely bent over, topless woman. The long, hot African sun had burned her skin to a dark brown—it looked paper thin and totally wrinkled. This brings to mind something about "... as a brand plucked from the fire" (Zechariah 3:2). She walked with much difficulty, stooped over to a point where she could barely raise her head. We picked her up from her hut. Earlier that morning she had gathered her witchcraft paraphernalia into an old burlap sack, and we had publicly burned it all. People in the village were amazed at this public declaration—defying the powers of darkness.

Hundreds of people lined the lakeshore, some even in dugout canoes on the water. After a lively four-hour service, the baptisms began. Each of the baptismal candidates publicly professed their faith in Jesus. As this woman's turn came to be baptized, she was grilled about whether she had brought all the witchcraft paraphernalia to be burned. In such circumstances, national pastors know the wiles of the devil far better than we do. Rather reluctantly, she admitted that she had kept a dik-dik horn, which held the ashes of every part of the human body.

Satan needs only one small, uncommitted thing to keep a foothold. We bundled her back into my Toyota Land Cruiser and headed to her hut. With help, she retrieved the horn from the grass roof, and we returned for her baptism. Following close behind were people dancing and singing, "The power of Satan has been defeated." Truly it had.

Years later I heard another account of our time spent on the island. Because adults tend to pay little attention to children, we made sure to give them something to indicate that they, along with the adults, were special. A young boy was given a used Christmas card sent by a women's ministries group in Canada. As an evangelism tool for children, the ladies cut out the picture on the card, pasted it on colorful construction paper, and to that glued a Scripture verse in Kiswahili. A tree twig kept the card on his parents' mud hut wall for many years. It became John's daily reminder to give his heart to the Lord. In time he did just that.

Now, many years later, he is the Mwanza regional superintendent. Pastor John personally recounted the story of his Christmas card conversion. Six churches have been established on the island. It was only a used Christmas card, but it continues to bear fruit through Pastor John's life. Who can tell how many lives he has touched?

All the Lord requires of us is to plant the seed. His watering of that seed produces fruit, which in turn yields fruit, which in turn … Who can tell what will come from a single seed?

They Got Him To The Door And Dropped Him

Monda was a young crippled boy who was an orphan. He was one of the 28 children we found when establishing Starehe Children's Home. Extensive tests revealed that his body did not absorb vitamin D, causing his bones to become weak. Certain tribes associate people having deformities with signs of their having extra spiritual powers. Killing the person and distributing body parts is the usual method of harnessing the power. Knowing that, we were always aware of the danger that Monda possibly faced. Caregivers at the orphanage were instructed to secure all doors each night.

Arriving at the orphanage in the morning, terrified night staff met us, their faces still clearly evidencing a horrible night's experience.

In the middle of the night, they began hearing noises of animals laughing. Looking out to the illuminated driveway, they saw two hyenas loping toward the home. Local people can read the signs of their culture, so they immediately knew that these were witch doctors supernaturally manifested as wild animals. That is a common occurrence in that area. Please do not ask me to explain that theologically as I have no clue how that can be. Prayer is the only antidote, and they began some serious praying. Even those who do not pray, quickly learn how to pray when faced with situation beyond their control.

These two "hyenas" managed to pry open a back door and went straight to Monda's bed. They picked him up and began carrying him out. As they reached the outside door, they were overpowered by prayer and dropped Monda. They fled into the night.

When we met Monda that morning, he was radiant and seemed to have had a huge burden lifted from his shoulders. In our way of thinking, it should have been the opposite, but obviously the Lord had liberated him in ways that we could not fathom.

There are certain areas in a city where witch doctors prefer to live. Unbeknown to us, the hill area just behind our orphanage was one of those places. A footpath that people used to get to the main road ran along the hillside of the orphanage fence. People walking along this path travelled the length of the orphanage property.

It was a normal thing for us to commit the property and the children to be under the protection of Christ. A volunteer had a vision in which he saw angels guarding the gate of our orphanage. This was very comforting!

A local witch doctor met one of our workers and confided that every time he walked by the property, he began to shake all over. He said he didn't know what power "these white people" have, but it was bothering him. He confessed that he took a long detour around the orphanage property to get to the main road. It wasn't long before he and many other witch doctors began to move away. Apparently, some had even died since our arrival at the orphanage.

Years later, with 57 staff at the orphanage, many issues arose. Silence, as in secrecy, is one cultural aspect that baffled us. Things were going on that we were not aware of. Seldom, if ever, did anyone come and tell us. This not only happened at the orphanage, but also in church work, as well as in our personal lives. When a worker was fired for stealing, after the fact, people would say, "Oh, yes. We knew they were stealing for some time." In church work, the church treasurer may have been "borrowing" money to cover his own emergencies. After the audit, people came forward with the long-known facts. For some reason, never figured out by us in 35 years, it seemed that a game was going on to see when the white man would catch on that something was amiss. Possibly they wanted to take credit for divulging the truth when it would no longer cost them culturally. A person cannot begin to be policeman while being involved in multiple areas of the work. One hundred and fifty churches were under our care, five missionary couples worked in the country, and we had yard workers and house help plus the staff at the orphanage and farm. We relied on the Holy Spirit to reveal things or to direct us even when we were in the dark as far as information goes. We were vigilant but not consumed by this one aspect of the work.

Oftentimes things came to light without any undercover work. These were the times when people experienced some fear of the Lord. Two lady staff at the orphanage had their employment terminated. This put fear into the witch doctors in the area. Without our being aware of it, these two ladies were bringing witchcraft items into the orphanage. A witch doctor asked our pastor, "With so many staff, how would those white people know exactly which two were involved in taking things into the home?" We didn't, but the Lord did.

A warning to tourists or people visiting countries where the powers of darkness operate more openly than at home: do not purchase any carvings with faces or the disfigured form of a person. Often these things are dedicated to some power of darkness, and it will accompany the article into your home. Do not throw them away; burn them. Second Kings 23:11 records that when King Josiah was cleansing the land, he burned things that had been dedicated to the gods of the people.

He Probably Came While You Were Sleeping

Witchcraft still holds a prominent place in many people's lives. When the "white man's" hospital or procedures don't produce expected results, the witch doctor becomes the go-to expert.

Our neighbor's husband did not return home one day. She was not too concerned as she knew he was husband to another family, and probably after a week he would return. When that did not happen, she contacted the "other wife" and was shocked to hear that he had not been with her. A police report was filed, but no reports of accidental deaths had been reported, nor were there any remains in the city morgue that matched her husband. After another week she became more desperate and decided to consult a witch doctor. His instructions were that she must sit by a certain window in her house, and her husband would appear. She must do this day and night until he appeared. Food would need to be put out as well to entice him. Her husband did not appear. Complaining to the witch doctor, she was told that he probably had appeared. The reason she did not see him was that she had most probably dozed off, perhaps even for a second.

Eventually she tired of paying and realized the witch doctor route was fruitless. Her friend said he would help as he knew people in the police force. Further investigation was carried out. In case it might trigger some recollection, the wife had informed the police that her husband loved to read newspapers. It was customary for him to be walking with a number of publications under his arm. Nothing showed up in any police files. An officer offhandedly mentioned that the traffic division might be a department to check. Apparently, no information is passed between police departments.

Sure enough, a fatal traffic incident report noted that the pedestrian walking along the road was carrying a number of newspaper publications. Many major roads are narrow and do not have sidewalks, so pedestrians would share the road with vehicles. A number of weeks had passed without the body being claimed, so the municipality buried him. Due to the lack of space, and in the interest of time efficiency, multiple corpses occupy the same shallow grave. There are no coffins, and the corpses are wrapped only in a cotton shroud. In order to positively identify the body, the grave would have to be dug up. A friend of mine attended the "opening," and apparently the sight was not pretty. He mentioned a few gruesome details, which were enough for me to switch the topic of conversation.

The wife did identify her husband's body, and she was able to give him a proper burial as her tribal custom dictated. Her mind was also at ease, which was a comfort for the future.

Evil Spirit, Come Out!

Every African is very familiar with witchcraft and the power of darkness. Africans are no different from people all over the world in that, when something unexplainable happens, it must be attributed to something that is known. In Africa, the unknown is attributed to some spiritual pen name. In the college class I was teaching, I tried to get them to think beyond the fact that the unusual always has a spiritual connection. A logical and physical explanation is a possibility.

One day, as the students were in the dining hall, I placed my remote-controlled car just outside the door. My perch on a rather large boulder under a tree offered the perfect set-up. The first student, comfortably fed and with a leisurely gait, ventured out. Within a few meters he noticed something moving. Snakes and reptiles instantly accelerate their reaction time with a double shot of adrenaline. This student had just had a good supper and was already looking forward to having it sit happily in his stomach, bringing a sense of peaceful bliss as it digested. The unrecognizable object moving toward him suddenly launched him into a high scream and shouts of "Satan, I command you to come out!"

Within seconds the dining room emptied and joined in a rabble of shouts and screams as this "thing" moved toward them then backed up (with them following), then forward toward them again, making them jump backwards, with some of them running, eliciting more screams and shouts of command.

Coming out from my hiding place, I calmed them, then explained, and demonstrated what the "thing" was and how it worked. Even then, when I remotely moved the car, some still began screaming and shouting in near terror. Even the brain finds it difficult to override ingrained reactions. My point in the lesson was that not everything that is unknown has a spiritual connection. Most probably we in the West are far too prone not even to entertain a spiritual connection. We need to be more aware of a possible spiritual lesson when something out of the ordinary, like a healing or a catastrophe, occurs. Our brains want to make an instant assumption based on something we already are familiar with. People living in areas of unrest associate a loud bang with a rifle going off, and they dive for cover.

Witch Doctor Apprentice

Samuel grew up in a rural area very near the Ngorongoro Crater. Samuel's grandmother was a witch doctor and was passing the mantle on to her grandson. The practice of skipping a generation is common. From a young age in his apprenticeship, Samuel was taken into the forests for long periods of time and taught which roots were antidotes to certain ailments. Sometimes these roots were from trees, and others were from plants. Certain plants were also identified as having healing properties.

Samuel was in his 20s when he heard about Jesus. Without hesitation he committed his life to Christ. His witch doctor grandmother was furious that he would throw his life away in such a manner. In Africa, witchcraft affords a good life. People always find themselves in trouble and are willing to pay to find out who or what is the root cause. Samuel eventually attended Bible school and became a pastor. He had found the source of true healing power.

You're a Witch Doctor![5]

Heathens, when confronted with a miracle of the Lord, see witchcraft as the only logical explanation. That is what many have grown up with, so it is completely logical that this would be the case.

It all began innocently enough when Pastor Julius, who was a caregiver at Starehe Children's Home, went to his rocky lot to break rocks (granite boulders) with fire.

Julius and his wife, Prudencia, worked as caregivers to 54 young boys and girls. While at the orphanage, he received a small piece of land from the government on which he wanted to build a hut for his family. The problem was the rock. To Westerners that might present a problem, but in Africa it does not. They use the heat, cooling and cracking method. Charcoal is piled on a rock, lit on fire, and covered with iron sheeting. As the rock heats and then cools, it cracks along tiny fault lines. That is chipped off, and the process is repeated.

Julius got one day off every week. Unfortunately, this particular day did not look promising for starting a fire sufficient to heat
the big boulders. Adding to the misfortune was the fact that it is
hard to find dry charcoal during the rainy season. We will let Julius
himself tell you the story:

*I was trying to outrace a government announcement that if
people had not started building by a certain day, the lot would be
given to someone else. With my last $10.12 before payday, I pur-
chased a bag of charcoal and endeavored to light it on the boulders. My young Muslim friend, Atibu, 22 years old, was with me as
I lit the charcoal, which was laid over the granite rocks. It was a
bright, sunny afternoon, promising to be a great day to finish this
task. Shortly after the charcoal was lit and burning nicely, a huge
thunderstorm of heavy rain and lightning erupted. Others around
me, who were also burning rock, ran for their lives to avoid the
pelting rain and the lightning flashing all around. All feared being
struck by the bolts of lightning. I stood in a stupor, totally bewildered by the turn of events. Literally, I was seeing my last money
being soaked up by the rain. In my great distress and disappointment, I began to cry (African men especially do not cry).*

*Atibu began shouting that the charcoal was finished and to take
cover with him before we both caught pneumonia or got killed by
the lightning all around us. Without a lot of premeditation, but in
desperation I said, "No! Wait! I am going to pray and ask God in
the name of Jesus to sustain this fire in the midst of soaking wet
charcoal because it was my last money and I have no more. Then
we will run."*

"That is ridiculous in this deluge of rain!" Atibu yelled.

Fierce lightning and wind accompanied the downpour, giving rise to the young man's remark stating the obvious: "We're going to get hit by lightning and be killed! Let's get out of here!"

By this time, both of us were drenched and shivering. I prayed, "Dear Jesus, light my charcoal. I have nothing else and no one else to lean on but You. Jesus, save me! Amen."

As we turned to run for cover, the charcoal burst into open flame. It was as though gasoline had been poured on the charcoal. It produced such a heat, in fact, that as the rain hit the boulders, they began to crack from the heat expansion and contraction in the cold rain. As the rain subsided, we returned to the rock. The charcoal was still burning.

My young Muslim friend, seeing this miracle, began to shake and said, "Pray for me. I accept this Jesus." Others, whose charcoal fires had been totally obliterated by the tropical downpour, had also come around to see the miracle charcoal. When they heard Atibu confessing Jesus, they tried to discourage him by accusing me of being a witch doctor.

He defended me, vigorously saying that he had heard me pray to Jesus to light the fire and nothing else. He said, "Friends, Jesus is the true Kiboko," meaning "He is the real answer." (Kiboko is a strip of hippopotamus hide made into a whip. During colonial times it represented authority). *Atibu went on to say, "From this moment I am a follower of Jesus."*

Julius was not always a follower of Jesus. In fact, he was far from it. His entire family were Muslim, and many were also into witchcraft. People of any religious persuasion can be into witchcraft. Julius used to help his aunt with witchcraft practices. During her "healing" sessions he beat the drums. Other times he went with her to collect herbs and roots in the forest. Julius was converted in a rather remarkable way. Living in Dodoma, he was going home from a bar. Drunk, as well as being on drugs, he followed a path that led past the Pentecostal church. Right at that time they were performing the drama, "Heaven's Gates and Hell's Flames." As he stumbled past the church, "Christian vigilantes," as I call them, caught him and dragged him, kicking and fighting, into a room in the church. There they prayed for him from 6 until 11 p.m. Miraculously, he sobered up. As he heard the gospel, he accepted Christ as his Saviour.

Shortly after that, he heard God's call to ministry. For Julius it was unique, as it is for many of us. The Lord is no cookie cutter, using the same methods with everyone. After becoming a follower of Jesus, he felt compelled to work with children. However, none of our churches had a children's ministry portfolio. In fact, children were hardly given any attention in a service; adults were the focus. A number of years passed before we started the orphanage, but there Julius's call to work with children was realized.

"Jesus" Respect

The misbehavior of children is as normal as a dog having fleas. Having 128 children in our orphanage brought 128 different challenges of character. To us it was just normal that we would call the child into the living room area, have them sit on a chair and speak to them. The goal was to help them change their actions, which would change their lives, and in the future help them to be stable, well-adjusted adults.

We had no idea how powerful this was until we listened to our orphanage pastor, Rev. Julius, preach to the children. His subject was "Examples of how Christ respected everyone, even the sinner." In his sermon he used us as an illustration. According to local custom, a reprimand was administered in a number of ways. Often the reprimand involved humiliation. Having them sit on the floor, making them take off their shirts, shoes and socks, severe beatings, and yelling were some of the tactics used.

From that cultural background, Pastor Julius had himself been a thief before his conversion and knew the ropes, as it were. He reminded the children that when *Babu* and *Bibi* (Grandpa and Grandma, as the children and the staff respectfully addressed us) called them in for discipline, they allowed the children to sit on a chair. To someone facing discipline in their culture, such treatment is unheard of. In our culture we would compare it to the police taking us for a steak dinner after catching us speeding and, in the course of conversation, gently telling us that speeding is not good.

Without knowing it, even in discipline we were mirroring Christ in a culture where mercy and respect are seldom offered to the accused. It is guilty until proven innocent or until you confess. Confession is painful but the quickest way out.

Don't Take Us to the Police!

Petty thievery is a given in Tanzania. Being white Canadians who operated the orphanage carried with it the fallacy that riches accompanied the skin colour. As a result, the orphanage could become a target for thieves, so we employed watchmen on our seven-acre site. Most important, we wanted the children and caregivers

to be safe at night. As the watchmen were watching their dreams—
sleeping in layman's terms— gun-toting thieves accosted them, as
they remembered it. They were made to lie face down and were
threatened death if they moved. Only when the sun came up did one
dare open his eyes and slowly turn his head to assess the situation.
Seeing no gun pointed at their head, they all scrambled to their feet.
Their own spears had been stolen, so they were forced to report the
incident. Then they became even more terrified because now an
investigation would ensue. I mean, you can't tell the director that
overnight termites ate the spears, steel and all, and expect him to
believe the story.

Because of our responsibility to the children, who were technically wardens of the state of Tanzania, I insisted that we report the
incident to the police. All four watchmen began to beg us not to
report it to the police. Through local "word of mouth investigation,"
they themselves would find the culprits. I insisted on going to the
police and finally found out from them why they were so adamant
about "self-investigation." They said that most crimes in Tanzania
are inside jobs, and that the first thing the police would do was to
throw them all into a windowless cell with 53 other suspects and
leave them there without food or a bathroom for three days. After
that they would be severely beaten with rods until they confessed.
I have walked past those holding cells. Holding your breath is the
best course of action if you don't want to gag on the spot. These
holding cells are always situated in close proximity to the desk
where police statements are taken. Why? It is just another thing
that confuses the Western mind.

I have never seen a more somber-looking foursome slowly
squeeze into the back seat of the vehicle. No one volunteered to

sit in the front with me. The tightness in the back seat must have felt like a cozy, comfortable blanket as they contemplated the dreaded destination's end. There was not a peep out of them the whole way—only the strong body odor connected to profuse perspiration that had been going on long before we got into the vehicle. I entered the police station and asked to speak with the police commander. At a time like this, you go to the top. Now, being Canadian had its advantages, and without question I was taken to his office. It took a lot of convincing for him to agree not to take the "normal" course of police action: the stripping of clothing, throwing them into a cell, taking a statement days later, and then a solid beating, which led to a quick confession—then on to the next case. To put it into perspective, one must realize that this is all done in public, in front of other police and prisoners.

I received the police commander's assurance of restraint in writing. Written documentation in Tanzania is akin to the Old Testament law of the Medes and Persians—unbreakable. Even with the re-assuring paper in my hand, the watchmen slowly peeled themselves out of the vehicle and, trembling, followed me into the police station. A washroom break at this time would have been most welcome by all of them, but washrooms are not there for suspects. They were thrown into a cell until their statements could be recorded. They were severely threatened but, because of the letter, were not beaten. No one was ever arrested, so whether their story was legitimate or not, I will never know. I doubt they ever slept on the job again. Well, maybe—but no doubt they took turns.

Respect, African Style

Respect is shown in different ways in various societies. Sitting on the floor during a discussion is one way respect is shown. Although it makes Westerners very uncomfortable, to reject its expression brings humiliation and grave disappointment to a national. Trefosa, a young woman whom we had working for us, wanted to share a desire she had. She also wanted us to bless her in agreeing to her request.

Very formally she asked to meet with us. At the agreed upon time, she arrived and greeted us with the usual Tanzanian greetings of respect. Being welcomed into the house, she proceeded to sit on the floor adjacent to our couch. This was very uncomfortable for us, and we asked her to sit on the nearby chair. She declined this offer, so we knew she was demonstrating the highest form of respect her society acknowledged. Without any feeling of superiority, we accepted it as such. She painstakingly laid out her request. Her parents in Rwanda were getting on in age and she wanted a month off to visit them, possibly for the last time. To us this might not seem like a huge issue, but for her it was connected to a deep respect for her elders and for us.

Was That a Rifle Shot?

Going to the police station was a regular occurrence for me. Sometimes it was to renew gun licenses or my driver's license or to obtain a permit of some kind or another. One particular day my visit was of a more serious nature, so I was taken upstairs to the assistant regional police commander's office for his approval of an application. It was no big surprise to hear from his secretary that he was out for a bit but would return shortly. I was ushered into his very large office and invited to sit on a chair near his desk. A secretary type policeman sat at another much smaller desk near the back of the room. The high authority of these men was reflected in their surroundings. His desk sat near a window overlooking the street. The room measured probably 20 feet by 30 feet. Being that the room held only two desks and the chair I occupied, a distinct echo was the only other furnishing. Important positions came with one private secretary sitting outside the office and another inside. These had the responsibility of quickly typing up a letter or being ordered to fetch a cup of tea for the visitor.

Upon being ushered into the room, I observed a young man sitting on the floor just to the left of the door. His dress, or lack of it, immediately indicated that he was a thief: no shirt, no shoes, no socks, and no belt—only a pair of summer shorts. The man I wanted to see was off somewhere for a long time, but because I needed his signature I hunkered down and waited for him. Every half-hour or so, a heavily built policeman dressed in full uniform would come in, look at the man on the floor and then at me, then exit. This happened four times. On the fifth entrance, two policemen stepped into the room and, ignoring my presence, proceeded to deal with the thief.

With heavy army type boots he was kicked and instructed to stand, then brutally kicked in the shins and told to sit down. Then the process was repeated a number of times. With open hand he was slapped across the face; the noise literally sounded like a rifle shot in the nearly empty room. It was obviously not the first time these men had executed this maneuver. Having completed their assignment, they kicked the man out the door and admonished him not to repeat his deeds of thievery again.

A-Hunting We Will Go

This was a big, important year for me: general hunting would be opened on July 1, 1975. It had been closed since September 1973. Sadly, upon the advice of a non-hunter, I had sold all of my guns in Canada. He assured me there would be no opportunity to use them. When I arrived, I realized that hunting was a pastime of many expatriates as well as some locals. Besides, the meat was very welcome. Contrary to what we hear in Canada, there are still large numbers of animals, and the government must cull as a result of overpopulation. Leopards, for example, are found close to towns. It is a regular occurrence for children to be eaten by leopards or lions as they herd cattle. Animals are a renewable natural resource that the government closely regulates with permits to harvest, just like we do oil, trees or fish.

Amazingly, Tanzanians do not eat much wild meat. Part of the reason may be that few have guns, and the old ways of spear or bow and poison arrow are too energy consuming, not to mention dangerous. They prefer the domesticated cows, goats, sheep and chickens. It certainly does not have anything to do with the taste since the vast majority of African animals are grass grazers, so the meat is very tender and has only a slightly different taste from beef. In the rural areas, wild animals live in relatively close proximity to people. Each year they invade farmers' fields and decimate crops of corn, spinach, peas and sweet potatoes. Baboons, although not seen as a trophy animal, are particularly destructive. They grab a corncob, take a bite out of it, and then take another. It doesn't take long for a group of 50 to destroy many acres of corn.

I did purchase a nice .30-06 calibre from a Swedish fellow leaving the country. Purchasing was not an easy thing as all firearms in Tanzania are registered and ownership is closely monitored. This means you have to go through a lengthy personal verification process. The first step is to request a letter from the Canadian High Commission verifying that you are a Canadian citizen in good standing. Then the local village chairman (yes, even in a city!) writes a letter of recommendation. The process then goes on to the district official, the regional official, the district police commander, and finally the regional police commander. All of those signatures appear on your application, which is then sent to Dar es Salaam, where the highest police commander in the country passes it—hopefully. That permit gives you permission to buy a gun. These committees do not meet too many times a year and, due to a long agenda, the matter of guns is pushed to another meeting. A year can pass even before the permit has gone through all of the committees.

Every gun comes with a logbook showing its history from the first person who owned it. Each year the gun registration is renewed. That receipt is pasted on top of all other previously issued receipts. All of this involves a lot of paperwork: filling out at least three copies of the yearly application stating your name, age, address, type of rifle, calibre, the number of rounds purchased that year, and the number of rounds in your possession. Each gun had a maximum rounds allowable attached to it. For example, for the .22 calibre, I was allowed to have only 25 rounds in my possession.

I enjoyed hunting as that was real Africa to me. I had hunted since I was boy of five. Many sparrows and gophers fell to my trusty BB gun. Sometimes I called it lightning because it never struck the same place twice. At that age, and even later, I found hunting to be relaxing and exciting. Now the enjoyment of simply being out in nature is more fulfilling to me.

Over my years in Africa, I harvested hundreds of animals. My license permitted me to take 24 animals a month covering 10 different species. Animals were plentiful, so this gave me an opportunity to bag some beyond the average size. Four of my animals are of sufficient horn size to be registered in *Rowland Ward's Records of Big Game*. Hunting in Africa is efficient in that every part—and I mean every part—of the animal, except the skin, is utilized for food.

Prior to 1973, hunting was open 12 months of the year. Everything could be taken, including the "Big Five" of Africa: the elephant, the leopard, the lion, the Cape buffalo, and the rhino. All were on one license, but within a set quota for each species. Due to heavy poaching for rhino horn, it was taken off the list. These poachers operate across borders and are hard to apprehend.

On September 1, 1973, the government closed all hunting while the Canadian Wildlife Service helped them evaluate their hunting laws and quotas. After two years new regulations were brought in, resulting in the hunting season's being reduced to six months of the year, July 1 to December 31, along with new district quotas for each animal. For example, if the quota for the region was 200 wildebeest then, when 200 licenses had been "cut," that was it. In November the Tanzania Game Department tallied the number actually harvested and opened the season again if the quota had not been reached. This was determinable because, at the end of each month, your license tags were returned to the game department indicating which animals you had paid for and which you had actually harvested. It was a very good system and helped in the management of a huge natural resource in the country.

To keep the skull or skin or teeth or tail, a certificate of ownership listing the particulars of the trophy was issued. On exportation of the item, the game department, upon surrender of the certificate of ownership, issued a "Permission to Export" certificate. Wild animals are a renewable natural resource, and its governance brings in huge amounts of money to the national coffers.

Another aspect of the hunting review was that a two-tier hunting system was established. Animal species were separated into "professional" and "resident" categories. All of the big desirable animals like elephant, lion, leopard, zebra, kudu, oryx and sable antelope could be hunted only by high-paying foreign hunters. When I say "high-paying," I am talking up to $100,000 or more for a three-week hunt. Rhinos are the only animal on the endangered list and are protected.

Residents could acquire a license for the "less exotic" varieties such as warthog, wildebeest, Cape buffalo, all varieties of gazelle, hartebeest, eland, bush animals, reedbuck, bushbuck, topi, and so on. Each category of animal had a monthly tag limit. For example, two warthogs, two Thomson's gazelles, one Cape buffalo, one hartebeest, two dik-dik, and so on. With the abundance of species, I was allowed 26 animals per month. Professionals could hunt any animal they desired, but it all came with a hefty price tag. Few professional hunters ever came without wanting to hunt the big game. This meant that they had to pay USD $1,500 per day for a minimum of 21 days. On top of that, there was the tag price for any animal they shot plus some extras. For example, for a lion and leopard there was the tag price plus so much per foot of its length. On elephants, there was the added cost per kilo of each ivory tusk. If a tusk went over a certain weight, there was an extra surcharge. Added to that are relocation costs to different parts of the country since not all animals exist in the same location. There are guides, trackers, cooks, skinners, drivers, people to set up camp, and tips for everyone. It does not take long to reach $150,000.

Why Take Water? We Do Not Intend To Get Lost!

Bob, a missionary hunter friend living in Kenya, wanted to come on a hunt. Immanuel arranged a hunt in the coastal Tanga area where he lived. Situated on the Indian Ocean, it is extremely hot and humid. Because we all loved to hunt, we decided to make a two-day trip out of it. Normally we went out only for a few hours— and came back with game. A drive in the coastal area over narrow, sandy roads and through numerous roadside coastal villages with high, lazily swaying palm trees cannot be compared to hunting on the African Savannah, which is found inland.

Since this was a "real" hunt, two African trackers were employed. Apparently they knew the bush in that area. This would be an asset since our time was limited to two days. Bob especially wanted to bag a Cape buffalo. We left the old German town of Tanga in the early morning and headed toward the forest, a bit inland from the coastal shoreline. Coming out of Tanga, we crossed a small bridge spanning a dry historical canal. During the war with the British, German soldiers had dug a deep channel from the port inland so that the British could not detect them disembarking from their troop ships. The book *Battle for the Bundu* by Charles Miller is an excellent read if you are interested in the fabulous history of Tanga and the British-German war in East Africa in particular.

Rains had started in Kenya only 50 miles or so from our hunting area. This meant that a lot of game had already headed that way to feed on the rich young grass shoots that pop up overnight after a rain. The vehicle was parked, and we followed our trackers into the dense bush. Suddenly they stopped and raised their hands, indicating they had spotted an animal. Cautiously we crept up to them and, following their pointing, saw a warthog. Bob downed it with a well-placed shot.

The crack of the rifle had barely subsided when the two trackers bolted for the warthog. One had his knife out, making a lunge toward the pig to cut its throat according to his religious custom. Bob gave out a huge bellow for them to stop. Momentarily he did, in mid-motion, then made another lunge as Bob repeated his command to desist. I never saw such two downcast men. Here, before their very eyes, lay a wonderful meal, but they were robbed because they were not allowed to butcher it as required by their faith. Bob, being a taxidermist, wanted to do a full mount so did not want the skin punctured with a lot of holes. One bullet hole was enough.

It was the only animal we saw that day. Driving in, we had found a nice camping spot and returned there to spend the night. It consisted of a level area in the shape of a cove by the forest. Close to the road there was a large flat outcropping of rock. On it we ate our supper. For dessert we had the biggest pineapple I have ever seen during my 35 years in Africa. The coast produces an abundance of fruit, and this was a beauty. It was all of 16 to18 inches long and round like a watermelon. It was a good thing there was water standing in a hollow in the rock. Cutting into the pineapple, juice ran down our bare arms. Because they are field ripened, these pineapples are sweet and do not have the acid that can cut tongues like in North America.

We were up early the next morning in the hopes that it would be a better day. Starting the day with a good hot cup of very sweet tea is the only way to go. Naturally our trackers were dual purpose so they prepared the tea. The payoff was that they could put as much sugar into it as they pleased. Immanuel and I had at least four cups, knowing it might be a long day. Bob was not that used to being in the rural area and had not acquired too great a taste for the sweet tea. He managed one cupful. Driving along, we spotted an area that looked like potential good Cape buffalo habitat. Our plan was to go in a short distance and skirt around what looked like a small marsh. The "short distance" part was a huge mistake. Just as we approached the far end, expecting to head back to the vehicle, the bush exploded. We had jumped a Cape buffalo. No doubt he had heard us coming and waited to see what the noise was.

It was fortunate for us was that he had decided to run and not surprise us in a charge. Cape buffalo are notorious for running ahead and then lying down so they can observe what is coming down the path and often charge. The small animals had grazed the bushes off, so the Cape buffalo had a good viewpoint. With great fury and noise, their surprise tactic is to charge at close range. A single Cape buffalo can create so much noise that it would seem there are 50 animals. I have experienced this on a number of occasions.

Our trackers examined the tracks and determined that it was a very old bull by the way he dragged his feet. That is precisely what we wanted. Leaving our good sense behind, we eagerly began to follow—the trackers up front, Bob behind them, and Immanuel and I bringing up the rear. After about an hour, we again saw the bull jump up. Bob raised his rifle to shoot, and in that instant the tracker stood up right in front of him, his head directly between the rifle and the buffalo—a moment that was scary beyond belief. Seeing Bob lower his rifle slowly brought the blood back into our faces. Off we went again, being led deeper and deeper into the bush. After three hours we admitted with reluctance that the buffalo was the winner today.

It was then that we realized we were thirsty. The hot, humid weather had sapped us. It was also then that the second realization hit us: we had left our water in the vehicle. It was only going to be a "short" walk into the bush and back to the vehicle, right? By now the temperature hovered near 90 degrees with 90 per cent humidity. Without delay it was time to get back to the vehicle. One tracker knew the way back; heavy guns in our hands, we began to follow him. Surely it wouldn't take us long. Dehydration was a real enemy now. Very concerning to us was the fact that Bob was on autopilot. Every time we stopped for a bit of a rest, he flopped down on the ground. Sharply prodding him was the only way to get him up and moving again. He no longer had the strength to carry his own rifle.

We came upon what can only be described as a genuine mud hole, probably no more than six feet in diameter. By the tracks and smell, we could tell that pigs and hyenas used it regularly. Water is water, and we were desperate. Holding handkerchiefs over our mouths to hopefully strain out the courser materials, we drank. Bob seemed to revive a bit but was still in bad shape. After another hour of walking in the bush, we came to the same mud hole again. Now we knew that the first tracker did not have a clue which way to go to get to the vehicle, and we were in trouble. Even with that same pig's mud hole staring him in the face, he wouldn't admit that his guiding days were over.

The second tracker assured us that he did know the way. Why had he not said so two hours ago? That period of silence gave us great comfort in his abilities. You know I don't mean that. We had no confidence in him at all, but he was number two on our list of two, and there were no others. Tracker number one engaged in a big argument that he knew the way, and tracker number two did not. His insistence was wearing us thin, so we gave him an ultimatum: "Shut up or we will shoot you and leave your body for the hyena." We would not have done that, but it did serve to persuade him to clam up and hand over the leadership duties.

Off we went with less than a thin slice of hope that the number two tracker would be better than number one, who was following us just far enough behind so he wouldn't lose sight of us. As unbelievable as it was, tracker number two did lead us to the vehicle. By this time Bob was not well at all. Fresh water did revive him after a time, though. Our hunting trip was over, and we were glad it was. The next morning Bob weighed himself. He had left 17 pounds of himself in the coastal bush of Tanga.

You might ask, "Why didn't you take water?" We did, but it was in the vehicle—and it was not in our plans to get lost! Ring any bells in your own life of unintentioned actions?

Wild Animals

Wild animals are a large part of the mystique of Africa. The varieties, sizes and temperaments are unbelievable. Their varied characteristics are intriguing. Wildebeest run in a circle of approximately 10 kilometers in diameter. Therefore, if you have hunted them in a certain area more than once, you get to know their circle. I spooked a herd of about 100 north of where I was. Without hesitation I drove straight south and waited for them. It was fortunate that I waited in my vehicle because, had I not, they would have trampled me. Nothing deters them from their path. They thundered all around my vehicle.

Warthogs explode out of a small patch of tall grass and run like blazes. Their tails are held upright like antennas. Only after they have gone a distance, when their curiosity overcomes them, will they stop for a split second to look back. Then they are off, and you will not see them again.

When a lion is killed, the pride will double back and try to attack. If a male lion kills another in a territorial fight, the new victor will attempt to kill all of the young lions. The females will then produce his offspring. Nature is cruel. At the same time, it knows how to have the strongest animals produce the next generation. A lion can drag an animal up to three times its own weight.

Leopards drag their prey into a tree and consume it there. Hyenas have the strongest jaw of any animal: they can crack an elephant bone and leave teeth marks in a piece of steel. Hippos have to keep their hide moist; otherwise, it will crack like a watermelon. That is why hippos lie immersed in water during the day and come out only at night to graze. Although they have very short legs, they can outrun a man. If you get between them and water, you are in mortal danger. The giraffe's saliva is the consistency of glue; this allows them to eat from the thorn bush.

Elephants are matriarchal. Depending on the food source, they may have young once in 11 years. Only when a certain grass is plentiful is their reproductive system triggered. A peculiarity with them is that the mother's breast is situated between her front legs. Wildebeest bear young every year. They are known as the clowns of the plains as they jump and prance for no known reason.

Reedbuck bed down in tall grass and, when spooked, jump up and run for another cover and immediately lie flat down again.

Cape Buffalo can reach up to a ton and a half in weight. They are one of the few animals that will hunt the hunter. When confronted, they will charge 50 per cent of the time. Of all the people killed by wild animals, the Cape buffalo kills the most. Buffalo are hard to kill by spear because their ribs overlap, making it impossible for a spear to penetrate. However, a Cape buffalo, unlike most other animals, cannot run on three legs. If one of his legs is injured, he comes to a complete standstill.

Shoot Them Both

While living in the Bukoba region, a pastor informed me that hippos were decimating farmers' crops. With permission in hand from the game warden, I proceeded to do some crop protection with two young men. The plan was that we would shoot a hippo on land in the light of the full moon on an African night. Being so near the equator, the intensity of the light of the full moon is such that you can read or even drive a vehicle. Shooting a hippo on land is a lot easier than shooting it in the water. Trying to shoot one in the water involves huge labor to drag it out, not to mention the danger from other hippos.

A hippo's leg is only about 16 inches long, but it can outrun a man. One chomp of its huge teeth spells game over. Extreme danger results when you are caught between the hippo and its safe haven of water.

Two and a half hours from home, we finally reached the rendezvous spot. Separated by a meadow and approximately 50 meters of scrub brush lay the hippo pool. Moonlight was already sufficient, so we could walk without any trouble. Within about a half-hour, we came upon an abandoned hut. It didn't have a door, so we could see a man inside cooking. Exchanging the normal African greetings, he asked us to come in and join him. The "trackers" explained that the hippo pool was about a kilometer away and that we would wait awhile until the hippos exited the water. There are times when you have to rely on wisdom that you know is suspect, but have no choice.

Suddenly, as we were chatting and drinking tea, the men put up their hands for silence. After some length of time they said, "We will wait another half-hour. The hippos are coming out of the water."

"*What?* How do you know?"

"Listen! You can hear the sacred white egret birds chirping. They always accompany the hippos."

Having waited the allotted time, we began to head slowly down toward the pool. Coming to the scrub brush, criss-crossed with paths, we had not yet seen any hippos. Thank the Lord we hadn't, as it would certainly have been in the hippos' favor had we met around a bend in that brush. We emerged at the pool, and the trackers began to inspect the mud with a tiny flashlight. There were tracks everywhere. After some time, both men assured me that two hippos had come out of the water and headed in the basic direction from which we had come. Seeing all of those tracks in the mud, my level of confidence in their report of two hippos did not register too high.

Gingerly we moved straight up through the brush again. At the edge of the meadow, we stopped to survey the area. After some time they excitedly pointed about 150 meters ahead and, sure enough, there were two hippos grazing on the lush grass. Their hides glistened in the moonlight. They were just beyond a small clump of brush. Hippos have acute hearing, but their chomping on the grass drowns out sound. Scouting for danger, they eat for a few minutes and then stop to listen. Arriving undetected to the cover of the bush would be the trick. As the hippos' heads were down, we moved slowly. Seeing them raise their heads, we immediately stopped dead in our tracks.

Upon reaching the bush, which was about three meters wide, we realized that the hippos were only 75 yards ahead of us. The larger one was to the right of the bush, and the smaller one to the left. That meant it would be impossible to move into a position where we could see both of them. "Which one should I shoot?" I whispered to the trackers.

"We shoot them both" was their reply. That sounded good to me.

I stepped to the right and quickly put a shot into what I hoped was a vital organ. We were only in moonlight, remember.

Without hesitation I jumped to the left and saw the second hippo already charging down at us. Within a second I had emptied a shot at him. Back on the right, the first hippo was quickly closing the distance to us. Another shot, then back to the left to engage the second hippo. Back to the right, and a final bullet to drop the still charging bull. Then back left, and the last bullet in my gun to drop the second hippo. Both of them lay not more than 10 feet away. Having previously shot hundreds and hundreds of rounds with my .30-06 rifle paid off in that I knew it like a well-worn hat. To my amazement, when I looked around, my spear-carrying trackers were still with me! When seeing danger, it is not unusual for them to exhibit self-defense. The best defense is to pull off their shoes with lightning speed and head in the opposite direction as fast as possible. During my short time in Africa, I had learned that many trackers were not all that brave when facing the possibility of spilled blood—especially their own.

It is not hard to imagine that all of that rifle fire in the still African night announced to villagers kilometers away that something good was going on as far as meat was concerned. Shortly, dozens of them showed up, creating a babble of excitement and anticipation. Seeing that it was about 9 p.m., we decided to return the next morning and divvy up the meat.

Meanwhile, we still had to report to the local police commander to show our permission for crop protection. His part was to fill out a police report, and then we could all go home. That is easily done *if* the police commander is not drunk and, by the light of kerosene lamp, is trying to actually arrest us. After a few lines of slow painful scribbles on the page, he would crumple it up and throw it beside his leaning desk. This went on for a number of hours until we finally convinced him that we were coming back the next morning. Arriving home well into the early morning, my wife was not impressed. She has a keen imagination and allowed it to operate to full capacity as to why I had not yet arrived home.

Before 7 a.m. we (my wife and I) were on our way back to survey things in the light of day. I am sure most of the crowd from the night before had not left not to mention many other hopefuls who had joined them. Butchering was a long process. Hippo hide is one inch thick. Using large machetes, the skinners opened them up much like a watermelon. As they hacked, the hide split open ahead of them. The teeth had to be chiseled out of the skulls, plus I wanted all eight feet and at least four pieces of hide measuring about three feet square. Seeing all of the butchering and smelling the freshness of it all, LouDell would not even think of cooking any of the meat. I, however, cut some steaks, and it was absolutely delicious.

It was at least a half-kilometer from the butchering site to the vehicle. I could not believe the weight of each piece of hide I wanted. With the promise of a good chunk of meat for payment, four men were chosen to carry everything to the vehicle. With the promise of meat, there was no shortage of volunteers for the job. It was a fairly steep incline, but they hoisted everything onto their soldiers and, with much hilarity and singing, literally ran to the vehicle. Keeping up with them was out of the question. All of the remaining meat and the good parts (which is everything) was divided up and given to the host of onlookers.

Heading to the police station again, to our relief we encountered a sober policeman to record our crop protection letter.

At home, it took most of the next morning to skin out the eight legs. Each of the 24 toe knuckles had to be removed so salt would penetrate. Later I took them to Zimmermann Tanneries in Nairobi, where they made nice candy holders out of them.

Hippos have eight ivory teeth that are designated as trophy and which must be surrendered to the game department. Upon doing this, you pay a set fee and they furnish you with a certificate of ownership for the trophies.

That Was Our Warthog!

Although not the wisest or the recommended thing to do, I preferred hunting by myself. Once I hunted with a group of three other men who had permission to shoot game for an upcoming Bible college graduation. We drove into the hunting area, about two hours from home, with three vehicles. As was my custom, as soon as we came to the hunting area, I stopped and prepared my gun. Since we were on grassland plains, animals could be spotted at any time. A vehicle was about a quarter kilometer ahead of me, and I was slowly catching up to it. Suddenly, I saw them put on the brakes and come to an abrupt halt. Just then I saw a warthog, with its tail erect, bolt from the right in front of their vehicle, then veer to the left. In well-practiced routine, I jumped out of my vehicle onto the front bumper, pumped a cartridge into the chamber, and got the warthog in sight. It was still motoring as only warthogs can.

Having hunted them many times, I knew their characteristics. They run like mad until they feel safe, then make a very brief stop to look around, and then head for the hills. Having been frightened so close to the vehicle, it ran for some distance before it stopped. That meant it was a fairly small target to hit from an awkward standing position. As quickly as the warthog stopped, I cranked a shot at him. Needless to say, the fellows in the vehicle were very upset and said, "Why did you shoot our warthog?"

I replied, "I came ready to hunt and you weren't ready, so it was free game."

Coffee Plantation Crop Protection

Animals and farmers are constantly at odds. Farmers plant crops to harvest in order to make a living, and animals harvest crops to keep living. Places where this is an acute standoff are areas where farms border on parks or conservation areas. We are used to saying "dumb animals," but nothing could be further from the truth. Animals living in the protected areas will stand within a meter of the boundary and know they are protected. Sometimes the boundary may only be a plowed furrow from many years ago.

The Ngorongoro Conservation Area is a haven for large wild animals such as elephant and Cape buffalo. As soon as darkness settles in, they move into the bean fields or coffee plantations. On bean fields they can be easily spotted, but the coffee plantations have only an occasional maintenance road, so they wander at will.

A friend invited me to his coffee plantation to drive off some bothersome Cape buffalo. A few of his workers were to be my guides. We waited until around 8:30 p.m. and then drove into the coffee plantation with a tractor and trailer. Night hunting is permissible when protecting crops. I was on the trailer with my 458 rifle, which packed a powerful kick. We weren't into the coffee very far when herds of Cape buffalo started running across the cut lines in front of us. Under the light of the tractor it didn't take long, with my 458 Winchester Magnum, to down a nice bull with a respectable set of horns. Everyone feasted that night, and I am not sure if they were much good for work the next day.

Since the Cape buffalo had been taken while protecting crops, the horns had to be lodged with the nearest game post, which was at the entrance to the Ngorongoro Wildlife Conservation Area. A certificate of ownership for the horns had to be issued in the main game department in Arusha. Leaving definite instructions that I would be returning to claim the horns at a later date, I returned home to Arusha. About a month later, I was travelling through that area and presented my certificate. They checked on all of the tin roofs of the buildings but could not find the horns. Finally, the game scout confided that they had sold the horns to a tourist. Such is the life of the hunter.

CHAPTER 15

Don't Stop! Don't Stop!

A number of desirable African animals are nocturnal.

Many are to be found foraging on large bean fields around Arusha. My Italian friend, Mario, knew a farmer whose crop was being ruined by bushbucks and warthogs. The bushbuck was an animal I had not added to my trophy list, so I was excited to have the opportunity to bag one. In late afternoon we drove out into the sparsely populated rural area. Food is an important part of life, and we couldn't get away from the farm manager until we had eaten roasted guinea fowl. Sweet African tea followed. Things are done in a leisurely fashion. Managers lead a spartan, lonely life, so visitors are held captive as long as possible.

We headed out through the large Saskatchewan-sized bean fields. It didn't take long to spot ostrich, warthog, wildebeest, hyena and gazelles of different varieties. Our African helper was posted on the roof rack shining my one million candlepower spotlight. Suddenly he tapped on the roof, indicating he had spotted a bushbuck just to the right of us. Climbing onto the roof rack, I had a good view, and with a well-placed shot got a very nice buck.

African animals have very heavy hides compared to North American animals. This means if you want to save the hide, the animal has to be skinned and the hide salted right on the field. In my case it was to be a full head mount, which required meticulous skinning of the hide along with the delicate work of skinning out the eye areas, the ears, the muzzle, and the mouth. This was being done in the middle of a soft bean field with the help of car lights, the only sound being a chattering Italian. Immersed in skinning, which I enjoy as much as the shooting, I was oblivious to what was going on around me. As I finished skinning, I suddenly looked up. There was our helper on the roof rack merrily shining the one million candlepower spotlight. In my excitement I had shut off the vehicle, so the battery was being drained big-time.

My worst fears were realized. When I tried to crank the engine, all the battery produced was a faint, weak growl. We could try to push the Land Cruiser and pop it into gear once it got moving a bit. It took a try at that to realize that pushing a big 4 x4 vehicle in a soft bean field was not going to produce any desirable results. Being miles from anywhere, our only option was to sleep in the vehicle and hopefully, somewhere in that huge, sparsely inhabited area, find another vehicle or tractor the next day. My Italian friend exhibited his true national characteristics by swearing and recounting to us what he could be doing at home — mainly lying in a warm bathtub drinking a beer. It was close to midnight, after all.

Reluctantly, we each nestled into a corner and really had no hope of much sleep. It doesn't take long for a vehicle to get pretty stale smelling with three men in it. Windows couldn't be opened as we didn't want to be dinner for the millions of sadistic mosquitoes.

My mind wandered back to farm days and what people did when engines, or even tractors, didn't have a battery. They cranked the engine! I had done that myself on a number of occasions. Toyota Land Cruisers came with a crank to drop the spare tire, and I had seen an opening in the front of the vehicle to insert it into the engine. Suddenly we had new life and piled out of the vehicle. The crank fit nicely into the engine; however, it had not been made strong enough to crank the engine. As I pulled up, the three-foot rod bent like overcooked spaghetti. No amount of trying different tactics worked. We dejectedly returned to the vehicle. As quickly as our spirits had been lifted, they were dashed. Mario was a constant reminder of our hopeless situation and what we could be doing if we were at home.

Mulling the situation over in my mind, I realized that the tire-changing rod went into the motor just above the bumper. *OK, let's give it another try!* Mario would stand on the bumper with the rod running just ahead of the heel on his boot. That would drastically reduce the rod length, which was prone to bend. With Mario still swearing and I saying a silent prayer, I give the crank a sharp pull. It was just enough to get the engine to cough into life. What a sweet sound that was!

The way out could best be described as a track winding through soft bean fields and scrub brush areas and across sandy, dry river-beds. Every time I slowed down to take a bend in the track, Mario had visions of the Land Cruiser stalling and would shout, "Don't stop! Don't stop!"

My bushbuck head mount holds a unique story—one which, after all these years, never fails to give me another belly laugh as I think of Mario and his bathtub and beer.

Shooting at the Eyes

Crop protection is an ongoing and never-ending struggle. Visiting a church amid a giant sugar cane plantation near the Uganda border, the dinner conversation turned to destruction by wild animals. Realizing that I loved hunting, the manager invited me to help them. Crop protection legally happens at night. Animals have learned that daylight is not a good time to be around. Sugar cane grows to 20 feet tall, affording animals total and safe cover. Every half-mile, 40-foot swaths have been cut through the plantation to facilitate the movement of large machinery from field to field. Typically, these areas are where game meanders as it forages for food.

At 8 p.m. we headed out into the plantation to intercept these marauders. Heading down a long open swath, my guide motioned to stop the vehicle. He was sure he saw shining eyes ahead. Slowly we drove forward and, sure enough, there they were—shining eyes. I asked him if he was sure it was an animal. He had done this more than I had, so I could trust his assurance that it was an animal. Taking aim and allowing for bullet drop over the distance, the still night vibrated with the thunder of the rifle. The eyes remained stationary! That was *not* normal!

Cautiously we drove toward the "ever shining eyes," and slowly the rear of a truck came into view. Eyes became taillights. As we approached on foot, we noticed the engine was still running and both doors were wide open. It had been a very long shot, so the bullet fortunately never reached its target. Being so close to the Uganda border at a time when Tanzania's experiment with socialism was draining the country, these fellows were probably trying to smuggle something.

Shutting off the truck engine, we shouted to see if anyone would return to the truck. Both driver and passenger were very likely both still going full tilt in opposite directions through the sugar cane. I can only imagine the story they had to tell the next time they sat around the evening fire.

Good Times and Bad Times

My friend, Karl, and I went on a three-day hunt—the longest ever for me. With so many animals around, an extended hunt was not needed. Rarely were animals in a location where it was impossible to go and return on the same day. The area we were heading into was very near the Uganda border. Because we were expecting to do some major hunting, we brought three men along as helpers. With all of our hunting gear, extra fuel, spare tires and the men, it was necessary to take two vehicles.

With a rough map in hand, we followed the main road (dust, stones and washboard) heading toward Rwanda. This part of Tanzania is exceptionally beautiful as the roads follow along the crests of high hills. The grass is very fine, and the landscape looks like it has been freshly mowed. We had been following hill-tops for probably three hours and finally arrived at a game warden station, where we signed in. The warden directed us to our area.

We took a sharp right turn off the road leading to Rwanda, and we were headed down a steep escarpment. Apparently, long ago, buses travelling into Uganda had used the road we would follow to the hunting area. Descending down the steep slopes of the escarpment was hair-raising and beautiful all at the same time. Washouts and deep ruts kept us in four-wheel drive and low gear. The fact that both vehicles had four-wheel drive certainly increased our level of confidence that we would ever make it up those slopes again. Eventually, the steep grade bottomed out into a mixture of plains and scrub brush.

From the ancient maps we identified hills, wound our way between them, traversed through small creeks, then kept another hill to our left and headed for one far off in the distance. That was our target area. Our plan was to hunt until dark (7 p.m.) on the third day and then head home, hopefully arriving around midnight. It had been a few months since the long rains had moved on. We weren't too concerned about finding our way out since the dry grass was three to four feet tall as we drove in. Following the track would be a simple thing. Besides, we went in with two vehicles, so the grass was well trampled down. Eventually, even the old bus track became invisible, so we needed to keep a visual of the topography in our heads. Some streams we crossed had water in them, and some were very simply soft sand. Past experience had taught us that getting stuck in that is worse than getting stuck in mud. Sand has no foundation: within a short time all the wheels have dug down, and the vehicle is hopelessly hung up. Taking a full run at the sand streams was the way to go.

It was nearing 5 p.m. when we arrived in the area, where hunting looked promising. Two men were dropped off to arrange camp. We went in search of an animal for supper. We got a hartebeest, which later thrilled our national men as it meant lots of meat for supper. Butchering and skinning does not take long when you have done it dozens and dozens of times. Returning near dark, a cozy fire with big dry logs in place was already blazing. It always amazed me how quickly they could get a fire going and have green sticks, acting as BBQ spits, in place. People knew that "whites" liked steak, so that was on the menu. Tongue, intestines, brain, liver, kidneys, heart and lung are among their absolute first choice. That bit of menu preference meant that everyone was happy. Nothing compares to eating in the African outdoors at night. Roasting meat over an open fire adds the cherry to the sundae. It was a leisurely feast that was still in progress when Karl and I called it a night.

Most people can only afford to buy meat once a month, so having this smorgasbord in front of them was unbelievable. It was definitely an opportunity not to be squandered. Erecting a tent seemed like a waste of time and, although possibly a bit less comfortable, I opted to sleep in my vehicle, parked not far from the bonfire. I was there to hunt, not to live in luxury. It was 2:45 a.m. when I woke and looked toward the glowing bonfire. Three men sat huddled around it, and in slow motion stretched out sticks holding more meat for roasting. In the same slow motion, after a period of time, they withdrew the meat and raised it to their mouths. Their stomachs could not convince their brains to call it a night. I can only imagine what shape their stomachs were in. Wow! This was a memorable occasion about which they would long boast to their meat-starved friends. The three men weren't much of an asset the next day.

Hunting the next day was successful as we took, hartebeest, impala and topi. Hunting under the hot tropical sun holds some challenges, such as how to keep meat from spoiling. Warthog spoils the quickest of all, so it is always taken last on the way out. Going into the area, we were not aware of the fact that the army hunted there for their meat supply. They use the chase and slaughter method. That translated into the animals hearing a vehicle engine and heading off at full speed. This was one time where we really had to actually hunt to be successful. Among other animals, Karl and I both had tags for topi. They proved to be the most skittish, and Karl was unable to reach them with his smaller calibre rifle. After a number of failed attempts, he asked me to shoot one for him. Topi typically like to stand on a termite hill with their two front legs. This seems to give them a good stance for spotting danger a long way off.

As we drove through the tall grass, we spotted one at least a half-kilometer away. Checking him through our binoculars, we saw that he was a fine animal. However, we knew there was no way we were going to be able to approach him in a vehicle. Trying to be as inconspicuous as possible, we climbed down off the roof rack. My friend was sure there was no way we could approach him. I had a totally different outlook on the situation. A plan was forming in my head. We would slowly drive off into an area with bushes some distance away so the topi would not be spooked. Coming back, my friend would drive slowly and I would jump out of the vehicle without him stopping. He should come back in an hour. He was most skeptical of my assurances that within that time I would have his topi. Contributing to that was the fact that between the track and the topi there was nothing but grass — no mounds and no bushes.

After we rounded a bend some distance away, I had Karl drive and we returned. I rolled out of the moving vehicle, and Karl kept on going out of sight. Belly crawling was my only option. As tempting as the topi was, it was tempered by the constant realization that snakes are everywhere in these parts. I was at their eye level should I meet one. But this was only part of the problem; fully 98 per cent of all snakes in Africa are poisonous.

Getting within range where my gun scope could clearly view the topi, I realized he was a huge old bull. True to my promise, when Karl returned in an hour, I was already skinning his animal. It was late afternoon, but the sun was still hot when we finished skinning it and headed back to camp. Our camping spot was well chosen, with a few large trees acting as welcome shade. The topi was hoisted into the tree from a large branch. It is important to get it high enough so lions and hyenas cannot reach it. At the same time, it cannot hang too close to the top branch, where a leopard is able to reach down, pull it into the tree, and begin having a feast.

Later, checking the horn length and circumference in *Records of Big Game* for African animals (Rowland Ward Publications), I was thrilled to see that it ranked in second place. However, Karl wanted to keep the horns, so we never had them officially measured. Later Karl returned to Sweden and was unable to take the horns with him. What a shame!

Coming from North America, hunting season coincided with cooler weather, which is not the case in Africa. There is a trick, though, for keeping game meat in the field, even for a number of days. Before pulling the carcass up into the tree, it is wrapped in cheesecloth to keep the flies at bay. It is amazing how much the tree in full leaf acts like a refrigerator. As the carcass cools, the thin membrane that remains turns into another very thin skin. This further protects the meat from spoiling. Once home, that skin is removed. The above pertains to all animals except warthog. Their meat begins to break down very quickly in any type of warm weather. The only thing to do is to harvest them on your way home.

Back to the bush. As planned, we harvested a very nice old warthog on the way. As mentioned, we reasoned that darkness would not be a big issue, it should be an easy thing to simply follow our tracks through the grass, even in the dark.

Things went exactly as planned, with our headlights easily highlighting the flattened grass. We were coming up to the tell-tale line of trees that always grow along a small riverbed of any kind. It was easy to remember that we had to make a sharp right after clearing the brush area. Our progress was good and soon we would drive through the seasonal riverbed, come up on the other side, make the turn, and be on our way again.

But it was not to be. Coming out of the riverbed, we were stunned by what our headlights revealed. As far as the lights shone, we saw nothing but black grass ash. Together with the grass, our road map also disappeared. Our guiding instrument was no longer. Since we had passed, the grass had been lit on fire. In many areas of Tanzania, this is an accepted practice. With the tall grass burned away, and at the first rain, lovely green shoots appear within a few days. No wonder animals head toward the smell of rain.

With the game we had harvested, we did not have the option of sleeping out another night. We had to get it home. What were we going to do? The plan we devised was to have two men stand on the front bumper and direct us, as best they could, where they thought the road was. National men are great trackers. Standing on the front bumper, they were able to decipher which way the track went.

Suddenly they waved us to a stop. The road had been lost; most likely there was a turn. They both got down on their hands and knees and began patting the ground to sense where it was harder. That would indicate the road. Sure enough, they were right, and off we went again. A few hours later we emerged from the burned area and, with relief, saw the road heading up the escarpment.

Gearing into four-wheel drive, we began the slow, steep climb. It was tough terrain, with many washouts and big rocks sticking out of what used to be a fairly good road, I presume. Although painstakingly slow, we were making good progress when, without warning, smoke began pouring out from under my dashboard and everything stopped. We had no lights, and no there was no power to the engine. My wiring harness had shorted out and melted together into a mass of plastic string. It had decided it was time for the end of life. With what we would call back-alley mechanics, we bypassed a few things and hotwired the engine. No dashboard lights or gauges worked, but that was no big issue. The headlights did. It was great to finally reach the summit and the main road, where we could make better time. It was getting on to around 4 a.m., and we were tired and excited to see the town lights in the distance.

Being Africa, we had not yet arrived. The unmistakable vehicle swerving and rumble under my vehicle indicated a puncture (flat tire). It is not what we wished for at this time of night and this stage of the journey. As unpleasant as it is to have to change a baby's diaper in public, it has to be done. Exhausted or not, the same goes for flat tires. Well beyond midnight, with the vehicle full of meat and equipment on a hilly, narrow, dusty road, it is always a challenge. It is also helpful if the torch (flashlight) batteries have some life in them.

In a situation like that, it is best not to begin to think too much — just get out and get the job of changing the tire done. That is not as easy as it sounds, however. The tire tools have to be dug out. *Where is the flashlight right now?* We always carried a 'high lift' jack, so that had to be unlocked, removed and placed under the vehicle. Toyota Land Cruisers had this terrible setup where the spare tire is underneath the vehicle—exactly where all the mud and dust accumulates. A long steel bar had to be directed through an opening just under the back door. Getting it in place so it would catch to begin unwinding the chain holding the tire was a challenge. Invariably it would jam and had to be rewound a bit to loosen the chain so the downward journey could begin again. When it finally came to rest on the ground, it was well under the vehicle, so half crawling under the vehicle was the only way to retrieve it. This would be an acceptable operation if things were dry, but when they were muddy, it was a dirty job. During such times it was obvious that the engineers who developed this system were sadistic.

Wheel lug nuts have a tendency to come loose on rough roads, so they have been tightened to within a few pounds of beginning to strip or snap the lug bolt. Getting those loosened is grunt work for sure. Tire rims are tough steel in order to withstand the pounding roads and deep potholes. That makes them heavy— even heavier at that hour in the early morning. Roads are not built to make changing a tire easy. The vehicle always wants to slide off to the side when jacking takes place. By the time we reached home, the roosters were crowing. We figured we might as well call it an early morning and stay up. Besides, we had to cut up meat and get it into the freezer.

No Way Am I Shooting!

Mario became my hunting friend as a result of the Tanzania socialism experiment. You may not believe this—and I wouldn't either if I had not been living through it—but literally nothing was available in the country. Even locally grown fruits and vegetables were in scarce supply. No screw, no nail, nothing that needed to be manufactured. Mario was an electrician who was able to do electrical work *maarifa* style. For the uninitiated this means, "Use your ingenuity to make it work somehow." In addition to having ingenuity, you need to be inventive. Mario fought in the war for Italy, was captured, and sent to Tanzania as a prisoner of war. After the war, finding the country and lifestyle to his liking, he remained.

It is not uncommon for an ordinary person to meet a high-ranking official, whether from Tanzania or another country. It seems the skin colour breaks down some "positional" barriers. Mario called in a bit of a panic because the Yugoslav ambassador had called him, wanting to go hunting. How they came to be acquaintances I do not know. Arusha was a good location for hunting. The problem was that Mario had no vehicle, which again put into motion another African comment: "You don't have to have a vehicle; you only need to know a person who does." I was that person.

These men do not travel lightly when it comes to firearms. The ambassador had a pistol hitched to each hip plus two other rifles. A Thomson's gazelle was what he wanted to shoot. Unfortunately, it was the dry season, and the area we hunted in had very little cover. Thomson's gazelles have good eyesight, not to mention being fast runners. Adding to their advantage is their small stature and shimmering heat waves on the African plains in the hot sun. Spotting a lone "Tommy," shooting by the ambassador commenced.

After many missed shots, the Italian kicked in for Mario and he started shouting at me, "You shoot! You shoot!" With the ambassador's blessing, I lined up and dropped the Tommy with a very long shot. You never want to show surprise in a situation like this. Eventually another Tommy was spotted, and the whole scenario replayed. *Bang, bang, bang, bang!* Tommy still upright. Mario shouting to me, "You shoot! You shoot!" There was no way I was going to ruin my reputation of being a good shot. I had retired my gun for the day.

King of the Beasts

Nothing can get your adrenaline pumping like going after Africa's "Big Five": elephant, Cape buffalo, leopard, lion and rhino. Rhinos are totally on the endangered list. Only once in my 35 years in Africa did I see one outside of a game park or reserve.

Common to all of the Big Five is their fearsomeness and unpredictability, making them very dangerous. Each hunt is unique. The cats are all baited. For our lion hunt we first shot a wildebeest. It was opened up and dragged back and forth through the bush and across the plain for at least 10 kilometers. Finally, the carcass was securely tied to a tree. Lions drag their prey, making it imperative that it not be moved.

Coming back the next day, a pride of lions was ferociously guarding what was left of their meal. Crawling up to them in the tall grass and open plains is not for the faint-hearted. One fatal shot and a nice male was my trophy. However, skinning and preparing the hide are nerve-racking. Lions have a tendency to retreat and then suddenly charge. This happened more than once, and we had to be on our guard. A shot fired in the air made them reconsider their action but didn't set us at ease — that is for sure. We kept the vehicle doors open just in case. This hunt took place no more than 25 kilometers from our home.

Wow! They Must Live in a Pretty Fancy House

We had just relocated to Bukoba from Musoma. Bukoba is a small town in western Tanzania lying about 10 miles from the Uganda border. Our responsibilities there consisted of being over-seer to the 26 churches, teachers in the Bible school, encouraging women in their struggles of survival, and visiting churches each Sunday.

On the first Sunday, we were invited to speak at a local church in the heart of a large sugar plantation—easily 3,000 to 4,000 thousand acres. During the service we observed a young African lady taking part in the service. What made her stand out was the fact that she wore a brilliant white dress similar to a grad gown.

I am not that familiar with the fine art of laundry, but it didn't take too much "upstairs" to deduce that in the African setting of a church in a mud-walled building with its dusty floor, people walking to church, and no paved roads, it would be impossible for such a dress to survive in a mud hut. She and her husband must live in a very European style home. Therefore, it was not a great surprise to us that we were invited to have dinner at their home. The majority of a meal is made from scratch in rural Africa, so again, we were not surprised that the lady of the house had left for home as we continued to mingle with people after the service.

Probably 47 minutes later her husband and the pastor joined us in our Toyota Land Cruiser to show us the way to their house. Sugar cane grows to about 20 feet, so seeing too far ahead on windy, rough dirt roads is not possible.

"Turn left at the next path," the husband instructed.

To this day I am sure my mouth dropped six inches as I looked ahead to the "house" and saw a mud hut with a lady, dressed in pure white, squatting in the courtyard nurturing a fire set between rocks. Now the mystery of the clean white dress was really intriguing. Clean clothes emerging from a mud hut was something we later witnessed numerous times. How they kept their clothes clean is still an unsolved mystery.

Seventeen years later we lived in the same city, Mwanza. Meeting these people again brought back many memories. An interesting side note: they were African, their name was Brown, and the husband had blue eyes. His father, we learned, was Irish.

CHAPTER 16

Wisdom

Twice in my ministry I specifically recall God giving me

supernatural wisdom in situations that seemed to have no solutions, even after hours of committee discussions. One that I recall specifically took place at a missionary committee meeting. A certain very serious matter arose for which we could not seem to find an answer. We shared our thoughts, opinions and suggestions; we brainstormed and we argued, but to no avail. It was dinnertime, so we "tabled" the matter—no pun intended.

During dinner, as our wives and others were having a normal "good" time, I was troubled and kept silently saying, *Lord, how are we going to resolve this issue?*

Suddenly, the Lord showed me the solution. Reconvening the meeting, I said, "Here is what we are going to do." Unanimously the missionaries saw it as the solution, and peace filled our hearts. I don't remember the details, but I do remember that the Lord helped us. That knowledge served as a foundation for future administrative responsibilities.

"Take As Much Money As You Want"

Due to the lack of finances, many children were not sent to school. Pele was Rwandan, but with Tanzania offering a better opportunity for employment, she was sent there to live with relatives while barely a teenager. Rwandans, well known for their industriousness, are sought after as housekeepers—known locally as "house girls." The term in no way carries a negative connotation. In fact, it shows that a person is capable, dependable and trustworthy. Trustworthiness is a huge factor since these girls have the run of the whole house. This is especially critical because working people are not home during the day. In addition, these girls are regularly at places of employment from 7 a.m. to 9 p.m., so they hear the conversations going on in a house. Visitors speaking about business ventures also come. In short, these house girls become fully aware of the personal and business matters in a household. To refrain from gossip is a characteristic much appreciated and sought after. Pele was such a young girl. Very quickly she found employment with a family. She worked hard and was pleasant to have around.

However, some dangers accompany a young girl working in a household. Pele was maturing and catching the attention of the man in the house. Fortunately, he never forced himself on Pele. Regularly, he tried to convince Pele to sleep with him. Pele's moral values were grounded in her commitment to Christ. Day after day she refused the advances. Preying on her lack of finances, he said to her, "There is a suitcase full of money. If you agree to sleep with me, you can help yourself to as much as you want." Having so much cash around was a common practice for businessmen in a country where the vast majority of transactions were carried out in cash. In my role as mission field director, I carried a lot of cash, to the point of making me nervous. The most I remember having on me was $50,000 U.S. I was travelling from Mwanza to Dar es Salaam to purchase a vehicle. Local bank transactions (cheques) could take a month or longer to clear the bank. I must admit it was one time I kept an eye out as to who was behind me on the street.

The process of dealing with foreign currency was cumbersome. A foreign currency cheque was deposited into the mission bank account. If your credit was good, they would instantly give you the money converted into local currency. Then the process whereby the bank actually claimed the money from the foreign country started. At one point I had $60,000 dollars in the mission account in Canada which actually belonged to the local bank. For a cheque to clear a Canadian bank could take up to six months. Now that is what I call being trusted.

Back to Pele and her open suitcase full of money ... She walked out of the house and never returned to that place of employment. Pele came to work for us and related her story. She felt that having the opportunity to work for us was God's reward to her for being faithful to Him. Besides working in a Christian environment, she also worked fewer hours and received her full salary on time.

Describing her as being faithful, dependable and cheerful only begins to tell the full story. She had a brother who was attending school in Rwanda. It was an opportunity that Pele never had. Her salary, although generous compared to others, barely covered her personal needs. Living very frugally, she was able to send money back to Rwanda to help her brother with school fees and living expenses.

Pele had worked for us five years when she received word that her mother in Rwanda had passed away. She made the difficult decision to return there to look after her brother. Going back to a small village where no possibility of employment existed was a huge sacrifice. After some time, she phoned us to ask if we knew of anyone in the city of Kigali for whom she might work. Without giving her any hope, we contacted a couple whom we thought might benefit from her work ethic and outstanding character traits. They, in fact, did need her, and she worked for them for many years.

We smiled with delight as she had obviously gained access to a computer and sent us an email. However, she did not know what the space bar represented. Her letter was one long sentence, running on and on. There were no punctuation marks whatsoever. It was fun trying to figure out where a period or comma should go.

When her brother finished school, she had the opportunity to spend more on herself. But that did not happen. She found a young orphaned girl, took her into her home, and treated her like her own daughter. What caring she portrayed!

Over time she found a fine husband, and they now have their own children. We vividly remember her calling us again and telling us about that blessing. Then, when her first baby was due, she excitedly called again, asking us to pray that she would be a good mother. She was a gem in God's kingdom, and we are grateful for the privilege of knowing her.

Creative Commerce

Things have not changed any since biblical times. God instructed business people not to use dishonest methods; obviously, they were doing that or no mention of it would have been made. East Africa has a vibrant small business sector. Produced in local small gardens, goods are sold at roadsides or in open markets. It is buyer beware. People can be very creative and innovative. Bags of rice that weigh 100 kilos (220 lb.) are set on large rocks so that the indentation reduces the volume and weight. If people do it enough times, some profit is gained. Some gravel mixed in with it increases the weight and profit nicely. Potatoes and onions are sold in *debes* (tins). Cooking oil used to be sold in them. Over time, the thin tin develops bulges on the sides. It may not be much, but the tin becomes larger, requiring more produce to fill it. To return the benefit to the advantage of the sellers, the sides of the debes are pounded in. Very large potatoes are always placed on the top to give the impression that the whole bunch is of that size. Charcoal is handled the same way. The trick is that you do not buy the debe—only what is in it. Emptying the produce into your own bag, the hope is that discrepancies will not be noticed until you return home many miles away. Mixing in rocks that are the same colour as the potatoes enhances profitability. Watermelon sold by weight is injected with water, carrying who knows what kinds of little creatures looking for a nice warm home.

Purchasing curios carved out of wood is also a challenge. Broken horns or cracked wood is so cleverly patched up and covered over with shoe polish that it does not become evident until much later. Cow bones are a really good deal for the seller as gullible tourists readily buy it, thinking it is ivory. Plastic elephant tail hair bracelets fall into the same bracket. Sellers love to hear buyers ask, "Is it real?" It is a wonderful question to hear.

"Naturally, it's real!" That is what the buyer was wanting to hear, right?

In some instances a person is willing to pay just to examine the cleverness of the hawkers. Don't think for a moment that these men are simply thieves. No, they have honed their skills to exploit human greed. *Get something valuable at a good deal!* The thinking goes that the seller doesn't really know the value. One example: cell phones are a hot item. No monthly plans exist, so phones have to be purchased and topped up with "pay as you go." The sellers have many of them in their pockets as they mingle with crowds at the bus station. How many of them have been stolen is anyone's guess. Confusion, general chaos, hundreds of milling people with loads of luggage, and dozens of buses all add to the mayhem. Amidst it all, the clamor of dozens of "tan boys" (bus conductors trying to get people to board their company's buses) shouting, cajoling, and at times physically trying to push potential clients onto their buses adds considerably to the carnival atmosphere. Mothers carrying loads on their heads and bags in their hands are there with screaming, wild-eyed children. Man-pulled carts are hauling construction goods for out-of-town businessmen. Luggage and supplies are all loaded onto the bus roof rack.

Then, as people are settled into their seats, the confusion has subsided and everyone begins to relax. Getting to this point has been hectic. The bus begins to pull out, and people lean out of windows shouting final goodbyes to friends. Now the real professionals spring into action. Loping along beside the bus, they engage the people at the windows, offering them really good deals on a cell phone. Everyone knows it is "hot." The bus is speeding up, so the transaction has to be made quickly and secretly. There are police at the bus station. Quick as a wink, money is exchanged, and the seller reaches into his pocket and hands the phone to the buyer. A great deal! Friends begin to laugh as the "good fortune" buyer, upon closer inspection, realizes that the phone has cleverly been carved out of a bar of soap. Master craftsmen! Master deceivers!

On one of my out-of-town trips, I fuelled up at a "reliable" gas station a few hours' drive from home. With that top-up I could make it to my destination. Within 10 kilometers my vehicle began to lose power, sputtered, regained power again, and generally convulsed. Pressing the accelerator to the floor and watching in the rear-view mirror, my suspicions were confirmed. White smoke coming from the exhaust pipe is not indicative of a diesel vehicle. The diesel fuel had been liberally mixed with kerosene. I arrived at my destination in about five hours, but that was much longer than it usually took. Removing the fuel tank, washing it out, and reinstalling it was the only solution. It was expensive and time-consuming.

Bottled water became the "in thing" a few years ago. It was especially handy when on a safari, which is always hot and long. Buying the water involved some detective work. Creative entrepreneurs collected empty bottles, filled them with water, "crazy glued" the security cap, and sold it at a high price. If you did not hear the distinctive snap at opening, you were best to take your losses and pour it over your head. The best idea was to open it right there at the roadside kiosk. That way you might have a chance of getting an authentic replacement.

Counting your change was always a good habit to get into, even when the shop owner was a "friend." A surprised "Oh, sorry" was the usual response if the change wasn't the correct amount.

The mistake was never in my favor. Check your change again next time too!

The gasoline Toyota Land Cruisers I drove could be a pain to maintain. Bone-jarring roads and fine dust conspired to create a lot of banging and sputtering. Cleaning and adjusting the carburetor was a regular thing. Going to a garage was done out of necessity with not a little apprehension as to what expertise the mechanic had. No African worthy of his country's citizenship will ever admit that he does not know what he is doing. Even if he has never seen the motor or whatever, he begins to dismantle and poke around. It is an opportunity for on-the-job training, except the teacher is missing from the equation.

The statement, "Oh, sure! I can fix that" raises red flags after a few unfortunate experiences. My feelings of doubt were well founded when the mechanic left my engine and headed for his workbench. He began rummaging through a variety of old bolts and parts strewn on the bench. Attempting to show Canadian politeness, I asked him if he thought he could locate a part on that bench which would fit my rather new Toyota Land Cruiser. He assured me he could. I was standing and peering into the engine compartment when he returned with a rather triumphant look on his face. He had located an old bolt!

"Where is that going to go?" I asked.

Pointing to a hole on top of the engine, he said, "See that opening? It held a part that you really don't need. The engine runs just fine without it. When screwed in, this bolt seals the hole and everything is OK." *Oh yeah?* I thought. *The Japanese Toyota engineers have included a useless "appendix"? I doubt that very much!*

Once the "appendix" was removed and the engine sewn back up with an old, rusty bolt, I didn't have much confidence when I turned the key to start the engine. It ran like a clock. Until the day I sold the vehicle, I never did replace whatever part had departed. Again, *maarifa* (improvise, make it work) had come through. There were times when I was duped, but also many other occasions when I was amazed. That made for constantly being hesitantly optimistic. There were enough failures and just enough successes for the bet to go either way. But bet one had to at times.

The country is awash with one-man industries, mostly working in the open. This can be alongside a road or just a bit off the road. Welding is a huge industry. Making cooking pots shaped like a wok is done by pounding old steel barrel lids into the shape of a wok. It is hard work, not to mention the noise of pounding on steel all day. No type of ear protection is worn. All day long they endure the harsh noise. These men are masters at their trade. You can literally give them a picture or drawing of what you want, and they make it. Because the vast majority of these "workshops" are located outside under the hot sun, they are known as *Jua Kali* Industries (Hot Sun Industries).

Extreme Dedication

Jon was a pilot with African Inland Mission with whom we flew many times in a cramped six-seater plane. One felt very close to everything, including the long drop down to earth. You only had to turn your head a quarter turn and you were looking straight down with nothing below you. Scary!

Africa is dotted with small, dirt/mud/gravel airstrips. Flying is the fastest way, and sometimes the only way, to get into many remote villages. Pilots of small planes have scrutinized their maps for every dirt airstrip in the area over which they fly. That information could come in handy: there could be sudden bad weather or a need to pick up a sick missionary or even a mechanical emergency.

As Jon was flying, he noticed a small airstrip that did not show up on his maps; no other pilot had ever mentioned it to him. From his lofty eagle position, the bush strip certainly looked well maintained. He asked the passengers if they minded if he landed to check it out. Everyone wanted to get to their destination without interruption. Jon did not land, but he made a mental note of the airstrip.

That unmarked airstrip niggled at his mind. About two months later, he had occasion to fly that way again and made a point of landing. He circled the airstrip to make sure no goats or crown birds were using it as a resting place. Though entirely preoccupied by the demands of making a smooth landing on the strip, something registered as being out of the ordinary. Yes, that was it. In this area covered by tall, coarse grass, the airstrip was clean. Whitewashed stones lined the runway. Contacting the dirt strip, billows of dust rose behind the plane. He taxied up to what amounted to a four foot by four-foot corrugated iron sheet guard shed. Out in this wilderness he never expected anyone to be there. One last thud of the propeller, and the engine came to a stop.

Imagine his surprise when a man came out of the guard shed carrying a registration logbook that every landing pilot must sign. With an air of professionalism, he opened it to the page of the next landing entry. Handing it through the small open plane window, he asked Jon to sign in. Jon's head literally snapped back, and his eyes did a double take! Good thing Jon was still sitting in the seat! He could not believe his eyes. In absolute, stunned amazement he looked back at the attendant. To make sure he was not missing something, he exclaimed, *"What?* The last plane landed here 14 years ago? Is this true?"

"Yes, it is."

Apologetically the man added, "Well, actually, since that last entry a small military plane landed, but we aren't required to record those."

Extraordinary! The man was not off digging in his small plot of land or herding his cows or napping under a tree somewhere. Nope, he was on the job.

Proverbs 20:6 says, "Most men will proclaim every one his own goodness: but a faithful man who can find?"

Now I Understand

Missionary organizations have for years been the backbone of supplying transport for missionaries living in remote locations that have no roads. No one would argue the fact that they were a true lifeline and an indispensable cog in the wheel of missions.

One such organization is Missionary Aviation Fellowship (MAF). Although they have career missionary pilots, there are also a number who work for large airlines and take a year off to contribute their skills to missions.

In our 35 years of missionary service, we literally flew dozens and dozens of times with them. Sometimes it was to go into an area to teach. Other times it was to take our children to boarding school.

The times we most appreciated it were occasions when we were faced with a medical emergency and had to get to a hospital fast.

Lake Victoria has many islands that boast fairly large populations, mostly of fisherman. Missionaries live and work on these islands, and the only travel opportunity for them is by plane. Delivering supplies to an island, an MAF plane crashed, tragically killing the pilot. His sister, who lived in Canada, felt bitter. She couldn't come to grips with the fact that her brother would dedicate his life to this "worthless" job. It seemed thankless and certainly meager as far as monetary rewards went. It was definitely not a high-profile job. Airline pilots had huge opportunities in the "real" world, and here he was squandering opportunities for personal advancement.

A number of years passed until she finally felt she could conquer her anger at God and her brother. She had to have closure and go see the crash site on the island. To get to the island, she boarded a small ferry type boat, which was a literal Wal-Mart as far as the diversity of cargo. People, goats, sheep, corn, bananas, iron roofing sheets, bales of clothes, and fuel were some of the items carried.

In some respects, Africa is a wasteful society. However, when it comes to fuel, the people are misers. Grab a cab and it immediately heads for the nearest service station to put in a liter of fuel. Why would you carry something valuable around without its being of immediate use? Spare tires also fall into that category.

Possibly because of the weight of the cargo or the headwinds, the boat operator had miscalculated the fuel needed. He opened the hatch to add fuel. A violent explosion rocked the boat at the same time, making it veer sharply. The unsuspecting sister, standing by the rail, was catapulted into the water. The boat, now having no steering capabilities but engines still going at full throttle, thrashed in circles. Unfortunately, the sister could not avoid it and was slashed by the propellers. Her head, shoulders and back were mangled. Eventually the boat was stabilized, and she was pulled aboard in critical condition. Word of the mishap quickly reached shore. By the time the boat limped onto the shore, an urgent radio message had already been transmitted to Nairobi informing them of an immediate medical emergency.

An MAF plane arrived within the hour. The sister was flown to Nairobi Hospital for treatment. Beyond anyone's expectation, she recovered. It was then that she finally understood the importance of the work her brother had been doing for years.

There were times when people challenged us, saying we were wasting our lives. Working at home, our salaries would be much higher, and certainly the living conditions could not be compared. Besides, you could be close to friends and relatives. There were so many advantages to living at home. Something we experienced many times was that God knows His math. He can add and He can multiply. He can still make something out of nothing. Suffering is a relative term. Suffering in happiness and contentment has no equal. We learned not to allow others to dictate what constituted happiness, contentment, fulfillment and a feeling of success. Being in God's will produces the ultimate fulfillment in life.

We Saw Your Phone Number on the Sign

Friends made plans to come and lend assistance at our orphanage, Starehe Children's Home. Part of the plan was to raise money themselves so the whole family could come. Creative ways were found to streamline their own monthly budget. Extra work was sought to help with their travel fund.

The husband received a phone call from a man, saying he had seen the business sign on a city road and needed his services. The husband replied that he did not know how he had seen his sign and phone number. He had never put out a sign. A competitor of his had. This happened a number of times throughout the summer. When people looked at the sign, they saw the phone number of our friend, not the one actually posted. God is creative!

When we do our part with a genuine motive to please the Lord, He steps alongside and takes care of the rest. He loves doing that.

Get Up! Let Go of My Feet!

Planting churches in a new area never loses its excitement and challenge. Rather than using a "shotgun" approach, we focused on opening new areas that were adjacent to existing ones. This gave pastors from the established areas an opportunity to lend support to the pioneers in the new area. Fellowship and a sense of belonging to a team are critical for moral support.

A pastor pestered me relentlessly to go with him to open a new area. I refused consistently since it did not fit into our proven criteria. It was at least a six-hour drive to Biharamulo from the central city of Bukoba. The distance was long and the roads were unfriendly. A big concern was that the area was on the other side of a vast area of nothingness. That in itself is rare in a country where people seem to be everywhere. That is especially true when your bladder had reached its max 20 minutes ago. Your eyes have been desperately searching for the right secluded spot. Wrong! You look up, and eyes are staring at you. We never were able to figure out how they materialized so fast.

Against my better judgment, I began to toy with the idea of going into the new area. Starting a new church was exciting. However, caution must prevail. I asked for all kinds of information about the area: how many villages were there, how far off the main roads these villages were, which church denominations were already there. Of special interest were Pentecostal churches possibly established by the Swedish Pentecostal Mission. I knew they were working close to that area. There was plenty of work, and competition was never on our radar. I was assured that no such churches existed where this pastor wanted to go.

I didn't have a good feeling even as we started on the weekend trip. Our methodology was always the same. We would arrive at a main market village on a Thursday evening and make contact with a few people. For the next two days we conducted meetings in the local fruit and vegetable market. The aim was to have enough converts to establish a church by Sunday. The plan never failed; we always established a congregation of believers.

Around 4 p.m. we left the main road and headed toward a village another two hours north. I am used to scouting areas, and I kept seeing mud buildings that looked like a typical rural African church. I was assured they were not of the Pentecostal denomination. Just as it was getting dark, we arrived at the target village, which also happened to be the home of my pastor passenger. A white missionary coming to the "bush" village was a big event. Visitors crowded into the mud hut until it was full, mud wall to mud wall. This large group was exactly what I had wished for. Interrogation couched in interest about the local area would get the answers I was looking for. My ulterior motive would not come into play while chatting with them as they did not know the reason behind my leading questions.

Nine o'clock rolled around, and growling stomachs were about to be hushed. Cooked plantains, rice and beans were supplemented with the normal fare of a tough old chicken. My mind was already made up; I had heard enough. After the scrumptious meal, I called the pastor to meet me outside, away from prying eyes, lip-reading experts, and eavesdropping ears. It was a short conversation.

"People have told me that there are a number of Pentecostal churches in the area that have been established by the Swedish Mission. You assured me there were none. We have driven six hours to this village, and we are now going to drive another six hours to return home."

The pastor fell to his knees sobbing and crying, grabbing my feet and asking me to please change my mind. Army commander style, I commanded him to let go of my feet and to stand up. Unlike a good army recruit, he did not respond, and I had to repeat myself a number of times. As I suspected, some time earlier he had had a falling out with the Swedish leadership and was looking for another group with whom he could work in his home village. I was not interested in beginning construction of God's work on such a foundation.

The only other surprise was the one registered on my wife's face when I was standing at the door of our home at 4 a.m. That was not supposed to happen for another five days.

CHAPTER 17

Blessings Left Hanging in the Rafters

Attention-getting antics seem to have no borders. They can be found wherever sales people ply their wares, when sellers of medicine assure immediate healing results, or when preachers take themselves too seriously. Whether he had seen it on some video or whatever, a pastor launched into the dramatic. In his "self-anointing," he proclaimed that he was throwing his jacket—no African pastor would ever preach without a suit—into the crowd, and that wherever it landed, a great outpouring of miracles would take place.

With fanfare consistent with the dramatic moment, he squirmed out of his jacket, and in one huge swing threw it into the crowd—well, nearly. He was speaking in a church where the rafters were just barely above his head. Whoosh! And there was his coat hanging in the black rafters. Rafters are always black, stemming from the treatment they receive to keep termites from eating them. A mixture of used oil and diesel is the recipe used. It seemed the blessing was for the rafters that day. This is what usually happens when we decide we are controlling the blessings.

Waiting to Be Raptured

As I mentioned in another account, bizarre behavior is not peculiar to any one culture or people. In the Arusha Region of Hanang, a group of "spiritually enlightened" people had received the inside track that Jesus was about to come back to earth. To be prepared for this, they all dressed in white *kanzu* type clothing (a long white robe or tunic) and trekked up the forested Mount Hanang to wait for Jesus to come. They spent days up there in the cold. Eventually they had to come down for food and to look after life's necessities. This went on for a number of weeks.

It all came to an abrupt end when, in the darkness, they descended the mountain and left a sleeping child there. The child suffered a horrible death as hyenas ate him. It did not take long for the government to demand a halt to this spiritual exercise. It is the spiritual "wet paint" syndrome. We don't believe the sign. Christ said that only the Father knows the hour of His return. We just need to believe that and leave it there.

Cut Down to Size

We had been living in Musoma on the shores of Lake Victoria for two years. Leadership, both in Canada and Tanzania, asked us to relocate to the very west side of Lake Victoria. Although rugged, it is very beautiful. Large plantations of bananas and coffee cover the hillsides. Receiving over 80 inches of rain a year, it is very fertile and lush green.

The Musoma area, on the other hand, is dry with very sandy soil, which supports drought-resistant crops like cassava, millet, coconuts and pomegranates. During the two rainy seasons, corn flourishes if the rains are spaced exactly right. People are always taking a chance that it will.

Road transport in those days from Musoma to Bukoba was next to non-existent. By road, a normal passenger vehicle took 20 hours to negotiate dirt road, dust, bone-jarring corrugation sections and potholes. Expecting our household goods to survive in a transport truck over such roads would be a giant leap of faith that we were not ready to test. Using a boat barge across Lake Victoria was the only alternative. Our goods had to be well packed. Freight handlers at the port did not wear "kid gloves." They were busy and couldn't coddle every piece of freight. Barrels, which had brought our meager belongings to Tanzania, were used once again. In total we had eight barrels plus a bit of furniture by this time.

Near the last day of our packing, LouDell twisted her ankle very badly, to the point where she could not put any weight on that leg. As she lay on the couch in pain, I continued to pack. There was no alternative: tomorrow the things had to be at the small port, ready to be loaded onto the barge. LouDell had just recently read the book *Prison to Praise* by Merlin Carothers. Contrary to her feelings of frustration and anger at not being able to help with the packing, she began to praise. In a short time she was back on her feet. A brief walk away was a medical clinic operated by Chinese doctors. They could not speak English, and LouDell could not speak Kiswahili or Chinese. Eventually, through sign language, they figured out what she needed. She was given a brown slug the size of a large marble. It looked too suspicious! Her sprained ankle got better without it.

The whole day, while we were packing, our local pastor sat in the yard watching us. He was ready to grab anything that we might be leaving behind. This was not something we were culturally accustomed to. To be truthful, being watched for hours, without receiving any offer of help, really grated on my nerves. Moving to an unknown area is stressful enough. Add Africa to the mix, and then add African culture, and the nerve strings get pretty tight.

Off our goods went to the port. We started the two-day road journey the next day. It was a long haul: five hours the first day to Mwanza, staying overnight in the African Inland Mission guesthouse, and then 15 hours the next day to Bukoba. A one-hour ferryboat ride took us over the Lake Victoria Inlet by Mwanza. From there it was dusty, bumpy and sandy. Bukoba is out of the way, so not much traffic makes use of the road. Villages are scattered sparsely along the way. For sure there are no eating-places, rest areas or hotels. Lunch was taken standing on the roadside. Bukoba town is only 10 kilometers from the Uganda border. Seeing that Tanzania was at war with Uganda, only local residents had any need or desire to travel there.

It was late evening, already dark, when we arrived in the wet town. We were hungry, tired and already lonesome. We didn't know anyone in the town and were aware that we would be the only white people there. Darkness elevates all of those negative feelings and emotions. Driving around the small town, we found the Bukoba Railway Hotel. Normally, the railway hotels did employ a cook of some sort—but this was 8 p.m. and well after dark. The dining room wasn't much brighter than outside. Being hungry, the menu offered some good choices. Good food may lighten our spirits. Choosing from the menu, turned out, presented no problem: the waiter emerged from the yet more dimly lit back and informed us that this late in the evening, only one item was available—so mixed grill it was. Being the sole customers in the dingy dining room had already dimmed our expectations. Turned out the "mix" was close to 100 per cent liver. My wife left still hungry.

Describing what it felt like arriving at our empty house in the dark and damp is hard to explain. Such times demand that a person become a realist and that your brain skip over the present and focus on tomorrow. Surely, with the light of day the worst would have come to an end.

Our house, although not a palace, was right on the lake. From our veranda a week later, we saw the tugboat pulling in the seven barges. On one of them, hopefully, were our barrels. The next afternoon found me in the huge marine godown (warehouse). I was taken a long way through an assortment of goods that had been off-loaded. Among them I eventually found our barrels. However, upon getting them all out, I noticed one was missing—not an uncommon occurrence. Going back to recheck, I still could not see the barrel. Suddenly my eyes saw a steel form that had the grey colour of our barrels. Closer inspection revealed our missing barrel—a miniature, now one-third the size of its adult life. A very heavy object had been placed on it and compressed it. Straight down.

Chinese exports were prolific during Tanzania's experiment with socialism. LouDell had purchased a full set of Chinese dishes. Moving the barrel, we heard only the *tinkle, tinkle* of fine china. Unpacking was not a joyous occasion. Only one cup and saucer had survived—my wife's emotions did not.

No One Is Alive!

It was shortly before 6 a.m. when the phone jangled. Groggily, I reached over and picked it up. Over the loud crackling and intermittent audible voice, I tried to decipher what the call was all about. It is not unusual to receive calls at all hours of the night from bars or wherever. Who knows how the African telephone system operates? Once I got connected to a gentleman in England with whom I had a nice chatty conversation about where we were and what it was like at our place. Crossed lines can be interesting at times.

Exasperation was beginning to become evident in the person's voice at the other end. That much I could tell but still could not make out the gist of the call. Finally they simply said, "Come. All dead." Immediately I knew that something must have happened to our friend. I also knew that someone must have been alive for him or her to get my phone number.

It was a mad scramble as I turned our Toyota Land Cruiser into a potential ambulance with back seats removed and foam mattresses thrown in. Meanwhile, my wife informed her friend that we were coming over. We were not sure, but it sounded like something may have happened to her husband. The day before, she had not had a good feeling about his trip.

Much of a missionary's time on the field is spent alone. Especially for those living in remote areas, this can become a stress point. However, even when living in a city it is not always easy to organize fellowship times. People are busy with their things, plus Africa tends to shut down after 7 p.m. Friends of ours in Arusha worked with The Pentecostal Evangelical Fellowship of Africa (PEFA) organization, and we forged a close friendship. Dieting and then making homemade ice cream once a week seemed like a good, healthy routine. Our times together were always filled with lots of stories and laughter. It was our regular afternoon for ice cream, and we again enjoyed one another's company. In the course of the conversation, our friend mentioned that he was heading out of town early the next morning to attend a mission conference. Remembering that bit of information was what determined which road I would take out of town that morning.

Being early, the road was clear as I headed in the direction I was sure my friend had taken. Over every hill and around every corner I was anticipating coming upon an accident. By this time it was getting light; living so near the equator, light just happens—there is no long-drawn-out dawn. Coming to the crest of a hill on my far right, I saw the Arusha International Airport. Its white buildings stand out like beautiful sentinels against the brown landscape. In the early sunrise it looked spectacular.

Just over another slight rise a few kilometers past the airport road, I was confronted with a grisly scene. There lay a scrap of unrecognizable metal with six bodies lying scattered on the pavement. My friend drove a small white Toyota pickup, but from the mangled mess I could not identify it. I inspected the face of each lifeless body to see if my friend was among them. He was not.

Shortly, I saw a bus come down the hill toward the scene of the accident. It was an unbelievable sight to see my friend disembark. He had recognized my vehicle and had the bus driver turn around to bring him back. His face was splattered with blood and bone fragments. A Third World country, a foreigner, and an accident involving death are not a good mix. The general course of action is to immediately put the person in jail. He remained in jail while a slow police investigation proceeded. "Slow" provides lots of opportunity for behind-the-scenes negotiation to take place. In this country you are first guilty; then you prove your innocence.

In the midst of a sudden catastrophe, keeping a cool head is imperative. Leaving the scene with bodies strewn around yet no police presence, I headed back to the city with a plan formulating in my mind. My friend was taken home to clean up. I chartered a small plane, and within an hour he was on his way to the neighboring country. Only then did I contact the police. It took months for the case to come before the courts. Things settled down so our friend was able to safely return to the country to attend ensuing court appearances.

My friend explained what had happened to cause the accident. As he came over the knoll of the hill and began heading down, his lights picked up a heavy steel trailer, loaded with sand, being pulled by a tractor—there were no lights or reflectors on anything. There was simply no time to react. Magically, by the time the police arrived, the trailer and tractor had disappeared. Amazingly, in his dazed stupor, our friend found his camera and had the presence of mind to record the scene. Court allegations were that he was driving without lights; however, the pictures clearly showed his lights still on in the mangled wreck. He was exonerated of all fault in the accident.

The phone call had been made by someone who was travelling to the airport about five kilometers away. Who the caller was, I never found out.

Corn in the Dung

In 1974 the rains failed. The Musoma area which lies around Lake Victoria consists mostly of sandy soil. Possibly this was due to Lake Victoria's being much higher many years ago than it is today. Agronomists have designated that whole area as semi-arid, which means that only drought-resistant crops should be planted. However, as we all are aware, there are certain grains we like and others that we use only when necessary. Cassava is an interesting root. It contains a poison, and if not properly dried or roasted, it can cause death. The cassava root is dried and pounded into flour, which is mixed with corn flour. Millet is another grain that is recommended for planting. Unfortunately, few people like to eat it every day.

Cooked cornmeal is the preferred food around the lake at Musoma. The drought to which I referred was so severe that people actually began to starve. Distributing government aid through aid organizations involves a lot of red tape, so by the time help arrives, a situation that is already dire has dramatically deteriorated. Transport into the rural areas poses further delays. The logistics of prioritizing the needy areas and getting it there is a nightmare. Prioritizing involves not only the general area but also people within that area. There may be people requesting food aid who don't actually need it. If they do receive it, they sell it at an inflated price. People are people.

A hand–operated ferry (a man pulling hand over hand on a cable) was the only way to cross the Mara River when travelling north. Its capacity was three cars and a bus. On one occasion, a herd of 16 cows joined the "river crossers." As the cows disembarked on the other side, we saw just how desperately hungry people were. An elderly woman began picking through the cow dung searching for any possible kernels of corn that the cows may have eaten and not digested. That is a picture which still vividly sticks in my mind.

God Does Not Tolerate Sick Pastors

At our Bible college, students were given the opportunity to pay part of their school fees by working on the college grounds. Saturday was the dreaded day. *Dreaded* because they felt it was below the dignity of a pastor to do manual labor, and culturally dreaded because men aren't involved in manual labor. After all, a wife at home might find out that her husband actually had a hoe in his hand. Invariably, Saturday saw a lineup of men at our door requesting permission to be exempted because of malaria, diarrhea, headache, a bad back, and so on.

A pastor came to the door showing us his somewhat mutilated toe. He had hit it with a large Chinese garden hoe. To this day I maintain it was intentionally self-inflicted although, knowing that the hoe was probably an unfamiliar weapon for him, it may have been an accident. LouDell has a soft heart and immediately began to commiserate with the man, who was in obvious pain. The pain wasn't too obvious to me! It was bleeding, covered in dirt, and did look a bit mangled, but not enough to open my tear ducts. LouDell wanted to get some antiseptic and a Band-Aid—really do a compassionate job. I had a look at it and requested a machete to lob off the toe so the man could get back to work. A look of consternation crossed his face as he processed that solution.

Our next announcement to the students was that any pastors showing up sick at our door next Saturday would promptly be sent home. How could a pastor be sick on Saturday and expect to preach on Sunday? From then on, only the genuinely sick dared drag their frame to our door on a Saturday.

Where Have These People Been?

"Oh my! The spare tire is also flat." That statement was pretty indicative of the missionary whose calling in life pretty well excluded any vehicle maintenance. Even the fuel tank was a hollow drum most of the time. The blue Toyota Land cruiser was left parked "as is" when the said missionary went on furlough.

Some missionary colleagues, LouDell and I had flown to Mwanza for our two-week teaching sessions at the Bible college. Upon completion we were thrown into a "let's see what options we have" situation. All small mission planes had been grounded overnight due to a border spat with neighboring countries. Unfortunately, our children, at boarding school in Kenya, were due for their mid-semester break. It was also a break for teachers and staff, so no students were allowed to stay at the school. We had to get back to our home in Arusha, where we had left our vehicles, and then on to Kenya.

With the deadline now staring us in the face, road travel was our only option. But how? Bus service was a Russian roulette game we were not willing to consider: accidents, dust and breakdowns were far too normal, so that option as a mode of transport was quickly dispelled. Before this, the "flat spare tire", "hollow drum gas tank" vehicle was a nervous adventure, even to go for an occasional short trip to the city vegetable market. Now it stood mocking us to take our slim chances with it. It seemed to have "revenge time for despising me" written all over it.

Between the next lodging place, gas availability and showers stood a 10-hour drive, not to mention bad roads, riverbed crossings, and menacingly sharp rocks that were eager to rip into the tires. Few other vehicles travelled that road, so in case of problems, roadside assistance was not really part of the safe arrival equation.

With good humor on our faces but hearts full of trepidation, we packed up and headed out early the next morning. Much to our amazement, the vehicle rattled and banged but did not give up. Allowing billows of fine dust to seep through the doors was its only drawback.

We reached the summit of the world-famous Ngorongoro Crater and veered off through a small rainforest to arrive at the lodge. The parking lot was full of vehicles looking, for the most part, as though they had come through a mud derby. We began to laugh and wonder where they had come from because our road was nothing but dry and dusty. Only too soon did we discover the answer. The road from there on down off the crater and into Arusha was virtually impassable. Most of the parked vehicles had attempted the road and come back. They had been stuck for a number of days.

Desperation to meet the deadline and see our children, whom we had not seen in three months, outweighed the words of the many naysayers ("You will never make it!"). After an early breakfast and a missionary's "It can't be that bad" mentality, we struck out. After all, we were only seven hours from home. We had barely turned onto the main crater road and descended a kilometer when reality—and mud—struck. Deep ruts, greasy mud, steep declines, and narrow roads threatened to deem the "prophets" correct. A tractor pulled us out of the ditch, and we were again lurching, sliding and spinning—the vehicle and the wheels.

Coming over the crest of a hill, our hearts sank. This was my work area, and I knew the road layout well. Looking across the valley, we saw the muddy snake of a long road down into the valley and then up the other side. Adding to the flavor of adventure, which we were not looking for, were lines of vehicles, some stuck, some frozen in indecision about what to do. There was no such indecision for this driver: we had an appointment!

Our four-wheel drive had brought us this far, and we were headed home, no matter what. Gingerly we slipped and slid past other vehicles until we reached the bottom of the valley, where a small bridge crossed a raging creek. From numerous past experiences of crossing the bridge, I knew it had a solid base. Now we were headed up the next hill and the stakes increased. The road definitely looked like it had the upper hand; however, a farmer's harvested cornfield offered a possible detour route to the next plateau. With literally the pedal to the floor, the engine screaming, the vehicle fishtailing, and all tires spewing mud, we inched our way up via the self-made detour back onto the road. Then it was time for a breather to determine the plan of attack for the next step of our journey. Getting out to assess the situation, my feet immediately rose to the sky and I was flat on my back in the greasy mud. There was no time to lament or clean up. I just had to get back into the vehicle, as uncomfortable as that was.

Another field detour again seemed like the only option. With the same procedure in place, we began our ascent. Emboldened by our progress, even though precarious and slow, a vehicle from on top of the hill decided to start their journey down—on the very same path we were taking! When you are moving uphill in mud, stopping spells sure disaster. It was inevitable: I hit the oncoming vehicle a glancing blow but kept the momentum up to again hit a plateau. There were lots of angry words, but we had taken another step. Anyway, we had the right of way as we were the first to enter the self-made detour route. I like the African term for detour: their signs say "Deviation."

It seemed we had conquered the worst of the road and were smiling as we hit more solid ground. Then it happened! We rounded a corner and, looking down over the valley at Mto wa Mbu (the River of Mosquitoes), we saw nothing but a huge lake. Lake Manyara normally borders back at least a kilometer from the small village, but now it had engulfed the village and spread far beyond it into the valley. Having come this far, we needed to get a close-up view and again consider our options. Snaking our way down the rocky escarpment, things began to look even more grim. The lake, covering the road, was at least two kilometers wide. I had never seen Lake Manyara that high.

My knowledge of the road base gave me some confidence. The ladies did not share the same confidence, nor did the other missionary, who normally did not back away from adventure. It would be a treacherous situation if we got stuck in the middle somewhere. We had no clue how deep the lake covering the road actually was. A trick, learned years before in crossing bodies of water, proved reliable: shift into a low four-wheel-drive gear and go slow, no matter what. At one spot the vehicle's nose just kept going down until the water rolled over the hood. *Keep going!* Slowly the vehicle began to rise and we could see daylight. A paved road was just 30 minutes ahead, and Arusha an hour after that. It was jubilation as we drove onto our dry yard. I had to be pulled out of the vehicle as my muddy back and posterior had dried to the car seat. I think we all gave that old blue flat-tire Toyota a loving "We didn't think you would make it" pat. Getting the vehicle back to Mwanza could be someone else's adventure.

A Serious Carnival

On a number of occasions, we got caught in the middle of a circumcision (rite of passage) celebration.

Hundreds of colorfully dressed people were dancing and in a shouting frenzy. It always reminded me of a carnival, except these exhibited both a serious and a jovial side. There are carnivals within carnivals. Some sing a tribal song; some wear satanic-looking masks; some are pointing spears; some are waving clubs; some are pounding drums; some are blowing homemade whistles emitting piercing shrills; some are making threatening gestures; some are banging on the car windows, and some are jumping on the hood of the car. It is best described as general mayhem amid the dust.

Those with painted faces (white, of course—what other colour would show up on a black face?) are the circumcision candidates. Others also have painted faces, but the design easily distinguishes them from those going through the rite of passage. The crowd slowly moves along.

Circumcision of boys and girls is still practiced by a number of tribes. The age range of the group can vary depending on the number of those entering the rite of passage stage. If a large number of young people happen to be in a narrow age range, circumcision takes place every year. If only a few, then a year or two may be skipped until the number increases.

This Is Your End

Any safari in Africa can have an unexpected ending. For roughly six months of the year, our schedule was pretty much established and predictable. On Thursdays we packed the vehicle with used clothing parcels to distribute to the people of our churches as well as needy people in the villages. Boxes of Bibles in the local language of Kiswahili were also part of the load. Tools, spare tires, spare vehicle parts, and personal effects rounded out the list. By that time the rear-view mirror, because of the piles of supplies, offered only a limited view of the road behind us.

Mondays saw us on the road heading back home again—dusty, tired, happy, and feeling fulfilled. Unfortunately, our vehicle did not always share the "happy" part. Bone-jarring roads, heat, dust the consistency of flour, jagged rocks, deep ruts, and the heavy load all took their toll. Tuesday became the day for the vehicle to spend in the garage to get ready for the coming weekend.

One particular Monday was not the same. We arrived home early, around noon. After unpacking, I took the vehicle straight to its favorite car spa, the garage. Thieves had been staking out our house to determine our schedule. No vehicle in the yard signaled a good time for a leisurely rummage through the house. Surveillance confirmed no vehicle in the yard, so Plan A was a definite go. Sacks to carry off the loot were assembled. A good strong steel bar was also needed. A wire in the chain link fence could easily be removed, so no wire cutters were needed. A corner spot for the fence opening is always preferable. Spreading the wire apart would give ample room to drag out the bags containing dishes, bedding, curtains, and even the odd chair.

It would be an easy job. Remove a glass window louver, reach through with a stick and pry open the latch on the steel doors, and viola!—all kinds of things could easily be carried out. An easy job and potentially good pay as the things were sold. Most probably some advance payment had already been received. Where were the guard dogs? We had only recently returned from furlough and brought two young German shepherd dogs with us from Nairobi. We found that if there was not an older dog with them, it took young ones up to a year to become effective guard dogs. Until such a time, most often their presence was not enough to discourage casual entrance into the yard.

It was the stick banging on the living room metal doors that jolted my wife awake and resulted in a hushed but fearful "RON!" together with a sharp poke in my ribs. "I think someone is trying to break into the house!" Now fully awake, albeit with a bruised rib, I crept down the hall toward the living room. Peering ahead through the semidarkness, I already saw the dining room curtains half-pulled out of the window.

Get the gun! Self-defense is uppermost in one's mind as thieves were known to carry guns; besides, police encouraged self-defense in such situations. Normally the thieves would have placed a sentinel to observe if anything was moving in the house, but why bother? There was no vehicle—clearly, no one was home!

Sliding along the wall into the living room, I observed two men. The outside light illuminated them as well as our living room. They were trying to dislodge the hook on the doors with a long stick through a nearby window. By this time I had gun in hand. Both of the men were working the stick when I popped into view and hollered, "OK, this is your end!" and fired my .30-06 rifle above where I knew their heads were. An ear-shattering blow for sure. Dust, jarred loose by the reverberation, slowly descended from the ceiling.

There exists a man in Africa who holds the world record for the 50-yard dash. It wasn't run on a racetrack with sneakers either. It was across my yard in Arusha to the corner of the fence. Shattered glass fell on the man nearest the window. Stunned by the shock of the explosion and the glass, he dropped down on his knees. It took a minute or so for him to realize that he was not yet moving through the Pearly Gates or walking on streets of gold. Expecting a shot in the back, he hunched over and bolted for the opening in the fence.

An elderly gentleman passed our house midmorning the next day as I was working in the yard and casually mentioned that he had heard I had shot at some thieves. I asked him where he lived and he said, "Way across town." Years later, the severed steel burglar bar and lead-splattered curtains still testified to the botched thievery. We never had thieves again.

Is It OK if We Play Now?

Music in Africa is not always soothing to our Western ears. The accompanying instruments can be twanging and banging with no apparent correlation to the song. Each just seems to vie to be heard above the other, performing as a solo part. Before a performance, a long time is spent just getting ready. This is done on the stage of the venue – always at full volume.

We were meeting our longtime friends and fellow missionaries, Dave and Betty Anne, for supper at the New Mwanza Hotel. Years earlier we had spent some interesting times together in the same town, where we were the only Europeans. Although they worked with the Southern Baptist Mission and we with The Pentecostal Assemblies, it did not interfere with our friendship. We all had a pretty good sense of humor, which went a long way when living and working in conditions that would be considered far below the church poverty line in the West. There were no nice church buildings, no well-lit offices, no support staff, and no forty-hour workweeks.

In general, one eats late in Africa. Seven o'clock found us at the restaurant waiting for the doors to open. Already the "musicians" were in full rehearsal mode with noise at a level where shouting was needed in order to be heard. We called the waitress over and asked if she could please ask the live band to stop playing. She did that politely, and a lovely time of quietness settled in. Many things happen to missionaries, and we had a delightful time of sharing stories and just enjoying being together. Laughter was a big part of the evening.

About one and a half hours later, the waitress apologetically asked us if it was OK for the band to begin playing. Looking around, we realized the dining room was packed with people. The band stood, instruments in hand, like a group one would see in Madame Tussauds wax museum, waiting for our approval for them to come alive. Most embarrassing! Again, this showed the gentleness and "willing to please" attitude of the African people. It was an attitude we observed many times. We never knowingly took advantage of it but appreciated it when it was shown.

CHAPTER 18

This Vehicle Belongs to God

During the war with Uganda, there were not many four-wheel-drive vehicles in the country. During wartime the army basically has the right to take whatever they have need of—buildings, vehicles and supplies for war. I discovered this when, years after the war, I tried to get another piece of land for the orphanage. All Land Department documents showed it as belonging to the Tanzania Milling Company. Every time I followed up on why we were not able to purchase the land, I was met by silence. Eventually I was informed that the army had taken it over during the war, and therefore it was still their property.

Since the front of the war with Uganda was in the western part of the country where we worked, we encountered soldiers on the road. While heading to a rural church service, we were flagged down by heavily armed soldiers. They demanded our vehicle, which I hesitated to hand over. Going to a church service always meant that we were loaded to capacity with people, Bibles and books. Staring into the barrel of a gun that was most probably loaded was a bit unnerving. However, we were a long way out in a rural area, and I knew that no buses existed. Certainly anyone with a roadworthy vehicle had long ago hidden it in a banana plantation. If I handed over the vehicle, we would be stranded on the road, not to mention losing all of our supplies. Showing respect but firmness, I declined to turn over the vehicle. It is not wise to debate an order from a soldier with a gun. My defense was that this was a mission vehicle, which meant it belonged to God. I was not at liberty to hand over God's property without His permission. After a tense time of standoff, we were waved on, except that one of the accompanying pastors had to disembark so a soldier could get a ride further down the road. No amount of negotiations altered their mind on that demand. My poor wife was a nervous wreck.

Out With the Table

We spent our first two years of life in Africa living in an area at the southern end of Lake Victoria. Tribes in this area had a very predictable diet which consisted of very sweet tea, cooked cornmeal, and beans. Only occasionally would that change by their using millet instead of the cornmeal. Once in a while, spinach was also served.

Relocating to Bukoba on the most western side of Tanzania also brought a change in diet. Here the people owned large banana plantations. These banana plants grow up to 20 feet tall and are normally planted right up to the road. It felt like driving through a tunnel. The Wahaya tribe is dominant in the area; their diet consists solely of cooked bananas (plantains) and beans.

The first Sunday we visited a church we were royally received, as was usual. Africans are renowned for their hospitality. Not only was the diet of the Wahaya tribe different, but also their lifestyle.

A very fine grass grows in this area; that is spread on the floor of the hut. It is a nice change from dirt floors. Again, as is normal, the "living room" has a number of homemade wooden chairs with backs and seats, each leaning at their own angle. One is either leaning way back and staring at the grass roof or digging one's heels into the floor to keep from sliding off the chair. A small hand-made wooden table sat off in the corner and would most probably be brought to the middle of the room when food was ready. Seldom are any two legs of either the table or the chairs of the same length. Having spent six months of total immersion at language school, and a year and a half visiting churches, we were able to converse relatively well in Kiswahili.

This new culture of the Wahaya people was interesting, but we were thrown into confusion when it was announced that food was ready. The pastor got up abruptly and took the table outside. Following everyone else, we got up off our chairs and they, too, were carried outside. We thought, *Oh, maybe we are going to enjoy the meal outdoors*. But African guests are never taken outside to eat. Again, taking our cue from others, we stood back against the wall. With some fanfare three ladies entered, each carrying a large banana leaf. The first leaf was laid in the middle of the floor; then another was placed on each side, overlapping the first one. They exited and immediately two other ladies struggled through the narrow door carrying a large, obviously very heavy and hot aluminum cooking pot—no oven mitts needed.

With a deftness that certainly would take us a very long time to perfect, they flipped the big pot upside down on the banana table. Who says a table has to have legs? A resounding *plop* indicated it had accurately landed on target. A *whoosh* sound of suction accompanied their lifting the pot and exposed a heap of steaming plantains topped by finger-searing beans. Being guests, we were instructed to be the first to recline on one elbow around the meal. Others then reclined in sequence the way we had started the circle. If you are seasoned at this, the slope of the floor has already been calculated. Gravity works the same in all countries: the gravy flows downhill! Reclining at the low side dramatically improved the chance of getting more of the desired bean gravy.

A prayer of thanksgiving has already been offered, and as the last person lies down, it is time to eat. This was true finger food, as everyone used their fingers to eat the meal. This is an ordeal for Westerners who do not possess "blacksmith" fingers nor the dexterity to gracefully corral cooked plantains, beans and gravy—not to mention getting it to your mouth without dripping. Because it is a communal, hygienic "pot," you must not touch your fingers to your mouth.

Less than 15 minutes had elapsed before all that remained were three banana leaves exhibiting little sign of leftovers. Everyone got up, and in came the kitchen brigade to roll up the banana leaves and exit. Then came the furniture, and within three minutes we were again back on our chairs and into full conversation.

"Lutherans"

Language school was tough, and our class formed a special bond during that time. The language school was called Makoko Language School. Makoko was the village four miles out of town where it was located. Ten minutes of English preceded our baptism of floundering in the language. "These are the last English words you will hear in this school" was not an idle statement by the head-master. He did encourage us by saying that no one had ever failed to graduate. We later learned that it had taken one student four sessions in the course to graduate. In between, he had very nearly burned the school down by trying to burn bees out of the overhang with a rag soaked in kerosene.

For the next six months we heard not a word of English, no matter how many times we asked for the teacher to explain some-thing in English.

The school, which had an excellent reputation for success, was operated by the Maryknoll Catholic order. Classrooms were eight feet by 12 feet in size, and there were large open windows at each end. Classes began at 8 a.m., went to 10 a.m. with a 30-minute break, and then broke at noon for a typical local African meal. One o'clock until 3:30 p.m. was spent in the language lab listening to a tape and trying to repeat what the "tape teacher" had said. This was the pattern from Monday to Friday, except on Fridays we were given a 10- to 12-sentence story to learn by heart over the weekend. We just learned the sentences without having a clue what we were saying. You can imagine that by week's end our brains were already mush.

Sitting three feet across from us was the teacher. He slowly said a word or short sentence, and each student repeated it individually after him. In order for our ears to get used to different tribal pro-nunciations, a teacher from another tribe took over each month. Our lessons introduced us to basic language used in religion, mar-kets, medical terminology, greetings, customs and culture. After six months of total immersion lessons, we were able to converse on a basic level about a number of subjects. It would take some time to add to our vocabulary. The local people were well aware of new students and were very kind in helping us as we struggled to use the language. The local market was a favorite place to try out what we had learned. It would take many more years for us to hear what was being said and then, most important, to understand what had been said.

One example: *Karibu* means "Welcome!" It was something you heard when announcing your presence at the door. If nothing else is said, you remain at the door. Only if the person added, *Karibu ndani* ("Come inside") would you enter. The same went for arriving at a house where the people were eating. Out of politeness they would say "Welcome," and even "Welcome for food." However, unless they insisted you join them, they did not literally mean you were welcome to join them.

Most of the 45 students attending the language course dispersed into various parts of Tanzania after completing the course. Being that we were also from diverse missions meant that we would occasionally run into the other students on some road or in some village market.

David was my classmate; we sat side by side in a class of three. Although Lutheran by denomination, he ended up working very near to us in a Catholic Mission Station. He was an agriculturalist who had a very good rapport with the people.

After his two-year contract was finished, he went to Nepal. While there he had a rather humorous experience that he shared with me years later. Nepal is not known for having many churches, so David was extremely interested when he a met a man from a village who said they had a lot of Lutherans in their area. David made arrangements to trek to the village a few weeks later. Meeting the man, he inquired as to where all of these Lutherans might be. The man waved and said, "They are all over the place!" This really sparked David's curiosity.

"Take me to one," he requested. Pointing, the man said, "There is one right there." This was very confusing for David as the man pointed to an outside toilet. The unraveled mystery was that the Lutheran Church from Germany had given money to the rural people to build toilets. Having never had toilets, and not knowing that Lutheran was a religious denomination, these became known as "Lutherans."

This Could Be the Trumpet of the Lord

The best instruction engages the reality factor. Bible college students can become a bit desensitized to the realities in their lessons. Classes morph together when sitting in a hot classroom day after day. These students are adults and, when at home in their villages, spend the majority of their time outdoors. At college, notes are taken with the good intention of actually absorbing them once the head has cleared from instruction overload. My teaching subject centered on the trumpet sounding the Lord's return, and our readiness for it. I thought that a practical demonstration would reinforce the classroom instruction.

Our house, being just above the college dorm, offered the perfect opportunity. At five o'clock in the morning, I tiptoed past the sleeping watchman at the dorm door and quietly opened it. I pointed my trumpet down the hallway and played a loud bugle. For good measure, I repeated it a few times. Cement block walls are great for reflecting sound. Things began to happen pretty quickly.

Some students started praying. Others—wide-eyed with a blanket quickly wrapped around themselves—bolted out of their rooms. Some students began to feel around in the bunk above them to see if their friend was still there. Others started feeling their arms to see if they were still flesh and blood, and yet others just came out into the hallway in a daze, thinking it was a dream. One student thought it was the foghorn of a ship on Lake Victoria. The exercise offered a good time for reflection in our next class.

Raking Avocado Leaves

Avocado trees can grow to 60 feet tall. Ours hadn't reached that height yet, but you wouldn't know it by the pile of leaves underneath. Actually, that is what had me out in the yard one Saturday morning. It was 10:30 when I heard the sharp *rap, rap, rap* on the eight-foot tall steel yard gate. Strange, how after a while one can tell what a "gate rap" sound means. It was a rap of urgency. A boy of eight had passed away. Would I do the funeral? Funerals in the tropics take place quickly, so I had attended very few. We spent much of our time in the rural areas, where death and burial are separated by a matter of only a few short hours.

"What time?"

"Eleven o'clock."

"Where?"

"By the river near the flour mills."

Before I could say another word, the mother ran off. That math I could do in my head. I now had less than 30 minutes. All I had to do was prepare a short devotional, change clothes, dash across town, and find the spot on the river. Obviously, some things were going to be left off the agenda.

Arriving on time was going to be a fight between the clock, human-powered delivery trolleys, bicycles carrying iron roofing sheets, the odd herd of sheep and goats, and cars. Back roads are the answer— if they are not being dug up for some long-stalled road project or obstructed by a big tree that has been felled so firewood could be chipped from it.

With only a few minutes to spare, LouDell and I had parked the vehicle and zigzagged through the maze of tall grass, arriving at the river. Sixty feet away, a small group of people stood on the riverbank. That was the funeral! Fortunately, a narrow path took us to the group—just in time! They were about to lower the shroud-wrapped body into the shallow grave. Had another couple of minutes passed, we would possibly have been in time to help throw in a few shovels of dirt. There was only one short-handled rusty, broken shovel, so everyone took turns. Seeing me arrive, they squeezed in time for a short devotional, "There Is a Time for Everything," and a prayer of comfort. Attendees were already getting impatient. Some needed to get back to work, some possibly had another funeral to attend, and some were just onlookers. The "moment of interest" had passed. Besides, flies were already buzzing around the body.

I wonder about things. Who was that boy? How did he die? Would anyone ever know where he was buried? Would the high water of the long rains wash him away? Would his mother ever find that grave again among the fast- growing grass and bush? How much time did she have to mourn? Was the boy's life important to anyone? Would he be included in the oral history of the family?

Then I think of Jesus, who said that He takes note of every sparrow that falls and every hair on our heads. Amazing and comforting!

In Mud and Cow Dung Up to His Knees

It was early in my missionary career, and new experiences were coming every day.

I, along with some national pastors, had travelled to a rural area with the intention of opening a new church. The local market is a hub of activity, and that is the place to be. People are milling about, shoulder to shoulder. Tropical heat generated stifling body aroma. The still air kept it there. Some had made the 10- or 15-kilometre walk carrying heavy loads or herding their ever-wandering live-stock. There were vendors hawking everything from self-made pots and pans to bush roots capable of curing the incurable. Just looking at the mixture would cure me psychologically, if nothing else.

Gauging by their loud, high-pitched bleating, all the goats seemed to be in distress. More ladies carrying large clay water pots kept joining the melee. Fruits and vegetables, which were sensitive to the hot sun, were under the low-slung, grass-covered branches of a long shed. They were all heaped in pyramid-shaped piles: peas, beans, corn, millet, and sesame seeds, along with a great variety of spices, made for a kaleidoscope of colour.

African people are curious and have no inhibitions about getting close to the action. Loud megaphones inviting people to join our crowd within the crowd were actually redundant. Anyone standing on a chair and speaking is an automatic magnet. Add a white face, and a huge crowd presses in. Our well-worn yellow gospel song-book in hand, we sang acappella through a half-dozen hymns. No one knew whether we were off-key or not—no doubt we were way off! Anyway, who cares? The tune "Are You Washed in the Blood?" would be unrecognizable at home.

Once the gospel was presented and sinners were prayed for, we were off into the surrounding village —hoping that every villager was not actually at the market. Not to worry: there are always people everywhere. Heading down a dirt street—every street is dirt—I find a man up to his knees in mud. In front of his hut, he has dug a hole six feet across by three feet deep. I am curious about the dance he is performing. Some spiritual ritual, no doubt. He doesn't seem to mind my staring at close range. I make the first move. Literally translated, the conversation goes like this:

"What are your matters, old man/wise one?"

"I have no matters."

"What is your news?"

"Everything is all right."

"What is the news of your wife?"

"She is in one piece—she is fine."

"What is the news of your children?"

"My children have no matters. They are fine."

"What is the news on your farm plot?"

"It is coming along nicely."

"How are your cows?"

"No problems."

Greetings out of the way, I question what he is doing. A silly question, but perfectly permissible coming from a white face. He has returned the dirt which he just dug out, poured water into the hole, added a good dose of cow manure, and now he is mixing it with his feet. It is plaster for the walls of his new hut.

"May I speak to you a little?" I ask. This is crazy because I have already been speaking with him, but my question indicates I have something special to speak to him about. He interrupts his mixing job and motions me to a homemade wooden chair, which sits at the angle of a lounge chair, under the mango tree. He takes the other chair. No yard is ever completely furnished without chairs sitting under a mango tree.

"Have you ever heard about Jesus?" I ask him.

"No, I haven't."

Not too long into the conversation, he quietly says he would like to have his family come. In a rather subdued voice, he calls, *Njooni!* ("All of you, come!")

Naturally, by now they have all been hiding just inside the hut door trying to catch what this white man is saying. They come—the wife, the five children, and the visiting neighbor's wife. As a sign of respect and also because of the lack of more chairs, they sit on the ground in a circle close to us. I love this about the African culture. Closeness. I hadn't gotten too far into my story of Jesus before the "old man" had called his family, so I didn't have to go too far back to get to the beginning again.

At the end of telling them the gospel story, I asked if they wanted to become followers of Jesus. As a family, in their own tribal tongue now, they began to discuss it. After some time of discussion—everyone seemed to have something to say—they agreed that they wanted to become followers of Jesus. I prayed with them and invited them to a service we had planned for the coming Sunday. Sunday morning arrived and there they all were, sitting at the front of the crowd, eager to hear more. Their decision had been serious; they were committed.

He Washed His Last Dish

Moses was born in Zambia. I mention this because African countries also have distinct cultures, just as we do in North America and Europe.

We first met Moses while he was working for fellow missionaries. They lived in the coastal city of Tanga, which in itself has a very interesting history dating back to German occupation in the late 1800s. Most of the city sits on a plateau a hundred or so feet above the warm Indian Ocean. The coast is associated with heat and humidity, and Tanga was no exception.

Our place of work was Arusha, 300 kilometers to the north. It has a very moderate to cool climate at times. How else would the snow remain on Mount Kilimanjaro all year round? As we drove down toward Tanga, the weather got hotter and hotter, and the humidity notably increased to the high 90s. Driving into Tanga, "Old Arab" influence is immediately thrust upon you. Houses and businesses are built close to the road, and you see open verandas and flat roofs. Brilliant whitewashed walls reflected the bright tropical sun. Driving along the oceanfront and over the narrow bridge, we knew we were getting close to the gate (most houses in Tanzania have gates). Giving the horn three or four medium-length toots, we knew that shortly Moses would come sprinting over to unlock the gate and let us in. By the time we got the car parked and the doors open, he was already there enthusiastically greeting us. "What are your matters, Grandfather and Grandmother?" The usage of the terms "Grandfather" and "Grandmother" show special respect.

"We are completely fine."

"I fall at your feet."

"Do it many more times, Moses."

"Are you both well?" "How was your trip?" "What is the news in Arusha?" "How are Stephen and Rhonda?" Stephen is our son, so, although the youngest, he is mentioned first because he is a male. Rhonda is a girl, so she was mentioned after Stephen.

If LouDell had been alone, the greeting to her would run something like this:

"How are you, mother of Stephen?"

Seldom, if ever, is the lady addressed by her name. Her importance lies in having a son.

Sorry, I have left you hanging on the "I fall at your feet!" phrase. Sounds rather slavish, doesn't it? Well, in fact, it is slavish. The history of the occupation of Africa is symbolized by slavery. Arab slave masters demanded that they be "bowed down to" and shown deep respect. Slaves were commanded to use the greeting, "I fall at your feet" to show that respect. In great humility (I am being facetious) when accepting the subservience and the respect, the slave master would say, "Do it many more times."

Many slaves developed unbelievable loyalty to their masters. The servant of the explorer David Livingstone is an example. However, the absolutely callous attitude of reciprocating in any way is clearly exhibited by British aristocracy. When Livingstone's dried remains were delivered to England by the "slave"—actually, his faithful servant—he was dismissed without even a perfunctory acknowledgment. He boarded the ship back to Africa without ever receiving a thank you for his long service and dedication to a famous British explorer. This is hardly imaginable, but that is what an air of superiority still breeds today: disregard for anyone else.

Today the slave greetings remain in use and hold special significance. Gone are the "slave" connotations. Their meaning has merged into an expression of deep respect for elders, people in position, or anyone above one's "class," as it were.

Resident missionaries were completing their time in Africa, and we needed a worker to help LouDell in the house. Moses indicated he was interested in working for us. Houses, even European ones, may resemble "home" but are subject to weather conditions. Tight windows and doors are not needed in the tropics. As long as snakes or rats have a problem getting in, you are OK. Dust and wind are common, so every morning a fine layer covers the furniture. Dusting is an everyday chore. Guard dogs need their food cooked, which takes time when only a small kerosene stove is used and the pot is large. Ironing, washing dishes, and regularly going to the market round out the normal things to do. Oh, yes! Floors need to be washed every day as well.

Besides, the local economy hinges on people having employment.

Moses came to Arusha to work for us. Workers, not unlike us, develop a certain loyalty to their employers. For them to switch that loyalty is very difficult. No matter how mama wants things done, he/she does them exactly the way the first mama wanted them done. This can develop into a bit of a tug of war. If you really want to exacerbate the challenge, come into a house where a worker remained at the house but had formerly worked for another couple. It will be a very long time until you become the boss. Get used to it! You are the visitor.

Moses had this interesting little quirk that when mama got after him to do things as she wanted them done, he would give a deep sigh and say, "Hallelujah!" It took a while to figure out that it really had very little spiritual connotation. It was his way of dealing with the stress of having to change his ways, which he knew were the best. The former mama had taught him, so they must be right. Our calendar pages kept turning, and one day it was blank except for the scrawled words, "Time to return home." It was a sad parting. We paid Moses his severance, holiday pay, and a good amount as a gift for being faithful. He was retiring, so our gift to him was generous. In fact, it was sufficient for him to buy a small house in his hometown of Mwanza. He did not want to do that, so we suggested he open a bank account and slowly withdraw the money as he needed it. "What if I die?" was his reaction to that! In other words, somebody else was going to enjoy his hard-earned money. That was not going to happen.

Four years later, we returned to Africa and lived in Mwanza. It was a delight to see Moses and his wife again. He had not changed. Each Sunday morning he was in church smartly dressed in his suit. We occasionally visited them in their very low-cost square compound home. They had two tiny rooms. It was a mud building with a tin roof structured somewhat like our motels. Each house, if you can call it that, looked out into a dirt courtyard where people cooked, washed clothes, and just sat around under the lone tree. As always, you could find lots of children kicking around rolled-up plastic serving as a soccer ball. The outside wall of the rooms served as a security fence. A large tin gate was the only way into the compound. Built mud wall to mud wall, the home lacked any privacy.

Early one morning we received the sad news that Moses had passed away. A four-wheel drive was needed to get to the house as we had to negotiate the washouts and overflowing drainage ditches filled with all kinds of unsavory things. We headed across the compound toward the open door of the room. Only a few people gathered. Crouching down to negotiate the low door, we entered the dark room. As our eyes adjusted to the darkness, we saw Moses, or what was left of him, lying on his single bed. He would remain that way until the next day. Throughout the day, mourners came and went to express their condolences to his wife.

Midmorning the next day he was carried, shoulder high, on a stretcher type of platform to the public city cemetery. A few church elders had secured a burial plot in the city cemetery and had dug a shallow grave, into which his body, wrapped only in a shroud, was laid. A few Christians from the church were there to sing a hymn to bid him farewell. There was no money for anything resembling a proper burial—that is for sure.

As Moses lived, so he died. Simple, yet dedicated to God. We said a sad farewell to a faithful servant. Our feet were heavy as we meandered our way through the helter-skelter graves to our vehicle. Moses was gone, but to this day our memories of him are filled with joy and laughter as we recall his "Hallelujahs" in the kitchen.

CHAPTER 19

Around the Dust Bowl

In Canada, "off-roaders" look for muddy roads on which to make use of their four-wheel-drive vehicles. During the rainy season, such roads were at our doorstep every day. They consisted of deep, muddy potholes and long stretches of soft sand that were determined to pull your vehicle off the road. Some roads morphed into fields with no ditches to separate them. In essence, the roads were mere tracks that could change during the rainy season. The change came about as vehicles cut new paths beside the road in order not to get stuck in the old ruts. Trucks never had that option, so they plowed straight through, enlarging the mud holes at each passage.

During the dry season, trucks on these roads sent up huge billows of dust. Trying to pass them was playing the lottery with your life. Through the dust any oncoming vehicle would never be seen. Not only was the fine dust thrown into the air by the tires, but also much of it became airborne through the configuration of the vehicle's exhaust system. The going was slow so trucks geared down, and in order to make any progress, kept the engine revved to where it was whining and screaming. That dispelled a lot of exhaust at a high rate. The huge exhaust pipes always faced down toward the dirt. As a consequence, the dust, having the consistency of flour, boiled up from underneath the truck. From a distance a truck looked like a huge, slow-moving dust storm. The driver looked like a coal miner; there were no airtight cabs here.

Our son, Stephen, and I were returning from visiting a church. I was doing the actual visiting; he was playing hide and seek among the sisal plants with his friends, whom he always seemed to quickly make. Chasing a little white boy around added greatly to the game. Looking ahead down the road, we saw the dreaded dust cloud from a truck. Closing in on it, I noticed that open fields presented themselves on both sides of the road. Quickly shifting into four-wheel drive, we cut out into the field and passed the truck, creating our own dust bowl. Pig holes and stumps sticking up 16 inches are always there to jump up at you in the fields, but this time we made it without breaking a spring or denting the roof with our heads.

Pedestrian, Lower Your Gaze

Living in a small town produced its fun times. I had purchased a small, beautiful white Honda 50 motorbike. It was actually for our daughter, Rhonda, and son, Stephen, but I liked to ride it. It only stood two and a half feet high and was easily maneuverable. Riding on the sidewalk was no problem and probably safest since vehicles could barely see it on the road. To the amusement of all around, I would drive right into a store with it. No one minded.

The bike was a hazard, however. Merrily driving down the street one day, I nearly went for a bad spill. Out of the corner of my eye, I saw a pedestrian come racing across the road right at me.

He was looking out for fast-moving cars so did not notice "little me" below his sightline. He very nearly ran over me. After a lot of swerving and careening down the street, I managed to get control of the bike again. After that I watched for anyone casting his or her eyes from the sidewalk to the other side of the road.

A House for Benjamin

Benjamin was an orphan raised by our pastor, Paulo. As he grew older, he felt God calling him into full-time ministry. As the time came for us to establish some churches in the coastal area of Lindi, he, together with his wife and family, were called upon to head up the work. As was normal, he rented a few small rooms to establish a home. Local authorities granted permission to use a schoolroom on Sundays to begin services. This was not always an easy feat in a small town that was predominantly Muslim.

Lindi was a town with "character" (this is how I describe anything that is not so appealing – to my wife). The main road leading into the town from the capital city of Dar es Salaam was dusty and rough. Although not consisting of too many kilometers, the seven hours it took to navigate the stretch speaks for the condition of the road. Making up for the slow trip was the magnificent view of the Indian Ocean with its hues of every colour of the rainbow. That was great if you were the passenger. Our booked flight had unceremoniously been cancelled, so road taxi was the only way.

Among a few other dozen things, road taxis are not known for having working shock absorbers. They are known for loose doors, fast drivers, seats with protruding springs (with no spring left in them), and potholes that invariably jump right into the way of the oncoming vehicle. The stifling interior served as a magnet for fine dust, which merrily swirled around in the vehicle. A reprieve usually presented itself as invariably the taxi ran out of gas or developed a flat tire. That reprieve served only to move you from a hot, dusty, uncomfortable vehicle into the hot tropical sun and humid atmosphere. No matter how much you think they will not be needed, never forget your hat or a bottle of water. A collapsible chair may also be a good addition to the emergency kit.

Lindi is a quaint coastal town. The road approaches it from a high altitude and descends into this town whose architecture is strongly influenced by the Arabs. Coastal houses are old and white-washed to reflect the searing heat. Roofs are made of woven palm leaves, making for an unappealing brown colour.

Pastor Benjamin had indicated that the best place to meet him was at the local fruit market. Although we were late, sure enough, there he was. He had booked us into a wonderful (read as having "character") local *hoteli*. Unless you have seen one or stayed in one, they are indescribable. A public toilet, Asian style (a hole in the centre of a small room), is regular fare. The hoteli room window usually has a screen with a flimsy curtain that seldom covers the whole window. The electricity will probably be off by the time you want to go to bed, so privacy is not an issue. These *hotelis* are the local watering holes, complete with indefatigable live bands. Fortune smiled on me (I am being facetious again), so I got a room with two windows. One faced the street four feet away; the other looked out onto the six-foot-wide long walkway connecting the street and the bar. That meant all-night free access to the thumping music, arguing drunken patrons, and strobe lights. Within minutes of turning off the lights—I actually had some this night—I heard a harsh rustling sound. Not to worry: I had heard the cockroach symphony before and knew that by now they were swarming out of the shower hole and covering the bathroom. That's when it's time to just pull the sheet up tight and go to sleep. Make sure your clothes are all in the suitcase and it is closed. These little critters are only too anxious to explore new country, and they look forward to hitching a ride right into your house.

Sunday morning we had our breakfast of eggs, a paper-thin slice of toast, and beans along with a nice cup of African chai. Gracing the little square table was always a chunk of very yellow margarine and bright red plum jam. It toned right in with the ketchup bottle of the same colour. *OK, push them aside and enjoy the breakfast.* Africa always has bed and breakfast, so it is a real deal.

Right on time—a bit early, in fact—the local town taxi pulled up to the *hoteli,* and a very proud Pastor Benjamin exited to come looking for us. Riding in a taxi picking up white people at a *hoteli* is about the highest rank you can achieve. No doubt people are taking note of it.

The church service was conducted in a large schoolroom. Students' school desks are built to accommodate two students; these served as pews. It was a Muslim town and the pastor had only been there for four months, so the attendance of eight people was rather impressive, even if four of them belonged to the pastor. Near the end of the service, a well-dressed visitor came in. That in itself is not unusual given that white visitors were there. Ladies from the church had brought food, so we enjoyed a nice local meal. The visitor had been invited to join us and, after the meal's casual chitchat, revealed why he was there.

He was a wealthy government official working for the Immigration Department in the capital city of Dar es Salaam. His family, especially his daughters, who had become used to "the good life" in a capital city, did not want to ever move back to a small Arab town. His house was for sale. That he was not really there for the church service could be surmised from his time of arrival. The first and firm answer was that we were not at all interested in buying a house. Admitting we didn't have the money was beside the point. No one would divulge that! However, this man was persistent that we look at the house even though we had no interest in buying it. Looking is the hook, right? We knew that, but we also knew there was no feasible way we could afford the house. Besides, it gave us something to do during the afternoon. We all crammed into the town taxi, and off we went. Lindi town, as I mentioned, lies right at the Indian Ocean. Pretty soon the taxi was taking us up the surrounding steep escarpment onto the higher levels of the area. Through familiar cornfields and down paths where no vehicle was ever intended to go, we drove and came to a house.

Breathless, actually, we stood and gazed out over the steep escarpment and out to the beautiful blue Indian Ocean. The air was actually cool and refreshing with the slight breeze, which is always present at the ocean. Two and half acres comprised the whole lot. On the property, a little down into a valley, was a cattle shed and a small flowing stream. The house was just in the process of being completed. Windows had not yet been installed, but they were leaning up against the wall at the ready. It was a four-bedroom with a nice local kitchen. As was customary, the actual cookhouse was outside. Another feature was a 20-foot-deep rainwater cistern which had been dug. Water in Africa is gold.

It was now just a matter of politely saying thanks and getting back into the taxi. "Don't you want to know the price?" No, we knew it was beyond our capability. For the sake of politeness, we asked what he was wanting for it. Our surprise was hard to hide. Together with a very reasonable price, he was willing to rent it to us for less than what the pastor was now paying. Besides, the rent would go toward the purchase price.

Our Canadian office does not get excited about the "good deals" that every missionary seems to regularly come across. They were not interested. Certainly Benjamin, already living from hand to mouth, couldn't afford it. It was too good a deal to pass up. I held the mission books so had some room for creative accounting. I borrowed some money upfront from some other accounts and purchased the house. Pastor Benjamin was getting monthly house rent, and I used that money to pay back the other accounts. Within four years we paid off the house by using Benjamin's monthly rent money.

The look on his and his family's faces as we later gave them the news that the house was theirs brought tears to our own eyes. This was something they could never have imagined in their wildest dreams. Sometimes chance encounters are not that at all. Reminds me of the Scripture declaring that the steps of a righteous man are ordered of God (Psalm 37:23).

You Are the Deacon

Church growth far outpaced the number of pastors we could train. Teaching personal evangelism was redundant in our 10 Bible schools and the Bible college. Telling others about their faith in Jesus is already within the spiritual blood of every Christian. There was a very sound reason for this. People who have been set free from the burden of spiritual darkness and the demands it puts upon them know what spiritual freedom feels like. As a result, Christians share their newfound freedom genuinely and with passion. Hearing the good news of the gospel, they receive the message, repent of their sins, and begin their life of freedom in following Jesus. On many occasions I heard a first-time visitor publicly state that he had come to church to give his heart to the Lord.

People are travelers. Even if there is a cholera or meningitis outbreak, they still travel. A ban exists, but paying some "checker" at the roadblock, a small travelling fee to help him with his child's school fees or to purchase a school uniform can easily circumvent that. Naturally, that greatly adds to the spread of a communicable disease. Spreading the gospel takes a similar route as the communicable disease — it spreads rapidly. In some cases the person who becomes a Christian suffers serious persecution, even from family members, who see it as a "white man's" religion or "new" belief when compared to their ancestral belief system that is possibly centuries old. Even a child may be disowned as a family member. The "new" Christian will continue to talk about their newfound freedom in Christ and be prepared to go it alone in life if necessary.

On a regular basis, as a direct result of this "gossiping the gospel," new gatherings sprang up. A person would be travelling, be invited to attend church by a friend, hear the gospel, give their lives to Christ, return home, and begin to invite friends and relatives to hut meetings. It might take only four or five months before the meeting has 40 to 50 new converts. Within a short time they are overwhelmed by the responsibility of running a church. They would come to us asking for a pastor to help them, but we didn't have any to send. When that happened, my way of dealing with it was simple. I would go to the church group, conduct a meeting with them, and at the end, pray for direction. Looking across the audience, I would point to a man and say, "You are the leader of this church." Then I would point to a lady and say, "You are the ladies group leader." I had no clue who these people were. My reliance was upon the Holy Spirit to do the choosing. More often than not, these people ended up attending the Bible school and becoming fully trained pastors. The process might seem random to us, but it is the Lord's work. He cares for His children.

I was amused by the ingenuity shown by the lady in another mission. Their denomination was opening a Bible school, and they needed students. Each missionary was required to send some men. This lady was in a new area and had not yet had many converts. But she needed workers. She went to the local market and asked men if they wanted to study. She found four and, without any explanation, put them on a bus and sent them to the Bible school. Within a week she received a phone call from a frustrated Bible school principal. He accused her of sending men who were not even Christians. She replied, "Yes, that is right. You make them Christians and send them back to me. I need workers." It actually worked. She received four Christian workers. It pays to do what God puts into your mind, no matter how bizarre it may seem. Most times we over think things. We try to fit inclinations into already existing shapes. God is still capable of making His own shapes.

Arrested Twice During Our First Week in Africa!

Justice in Third World countries can be instantaneous and swift, if not also negotiable. We had arrived in Tanzania on a Thursday. We picked up our Toyota Land Cruiser on Friday and drove from the city of Mwanza to the town of Musoma. Both lie on the shores of Lake Victoria, which is the second largest freshwater lake in the world. Saturday, I went hunting with two missionaries. We had shot one animal when a game warden arrived and told us that we were under arrest for illegal hunting. We were shocked and stunned. Why? What had we done wrong?

"Did you not hear the news last night that all hunting in Tanzania has been banned with immediate effect?"

Listening to local news in Kiswahili was not a pastime most missionaries engaged in. Pleading our genuine innocence and with much humility, we begged for forgiveness. We were let off and allowed to return home. The game warden also allowed us to take our animal, which was nice of him. It was the last animal we would hunt for the next two years.

On Sunday I attended a four-hour church service that was a two-hour drive from Musoma. Monday was my first day of total immersion language school to learn Kiswahili, the business language of Tanzania. Kiswahili is somewhat foreign to us in its usage. It is a language that employs many commands: "Do that," "Go there," and "Get that." Using "please" is considered a sign of weakness. As polite Canadians, that was a difficult transition to make.

Coming home from my first day of classes, I was met by devastating news. My wife, LouDell, had been arrested. She was in tears and severely traumatized. All she wanted to do was get on the next plane and return home. In this town, though, there were no "next planes." She had never wanted to come to Africa in the first place—that was no secret.

A fellow missionary's wife took LouDell for an orientation trip of the town in which we would spend at least our next two years. As they were walking down the street, a crowd began to form, shouting, pointing and jeering. As soon as they could, LouDell and her friend ducked into a small street side shop, but the crowd jammed in behind them. Eventually, the shop owner informed them what was behind all of the commotion. LouDell and her friend were forced out of the shop and marched down the middle of the street. A man grabbed LouDell by the neck of her blouse and yanked her to a stop as the jeering went on. Suddenly, they saw the missionary husband coming toward them. He was fluent in Kiswahili, so found out what was going on. LouDell, together with some East Indian girls, were marched through the downtown and into the Regional Commissioner's office compound. The missionary reasoned with him, saying, "This is this lady's first day in the country. What is she going to be thinking about the country?" The others who had been nabbed were marched to the local open vegetable market, where they had to sweep it clean. This was a great shame for them.

Apparently, the same day as the hunting ban, it had also been announced that in order to reduce Western influence, women's skirts had to come at least one inch below the knee. LouDell's skirt was at the knee. Incidentally, the tightness of men's pants was measured by passing a Coke bottle down inside them. If it failed, then the pants were too tight – too western. This was actually the beginning of the social revolution experiment by President Julius Nyerere to turn the country into a Chinese model socialistic society.

This traumatic experience stayed with LouDell for a number of years. She was totally fearful of any crowds, and especially soldiers with guns. Over time, with the Lord's help, she was able to interact with the local population in a natural way. Thirty-five years later, she was the one who found it most difficult to say goodbye to Tanzania.

Are You Sure This Is the Road?

Exact details are not always at the top of a national's list of priorities when it comes to giving directions. I had been invited to visit a church near the Uganda border. The directions were simple and straightforward: go to such and such a village, and anyone there can tell you how to get to the church. That village happened to be a three-hour drive from Bukoba town. Three hours sounds pretty straightforward when sitting on your couch at home. However, the road has long stretches of teeth-rattling corrugated roads. All the bolts on your vehicle had better be pretty tight! My Toyota Land Cruiser failed when it came to the exhaust system. Other stretches had rock hardpan—again, very rough.

Nothing can ever come close to duplicating the feeling of driving into an African village where the only regular guest vehicles are buses belching black diesel smoke. Indy 500 cars are put to shame when it comes to the noise emitted by their engines, not to mention the drivers who are constantly revving the engines to keep them from stalling.

The arrival of a strange vehicle immediately draws people, akin to a crowd of children being offered free ice cream. The fact that a white man was driving did not detract from the curiosity either. Here, and elsewhere in Tanzania, the word "crowding" does not exist; it is a European expression indicating acceptable space. Indeed, "crowding" to the Tanzanians might possibly mean actually sitting on your lap. In lineups at the post office, bank or grocery store, it is commonplace to have the person behind you pressed tight against you. The long line is not going to move any faster! More than once I have turned around—or tried to, at least —and said, "It is stifling hot in here, and I am perspiring. Would you please not press your body up against mine?" They never even realized they were doing it.

It didn't take long to find people who knew the village that was my intended destination. My original "direction giver" had indi-cated, or so my Western mind had assumed, that it was just a short distance from this present village. Once again, that assumption missed reality by a lot. Clearer directions now indicated that was a pipe dream. After some negotiations I managed to talk a young lad, probably about 12 years of age, into being my guide. Repeatedly he assured me that he really knew where the village was. It is not out of the realm of possibility nor probability that he just wanted a ride in a vehicle with a white man.

The safari started off innocently enough. Three kilometers out of the village, already travelling on a dusty dirt road, we turned off onto a secondary road. From that so-called road, the way narrowed into a path, and then into a double cow path. Within a short time, we were surrounded by bush, where only the far horizon ahead of us was visible. The "road" became an overgrown jungle of elephant grass that was five feet high. Even the comforting sight of a horizon was now lost. Had it not been for the tunnel between the trees, I would not have known that a road actually existed. The fact that we were not passing any signs of civilization was not comforting. Not only is it disconcerting to head in a direction further and further away from civilization; I was also concerned that this elephant grass could be concealing some lethal weapons. These can consist of tree stumps left 16 inches off the ground or large rocks, which just love to kiss the bottom of the vehicle's oil pan, or the proverbial washout waiting to launch your vehicle skyward and your head into the vehicle's roof. A hard hat would have been a reasonable addition to that day's attire.

The dilemma was a choice that had to be made. The first choice was to drive slowly enough so that if I felt that telltale bump, I could stop before I heard the rumble under the vehicle. The outcome of that would more than likely see us on the "road" long after dark. Choice number two was to hope that no hidden obstacles existed and we could possibly arrive before dark. So much for my optimistic foolishness that within 30 minutes the way might suddenly open out into a village marketplace. By now I at least had the assurance that my boy guide did not just want a joyride. He might actually know where this village was. Something else became very evident: he was patient. At least every 10 or 15 minutes, I repeatedly asked him two questions: "Are you sure this is a road? " and "Are you sure you know where this village and church are?" He always answered yes. However, having lived in Africa three years by now, I realized that a house could not be built on the foundation of "yes."

After just a tad over three hours, we did emerge from the forest into a village market. To say that it was far beyond my expectation is a huge understatement. It was beyond my expectation and experience of reality in Africa that the road actually led to a place resembling civilization. In spite of the many twists and turns, on my next trip into the area I was able to find my own way. I was beginning to acquire the knack of reading signs on the sign less roads.

"Pull Down Your Pants"

Those were the words of command I gave to our pastor. Being white and mysterious drew various reactions from people. Together with those two factors, Esau was also a bit devious. He liked to sneak into our yard to spy on what we were doing. The main road to our house passed through banana plantations. Our house, shielded by more banana trees, lay about 15 feet below the road. A small footpath led from the road, through bananas, across a small creek and into our backyard. The vehicle road was a longer round about route.

I was calmly sitting in my office, which was located in an outside storage room building. Suddenly, I heard this unbelievable commotion: a sound like the bellow of a huge distressed bull mixed in with dogs barking and snarling.

Bolting out of the office and around the corner of the building, I found Pastor Esau cowering with dogs nipping at him. Our female dog had a litter of pups, and this intruder was a threat. I called her off and went to see Pastor Esau. He was normally a brusque fellow, but now he was as mellow as an over ripe squash.

He was shaking with fear as he turned around to show me his ripped pants. In fact, it was his initial turning around to bolt that gave the dog a good target to sink her teeth into. It didn't take long to realize that not only were his pants ripped, but also his butt had not escaped the bite. That is when I had him take off his pants so I could administer the necessary first aid to both cheeks. He never again attempted to sneak into our yard, but at the edge of the property followed the accepted custom of courteously calling, "Hello! Is anyone home?"

I Didn't See Anything

While teaching a course at the Bible college, my wife, LouDell, ran a practical exercise. She had the students go outside the classroom for five minutes and look around. Coming back in, they were asked to write down what they had seen. For some, literally not one thing had registered on their minds. This is very amazing because Tanzanians are generally ultra-observant and have recall that is unmatched. An example of this stunned me. A pastor from Canada had been visiting us. When he left, he wanted his suit given to a pastor. I had placed the suit on the floor just to the left inside my office. A few days later, a pastor walked into my office and said, "I see the visiting pastor left his suit." He had walked straight into my office, and I had not even seen him look down.

Back to LouDell's assignment ... Because we were in the tropics, the classroom door was always open so the students could look out, even absentmindedly, and see things. Standing just outside the door under the veranda there were wonderful sights.

There was lush green grass which the recent rains had coaxed out of the dry ground. The sky was a brilliant blue. Within 40 feet stood huge mango trees with fruit hanging from their branches. Just off to the right, papaya trees were bending with the weight of their fruit. Down the hill not seventy meters away, a busy road passed the college. People pulling carts laden with cement bricks, iron roofing or bananas were passing. Women carrying babies on their backs and water on their heads were also passing. Men were carrying baskets of dried fish. Who could miss the thunderous trucks, overloaded with charcoal and belching smoke, akin to an old-time steam locomotive?

Houses were visible where children were playing soccer outside with the proverbial African football—plastic rolled into the shape of a soccer ball and wrapped with string to keep it together. Colorful flowers were blooming right outside the classroom. Lizards scurried up the sun-soaked walls munching on the ever-present and abundant flies. To the left stood a cornfield in which a lady, swinging a huge Chinese- made hand hoe, was getting it prepared for seeding. On her back she had a baby strapped with a *kitenge* (a wraparound which serves as a cradle, a diaper, and a basket). Scurrying around her were three other children playing in the dirt. They were only wearing oversized shorts, exposing too much of their lighter skinned butts. Over the road lies Lake Victoria. On it are sailboats with massive, handmade sails of cloth. Not far from shore are some local fishermen paddling in one-man dugout canoes. Not more than three-quarters of a mile away stands an island renowned for its variety of exotic birds. And they saw *nothing*! Hearing about the results of this exercise encouraged me to be more observant of the beauty and the variety of things I see around me every day. This especially holds true for God's continuous blessings.

Just Say a Few Words in English

One erroneous, deeply "spiritual" thought in some missionaries' minds is that in order to have an impact, a missionary needs to live with, and be like, the local people. In the 1970s the Rwandan

Hutu and Tutsi tribes had another of their disagreements, which led to killing and people fleeing to Tanzania as refugees. Most of them were from the Tutsi tribe, who are among the tallest people in the world. I would add that they are also the thinnest. Most Hutu are short in stature. Those who are taller are distinguishable by their small hands. In later years, many of these Rwandan refugees became citizens of Tanzania. In the area near the Rwanda border, we had 25 churches, all pastored by Rwandan pastors. It was in my area of work, so I regularly made the five-hour trip to spend a week with them. Pastor Paul met me in his village. Mixed in with the regular conversation was a "cultural" talk. That is not a common occurrence. Pastor Paul very honestly shared how respect is perceived in their culture.

In Kiswahili, he said to me, "You know that when we walk anywhere, there are always children all around us. Touching you, a white man, is something they all want to do. Those who are a bit braver will take hold of your hands as we walk down the path. Out of mud hut doors, from a safer distance, adults will unabashedly be staring. I want you to walk with me down the path from my hut through the village. In the hearing of the children, say a few words to me in English. This will really impress them. I have just returned from a semester at Bible college, and they will be in awe that I know English. I have been away and really received some education!"

He went on to explain that just being seen with someone who is perceived to be in a higher class elevates his standing in the community. He said that he does not need a vehicle; just being seen with me in my vehicle is enough. The same goes for a house. Having a

"European" style house is not important; all that is important is that he has European friends and is invited into their house.

"You see, being friends with a person who is on my social level offers no advantage. To gain prestige, I need to associate with someone who is on a higher social level. I don't need what you have; all I need is to be seen with you." I had never heard it described that way. In a strange way it made sense.

We actually saw this in operation in a town where we lived. A missionary couple came and lived in a hut and basically lived and ate like the people around them. Few people joined their group because they never believed that an association with them would in any way elevate them. People said, "How can a poor person help another poor person? They are like beggars—poor, just like us."

I know it doesn't sound especially spiritual, but we are called to be what we are without putting on any false airs. Seems to me I have heard a question asked about whether a leopard could change its spots. We are who we are. Understanding the nuances of culture is of immense importance if we want to make an impact. The things that we think make an impact can be way off the mark. Being practical has a bigger impact than our masked intentions. False humility, especially false spiritual humility, is detectable from a long way away.

In another instance, a couple went into an area that was new to them to establish a church. They took with them a trusted worker from the former area in which they had resided. A number of years went by, and they never saw anyone respond to their message. Finally, they gave up and moved on. It wasn't until later that they found out that the "trusted" worker was from a tribe that was totally despised by the target tribe.

In a further instance, another couple also attempted to live like the local populace. They even pitched a tent in a pastor's yard. The husband was a hard worker and helped people in the area with building projects. Within a short time, government officials forced them to move into the city. They said, "We don't know what you are doing, but it raises suspicion. You do not live like other Europeans in our country, and that is not normal." Showing sincere love for the people is far better than trying to live like them.

That Is Irksome

Nothing blasts our ego more than when we have done the work and another gets the credit. For three years I was the director of a family camp. It was a two-week event involving a myriad of details. On weekends the attendance reached 1,200 people. To show appreciation for all of the hard work put into preparing for the camp, a senior pastor decided to have everyone pass .25 cents down to the end of the aisle. That amounted to a lot of quarters. He had ushers collect them and then announced that the donation would go to the district leader and his wife so they could go out for a nice meal! Now that irked me! I was the one who had spent many hours making sure the camp ran smoothly. Fortunately, I had enough sense to swallow my pride and never complained to anyone, nor did I ever try to set the record straight. I only wish that in other circumstances I had exercised the same restraint.

It was a good lesson. Later in my ministry and missionary career, there were times when other people received credit for something I was actually responsible for. It rarely bothered me, nor was I tempted to set the record straight. Sometimes it is the hard lessons that stay with us the longest and end up being the most profitable for us.

Don't Talk to Me

Trina was a Bible college intern from Edmonton who came to work with us at the orphanage for six months. She was a feisty little gal, and it took her all of an hour after she arrived to inform us that she was not a morning person. "Don't talk to me at the breakfast table. I eat slowly and in silence. I wake up slowly" is how she put it.

She had no idea that she had just baited my trap. Three feet outside her open bedroom window was our African grey parrot's large cage. Talking is their thing. Every morning, just before sunrise around 6:30 a.m., they wake up and merrily go through all that has stuck in their brain. Talking and whistling are the two things they love to do.

When Trina was not around, I taught the parrot to repeat, "Trina, talk to me!" Trina, talk to me!" Groggily, at 6:30 a.m., three feet away through her open window, Trina hears, "TRINA, TALK TO ME! TRINA, TALK TO ME!"

By breakfast time at 7:30 a.m., Trina was usually awake! Mad, but awake.

Now That Is Appreciation

LouDell and I have been the objects of many celebrations which were extremely humbling for us. As special guests we were honored with songs and often gifts as well. Many of the gifts were hand made, making them very special to us. One stands out in my memory and in my wife's nostrils. I was given a very old leather skirt by the Mbulu tribe. It has an exquisite design of beads sewn onto it. To cure the leather it was soaked in cow urine – thus the connection to my wife's nostrils. I had to smuggle it into a suitcase to get it home. We also had to throw the suitcase away. Unfortunately now it is wrapped in a bag in my garage. If I hang it out my whole garage reeks. Speaking of cow urine, it is a wonderful additive to milk to keep it from souring. It may even hold a link to preventing tooth decay. In eight years my wife and I never had one cavity. Always there is some silver lining.

In our work we looked after 147 churches. We went into an area on a Thursday, visited two or three churches a day, and didn't come home until Monday morning. Sundays were a one-service event. This was due to the fact that congregations from the surrounding areas attended. They didn't just attend; they came with choirs, dramas, and other special contributions. Four- or five-hour services were the norm. In spite of that schedule, we only had the opportunity to visit a church once in three to four years. Numerous special occasions cut into that schedule. Most of these churches were rural churches, which were seldom built close to a main road. In preparation for our arrival, people brushed down all of the tall grass and filled in holes, even up to a couple of hundred yards from the church. They lined the cleared section with palm branches stuck into the ground. Red, orange and purple bougainvillea petals were spread on the road right up to, and inside, the church.

In the quietness of the rural setting, our vehicle could be heard from a long distance away. Excitedly, people lined the prepared "driveway." It was a sight to behold as they sang and danced, stirring up the dry dust. Waving branches and wearing brightly colored wraparounds, the ladies ululated and jumped around in a high-spirited manner. As we stopped the car, their faces were plastered against every window. With difficulty we slowly managed to open our doors and were immediately and enthusiastically engulfed in the melee.

All part of it were the prolonged handshakes, the grabbing of our bags, and leading us into the church and down the two-foot-wide aisle as the singing, dancing, ululating, and waving of branches continued. The more observant people knew that the church would be packed. With seating room at a premium, they forwent the festivities and forged into the church ahead of us. Unlike our Western church culture, front rows always filled up first. With much fanfare we were seated on chairs decorated with flowers. Those scenes and the accompanying emotions, not to mention the smells and the sounds, are forever engraved on our minds.

Genuine appreciation has a way of humbling one to the core.

CHAPTER 20

Release Those Spirits

I first met Faustini while he was building a retaining

wall at the Bible college. Every rock placed in the wall had a perfectly flat surface. That was intriguing. We lived in an area where granite rocks were plentiful, to the point of being a nuisance. Forget trying to dig a posthole or plant a banana tree! The telltale dull thud or twang of a shovel was always a clear message to dig somewhere else. If the location was critical, the rock would have to be dug out. The hope was always that the rock was not actually a boulder.

Faustini was a tall, slender man who had a muscular build. Constantly lifting heavy granite rocks produces stature. He didn't have many tools—only the few he needed for his trade. They consisted of a "wok" type of hand-fashioned metal bowl, a cement trowel, and a small, insignificant looking hammer. From a discreet distance I watched him. His helper had brought a stockpile of "wall" candidates, rock by rock. Each had been fetched from the nearby hillside in a burlap sack and added to the pile. Now a substantial stockpile of potentially useful rocks had accumulated. Faustini had not yet realized I was watching, so he continued on with his task. In what seemed like a calculated, slow movement, he bent down and, critical eyes searching, picked up a rock. The one that caught his eye was 12 to 14 inches across, of irregular shape, and most likely weighed 15 lb. As an individual rock, the shape was irregular, but it would fit right in with the space available. In life, I have seen irregularly shaped people/characters who were suited for an exact available space. A space was left for them, and they were the only ones with an exact fit.

With the patience and intent of a stork fishing, he gazed at the rock. Not finding what he was looking for, he slowly rolled it into a new position. Again, it did not satisfy his requirement. The next roll seemed to fit him better as he took more time in assessing it. Reaching into his back pocket, he took out a small hammer. With a gentle tap, like a mother on the head of her son, all I heard was a faint *ping*. The tough granite rock split into two pieces, each having a flat side. The mystery of the flat rocks on the already partially built retaining wall was solved. I had just witnessed a profound skill acquired from generations past.

Walking closer, Faustini recognized my presence and in true African fashion acknowledged that interaction with a person is more important than the task at hand. Even if that task was extremely important, it did not supersede the time taken to talk.

Realizing the uniqueness of this trade with rocks, I was disappointed to learn that Faustini was not passing it on to a younger generation. Most probably the speed and efficiency required by "modern" methods rendered his trade of little value.

During our conversation I commented on seeing evenly spaced scars on his bare arms. He said, "Oh, those are the scars left due to my sickness." He proceeded to open his shirt and show me more. In fact, he said they covered his whole body.

What kind of sickness would leave scars like that? I wondered. His story continued. He had not been feeling well, so he consulted a "local" doctor. The doctor determined that the sickness was not physical in nature; rather, it was spiritual. The only cure for that was to make cuts all over the body so the evil spirits would escape through the bleeding. From a distance Faustini's body looked slender; now, at close quarters, I realized he was gaunt. Within a few weeks Faustini never showed up for work—he had passed away. It could have been tuberculosis or any of a number of other diseases which are common in the tropics.

Living close to Lake Victoria, it could have been bilharzia. This is a disease that attacks the liver and is brought on by the tiny parasite that lives in the water. It is so tiny, in fact, that it enters the body through the pores of the skin. It lodges in the liver and slowly drains the strength out of its host until the final result—death. Its life cycle is not complicated: from the snail to the water to the host, out through the feces and either into another host or back into the water and a snail.

Where Is the Washroom?

We were visiting a rural church, wonderfully set in the midst of a banana plantation. Upon our arrival, we had been welcomed into the church and served tea along with rice cakes. People discreetly waited outside until we were finished. That part was obvious to all, as few churches in the tropics need glass in their windows. So, although not present in body, they were watching through the windows. People began to filter in, and the tea began to do its work. It is a great diuretic.

376

My wife's system works quicker than most, so she made her way outside and inquired about the location of the bathroom. In a rather panicked response, she was led into the pastor's mud hut and given a chair to sit on. That is not what she needed! Desperation was beginning to surface when, through the open door, she noticed a hole being hurriedly dug. Talk about "made to order"!

The hole in question was not being dug on the banana plantation somewhere out of sight, but within sight of the people sitting in the church watching through the open windows. As quickly as possible, four sticks were driven into the ground and two ladies' wraparounds became the walls. Just in time!

The normal washroom was big—anywhere on the banana plantation.

Open Door Policy

Our orphanage farm was six kilometers from the main orphanage. Getting there required a drive of two kilometers on the main paved road and the rest through a rural area. That was always an interesting drive with the tractor and trailer. Most times the trailer had kids in it—from a few up to 20. I use the word "road" rather loosely; in actuality, it was a narrow path barely wide enough for the tractor. In the vicinity there were always children playing, women digging in their garden plots, men sitting under trees, and young children selling roasted peanuts on a small home-made square wooden table that measured two feet by two feet. The peanuts were tightly packed into plastic bags holding barely a half handful.

The process was ingenious. Peanuts were soaked in salted water and then roasted in a pan on an open fire. The peanuts were then packed into long plastic tubes that were three inches in diameter. At about three- inch intervals they were separated, and the plastic tube slowly passed over a burning candle to create a small separate sealed bag. Nothing can beat the flavor of those crisp, fresh peanuts with a slight smoky taste.

Driving along the road, I saw "British long drops" (outdoor toilets). The sight raised questions in my mind as most were in disrepair in spite of having been built with cement blocks. Mostly the roof was broken or non-existent. Doors were either off the hinges or hanging by one hinge, if not completely gone. Each prominently sported the name of a well-known aid organization. In a leisurely talk about culture with the neighbor under the clan tree in his dirt yard, I asked him about those "long drops". It was an interesting insight into how one cultural value is valueless in another culture. Here is the story.

Personnel from the aid organization "happened" into this rural area although it was, in fact, only 15 kilometers from the major city of Mwanza. They noticed that the people did not have any outdoor toilets. A survey was carried out, which established that the people were in favor of having the organization build toilets. So they did.

Knowing that people kept the responsibility of maintenance tightly tied to the builder partly explained the dilapidated structures. I once visited a church that was built by a former missionary 28 years previously. During the speech time, which is another "must" in any gathering, I was informed that the missionary had not completed his project. In fact, it was causing them constant problems and they wanted it completed. Goats, sheep and cows were coming into the building at night and leaving their droppings. They wanted the missionary to finish the building and put in that door! You see, it was not "their" building; it was the missionary's building. They never had been required to contribute anything to its construction.

Back to my British long drops ... Another thing remained a puzzle: not every yard sported a toilet. Slowly, the true story emerged. African people do not like to disappoint, so when the aid organization asked if they wanted toilets built, they agreed so the agency would be happy. As to why they were in disrepair—it is not culturally acceptable for a son-in-law and a mother-in-law to use the same facilities, so they weren't useful or used.

But why were there toilets only on some properties? They were built in yards only where the people had children. Sound logical? Does it take a foreign aid agency to tell people who have lived there for hundreds of years that they need toilets? Wow, that must have been an earth-shattering revelation! Westerners are so smart and observant (note my sarcasm).

Speaking of toilets, in areas where tribes actually make use of them, they are as varied as the colour of people's skin. In the interest of hygiene and in an effort to stem the spread of certain diseases, the government health department demanded that people construct toilets. Obviously, all consisted of deep holes. The interesting part related to what was built over the holes. Sometimes trees were laid across the hole, and an iron roofing sheet, with a convenient hole chopped into it, was laid on top of the trees. Stepping on that created a lot of suspicious crinkling noises. Probably not too many who are reading this need to be reminded that trees actually rot over time. In Africa, termites like certain trees for food too.

Often doors, if present at all, faced the road so a "user" could see anyone approaching. That was the explanation I was given when I asked about the open door policy. Then there were the ones built of reeds or willows. They were "see-through" but at least had ventilation. The scary part was when willows were also used as the floor. There are some stories about that! Willows too are a special delicacy for termites. There is always the small can of water for cleansing. Western type toilets in commercial places like airports, rarely have a toilet seat as people stand on them and break them; they are used to squatting, not sitting. Describing in detail the condition of public toilets is no subject for a nice anti-septic book like this.

This Is a Tough One

Rainmakers in Africa represent a serious business. In the West, forecasters have cushy jobs; they only have to tell people when it is going to rain. No one throws tin cans at them even though they may be wrong 60% of the time. In Africa, they have to *make* rain. Statistics on Tanzania indicate that between 80 to 90 per cent of the population are subsistence farmers. That is the percentage of the population whose livelihood comes from food production, not to mention whose physical existence depends on it. Even one season of failed rain can result in starvation. Looking at the average rainy seasons, one sees that they can vary up to six weeks on either side of the normal start of rains. Over the years I kept my own rain record for a very simple reason. Getting close to a rainy season, of which Tanzania has two, no one could tell me when rains usually begin. Short rains normally occur from September to November, and the long rains begin in February and go through to May. Every year is slightly different.

I would hear farmers say, "The rains have not come. We are going to experience starvation this year." Then, within a few weeks, it would begin to rain. In frustration I became my own forecaster, and for years recorded everything connected with rain.

Rainmakers thrive on late rains. When rains are late, people begin to consult the rainmaker. There is no such as thing as going empty-handed. At first, payment may be a chicken. The rainmaker lives in the neighborhood, so he has had his eye on a particularly fat bird. During incantation he "sees" the rooster; having the most power to bring rain is a multicolored, tall bird. It has feathers growing down both legs. If the person requesting the rain does not have such a bird, he will find it in the neighborhood and pay whatever price is requested to get it.

Within a week the rains still have not come. A black goat with white feet is the next powerful weapon on the horizon. Still no rain and a brown, hornless cow with white markings on her rear is the antidote. But the rains still fail to come, so the rainmaker says, "This is a tough season. I need to go away to the hills and consult with the spirits." Every evening he has seen clouds beginning to form and knows that they are the sure signs of the weather building up to rain. When it begins to rain in 10 days, he triumphantly re-appears, having accomplished his task. The local people are aware of this scenario and joke about it, but are still afraid not to engage in the process.

Local people are very observant, so I was always curious as to why they couldn't read the skies. Another example of African logic confusing the Western mind. Speaking of reading the sky, for many years people, especially in the rural areas, did not have a wristwatch. Yet they agreed for appointments to take place at a certain time. Occasionally, out in a rural area, I would test their "time" skills. "What time is it," I would ask. They studied the sky and invariably were never more than 15 minutes off in telling me what time it was. This fascinated me.

Give Me Steak

Ah, meat markets. The first thing to do is figure out what time of the day your "smeller" is the least sensitive. Normally, the meat section is located at the far backside of the fruit and vegetable market. City markets are open but under a roof. Technically, it is not only a fruit and vegetable market, but because it started out as such, the designation has stuck. In it are handmade things like aluminum pots, kerosene lamps, wooden stirring sticks, swords hidden in walking sticks, clay pots and carved animals. Another section contains spices of every kind, bananas, oranges, papayas, peas, varieties of beans, cabbage, lettuce, beats, potatoes and more. It's like a big general store. The whole arrangement is very colorful.

We were on our way to the meat section, having passed the items listed above. Aisles don't bring in money, so they are only wide enough for one and a half people to pass each other. Taking into consideration that people are carrying large woven sisal or reed baskets, that space is further reduced. Nearing the set goal, one must still pass the heaps of dried sardine-like fish called *dagaa*. The name does not change the smell.

Then coming into sight is the meat market section. The odor has changed, but not the intensity of it. First, one must maneuver past low pens holding squawking chickens, roosters, ducks and geese. The pungent odor of their accumulated droppings makes paramount the taking of another quick breath. When they were last fed is questionable. No doubt before arriving, they were already chicken mannequins covered with feathers. Their long legs add to the accurate perception that this chicken is more bones than meat. It may barely suffice for one meal for a person who is not too hungry.

Openly hanging on steel hooks are long strings of intestines from who knows what animals—most probably sheep, goats or cows. In close proximity hang other delicacies such as stomach, lung, liver, kidney, brain and goat heads. If the aforementioned items are still available, you know you have arrived early because they are the first to be sold. Local wisdom says they contain the most vitamins, so why would anyone buy a "lesser" grade of meat?

Old blood in a hot climate has a distinctive overpower smell and hits one square in the nostrils. It is best to concentrate on what you want instead of the meat market décor. In addition to the smells are the ever-present hordes of flies. Remember, this is an open market, and flies have a keen nose for ripe, easily digestible food. Should your eyes dare to wander, they will take in the walls splattered with bone shards, blood, and a few other things. Intestines and stomachs are not always fully cleaned to Western standards before hitting the food shelf—sorry, the food hook. Within easy reach there is always a dull axe and a chopping block that look similar to the walls.

You ask for steak. No problem: out comes the knife, the hanging beef quarter is turned so you can see that you are actually getting steak, a neat chunk of meat is sliced out of the back leg area, and there you go. The chunk is dropped into your basket. Done. Flies, after having been so rudely dislodged, return to take up their perch. It is best to make a quick exit and not dwell too much on the origins of the meat as you drop it into your frying pan at home. It probably doesn't need a lot of tenderizing salts as it is pretty well on its way there already. Invited dinner guests need not know the details of the steak's journey.

I'm a Man—I Can Lift That

I have read that African women are able to carry up to 70 per cent of their body weight on their heads, plus carrying a child on their back and toting a basket of bananas in their arms. They can make it all look so effortless. I was parked on the side of the road when a lady carrying firewood came along. The tree branches and small trees from which this was cut probably measured eight or nine feet in length. The diameter of the bunch easily reached 24 inches. Politely I asked if she would mind putting it down as I wanted to try and lift it. She obliged.

Grabbing the twisted strand of bark with which she had it tied, I was not even able to raise it two inches off the ground. With ease ladies carry five-gallon pails of water (roughly 50 lb.) filled to the brim. Never a drop spills over the rim. Sometimes a small metal bowl is placed on top to keep the water from sloshing around. Something that never failed to bring an "ouch" to my system was seeing a woman carrying water and effortlessly turning her head to look the other way. I have no doubt whatsoever that if it was my head, it would have kept going, with my neck twisting off. To stop that motion, their neck muscles must have unbelievable strength.

In a local newspaper article I read of an experiment the U.S. army carried out. It dealt with women carrying loads on their heads. The army is constantly looking for ways to increase the load their soldiers can carry while expending the least amount of energy. The results showed that as the head-load increased, the walking gait of the women automatically changed. It became a rocking-like gait that actually propelled them forward. With the heaviest head-load they only used approximately 10 per cent more energy than for normal walking.

I observed a young girl, no more than eight years of age, carrying a full pail of water. Coming to a drainage ditch two feet wide, she walked to the edge, stopped, and got her bearings. With one smooth motion she jumped over the ditch without spilling a drop of water. I couldn't even do that carrying a pail in my hand.

It's the Bloodiest There Is

While ministering in an area known for its fierce warrior tribe, I slept in my vehicle at night. It was rigged with the seats removed and curtains sewn for the windows. Members of other tribes felt that it was a dangerous practice, but I never felt uneasy.

A doctor from a bush mission hospital invited me to stay overnight with them whenever I was in the area. The next time I was there, I did. While at supper, the doctor asked if I had ever attended a caesarean delivery. I hadn't. He then asked if I would like to, but immediately warned that it is one of the bloodiest operations there is. Normally, blood does not bother me, so I consented. We would wait until 9 p.m., and if the woman hadn't managed a normal delivery by then, an operation would be necessary.

At 9 p.m. we left the house and made our way in the dark across to the hospital. Our flashlights were glued to the path to check for any snakes that might be out looking for a rat meal. Most snakes, as they sense ground vibrations, slither off; however, puff adders are sluggish, so stepping on them is a real possibility. The scary thing about them is that their venom injection system is also sluggish. Therefore, if they strike, they need to hang on to the victim for about a minute until all the venom is deposited. Scary stuff! Other snakes strike and head for the hills to recharge their venom sack. An exception is the black mamba, which strikes and then comes back to strike again if necessary.

A diesel generator hummed in the background, assuring that power would be available. The first order of business was to scrub up: our hands, right up to the elbows, were thoroughly soaped. Clean fingernails were a priority. This brought back memories of a harsh schoolteacher. I never did figure out the reasoning behind a farm boy's need to have clean nails. Holding our arms straight out, nurses dressed us in gowns and placed surgical skullcaps and masks on us.

Entering the operating room, all hands were on deck with the patient in position as well. The anesthesiologist stood at attention at the head of the patient, holding the "pounce" stance. I was being repositioned to make sure I could see, at close hand, what was happening. At the doctor's nod the chloroform mask was clamped over the patient's nose. Within seconds she was in dreamland. Her stomach was exposed, and the doctor deliberately pressed the scalpel to her. As he pulled it down, the woman's skin began to peel back like a watermelon cracking open. That gave me the heebie-jeebies, but I quickly recovered. Like a sports announcer the doctor explained, "This is the first layer of skin, the epidermis. This is the second layer, the dermis. This is the third layer of skin, the hypodermis."

Doctor to nurse:

"Towel number one into cavity" (towels to soak up the blood).

"Hemostat number one" (instrument to clamp a bleeding vein).

"Towel number two into cavity."

"Hemostat number two."

"Towel number three into cavity."

"Hemostat number three."

"Towel number four into cavity."

"Hemostat number four."

Turning to me again, his voice muffled by the surgical mask, he said, "Here we are opening the uterus." A very tiny cut was made. Then the baby boy was pulled out and held upside down by his ankles until he began to cry. Thinking back on that, I marvel that this was his very first time using his vocal cords.

Doctor to nurse again: "Scissors."

The nurse begins her slow count:

"Removing towel number one."

"Removing towel number two."

"Removing hemostat number one."

"Removing hemostat number two."

"Removing towel number three."

"Removing hemostat number three."

"Removing towel number four."

"Removing hemostat number four."

Doctor: "Are all towels and instruments accounted for?"

Nurse: "Yes."

Doctor again: "Needle and thread."

Precisely the moment the chloroform mask is removed and just as the last stitch is tied off, the new mother is awake. Gas is expensive, so there is no reason to waste it.

A stretcher is wheeled up close to the operating table, and the woman is directed to get off the table and onto the stretcher! I could not believe my ears and then couldn't believe my eyes when she actually did—under her own steam. She was wheeled off to her bed and told to get off the stretcher and into her bed. Talk about not pampering anyone! Back in the house, the doctor explained their motto: See One, Do One, Show One. In an extreme life or death situation I might have graduated to Grade 2, but fortunately for any patient, I was never called upon to act on this motto.

Your Name Shall Be ...

As a child growing up in Alberta, Canada, we played many games outside. One of those games included "playing church." For some reason I always ended up being the pastor. We did mock baptisms (we believed in total immersion). We lacked water, so just went through the motions, saying in German, "Ich taufe dich in Kaffee Grund und dein Name soll sein Pudle Hund." ("I baptize you in coffee grounds, and your name will be Poodle Dog"—in German it rhymes nicely.) Don't ask me who thought up that rhyme. I had no idea that baptisms and baby dedications would become a huge part of my life. Throughout my years of missionary service, there were sure to be water baptisms, child dedications, and even the odd wedding that I found out about upon my arrival at a given church.

Very early on in our service as missionaries, we were invited to be at a church service about three hours from our home. Pastor Stephen was our guide, so we didn't get lost this time. Turning off the main sandy road, we followed a cow path. Around the corner, hidden by tall grass, a mud-walled church emerged. As we would find out was common, the place was already packed, with people standing outside. Glassless windows served as good overflow vantage points. Inside, exuberant singing was in progress, together with a lot of dancing and stirring up of fine dust off the dirt floor. Looking toward the open windows with the sun streaming in, that dust was obvious.

Part of the program complementing the many choirs, testimonies, and greetings from numerous pastors was the dedication of children. We follow the Lord's example of blessing children rather than infant baptism. No one possessing the proper church credentials to do this had visited the area for some time. The announcement was made that all parents wanting their babies to be blessed were to assemble at the front. That in itself is quite an exercise. First, the 'mamas' have to shuffle around on their backless seat while they unwrap the baby from the wraparound—which the baby may have soiled. A clean wraparound is found, and the baby is wrapped and prepared for public viewing. By this time, the baby senses that something out of the ordinary is happening and he/she is not being repositioned for breastfeeding. Some babies have that idea and, along with others who have just been awakened out of a sound sleep, begin to join the screaming choir. By the time everyone was assembled at the front, the screaming choir had swelled to 28 in number. As it turned out, some of the babies were walking already. They came wide-eyed as they saw the "white man." Their grip on the legs of their mothers nearly cut off circulation, I'm sure.

Another of the hundreds of lessons about culture was learned right here. "Parents" meant 100 per cent mothers. I never did find out why the fathers didn't accompany their wives. I was aware of the fact that women respond much more readily than men to spiritual things. I am sure many of the ladies did not have Christian husbands. Certain cultural practices are taken for granted and don't necessarily mesh into our antiseptic thinking of a Christian culture. Part of the reason is that many people are first-generation Christians, and maybe only newly converted.

The cultural norm in this tribe is that, when a child is born, certain things must immediately be done to ward off evil spirits. Strands of strings are wrapped around the baby's wrist, waist and neck. Sometimes these strings have things tied into them like beads or pieces of wood. I never did learn the whys and wherefores of that. Now that I think of it, this would have made an interesting study. As Christians, we do not believe in the power of string—no matter who has spoken anything over it—to ward off evil spirits.

It was our custom, before we prayed a prayer of dedication over these children, that we first cut the "cord." Some of these children were old enough so that the string around their neck was already tight. Imagine yourself being taken captive, even if it was by your mother, staring into a face having a colour you have never seen, and being confronted with a Swiss Army knife. For those of nursing age, the most comforting place was on mother's breast. Being the "honored" pastor, I was given the task of dedicating the children. It would be a lifelong honor for the mother and the child in later years to boast that a white man had performed the dedication.

It was already with some trepidation that I looked at the line and realized that each child needed a prayer of dedication. It was going to take some time, not to mention ingenuity, not to repeat myself 28 times. I was beginning to see the wisdom of mainline churches, which read the liturgy. Something akin to a mass marriage ceremony would have been the answer. The first mother I approached threw an even bigger wrench into the process.

I like to know the name of the child I am dedicating. "What is the child's name?" I asked. Blank stare. No reply. Knowing that sometimes a European's Kiswahili is not as clear as what they are used to, I repeated, "What is the child's name?" Again, blank stare, mouth not moving. No reply. I turned to the pastor and asked if the mother did not understand Kiswahili.

"Yes, she understands, but she is waiting for you to give the child a name—a Christian name." Now my brain really did a burn as I looked down the line extending wall to wall. Desperately, I mentally accessed my encyclopedia of appropriate biblical names.

As the dedications progressed, a routine emerged. While the mother held the screaming, squirming child, I tangled with the string around its neck, holding the knife at the ready to facilitate, as smoothly as possible, the removal of it. Within a split second the child was at the mother's breast again. Pulling the child from the mother produced a distinct sound as the child tried as hard as possible to remain attached to the mother. Every once in a while, I encountered a mother from another tribe who did not have the string custom. That was a relief. I would have liked to simply lay my hands on each child and pray, but I had to hold the children in my arms. After all, that is what I always did at home! By the end of the line, my front was pretty wet from little African streams. To this day I do not remember how many Johns, Peters, Pauls, or Marys, Esthers, Eves and Elizabeths I named that day.

Not to worry too much: within a few years their names might change anyway depending on whose ancestor they might call themselves after. Although mind-twisting for us, we accepted the fact that a person we had come to know would suddenly present themselves by a new name. This was especially convenient if a worker for a missionary was not well thought of, and over the course of time, that person applied for a job with another missionary. In these instances, a few years might pass. While you visit with a missionary, they proudly introduce you to their worker. It may be a new name, but it's an old face with a slightly embarrassed look. Sometimes it was more a look of triumph ("See? I was able to find another job."). There is no legal requirement to have a certain name registered.

The first assignment when teaching at the Bible college was to announce that every student must pick their name and stick with it for their whole time at the college. Some would forget; when marking an exam paper, suddenly you had six new students you didn't know you had. Or if a student failed his high school entrance exam, he would simply retake it under a different name. Best for the Western mind not to get bogged down in an analysis of logic. Accept the fact that it makes sense – to someone.

Is That The Mother From Whose Belly You Came?

"Every worker is allowed to attend only one mother's funeral." This was the confusing sign I saw posted at the entrance to a steel manufacturing shop. What lay behind it is the custom of generalities.

Kiswahili need not be a language of exact explanation. More than once a worker would come to us in obvious sorrow and share the news that their mother had passed away. No matter how hard-hearted you might be, that news is always met by long-faced condolences and your saying to the bereaved person, "Take as much time off as you need. No doubt it is a very difficult time for you."

As painful as it might be, you also promise to pay them full wages. To your utter surprise, four months down the road, this same person stands at your office door, again saying their mother has died. *Wait a minute!* As quickly as possible, you rewind the tape in your mind from four months back. A lot has happened, and you are not sure if this is actually the same person who asked for time off way back then. Then, yes! You are sure it is. So you say, "Four months ago, you asked for time off to attend your mother's funeral. Did your mother come back to life, and has she now passed away again?"

"No, that was not my real mother; that was my mother's sister, but we call her mother too."

If you were fooled at the first go-around, the next "mother" may even reach as far as a second cousin of the mother. So each time a worker came with the "funeral" story, an ancestry link had to be established. The simplest way was to ask, "Is that the mother from whose belly you came?" It might take a few shots, but eventually a yes or no would emerge, especially if you used the heavy hammer of allowing attendance at only one mother's funeral and told the person in question, "I am noting it in your file."

Every national was extremely fearful of two things. One was getting a written "letter of warning" for something they had been verbally made aware of in the past. The second was the statement that the letter would be placed in their file. I never did figure out what black, foreboding cloud those two things conjured up in their minds, but I do know it worked. Workers begged me, "Please do not write me a letter! Just talk to me. And please do not put any-thing in my personnel file!"

CHAPTER 21

Guess What?

Many overseas volunteers came to help us at Starehe

Children's Home. *Starehe* in the Kiswahili language means to be at total peace, to have no worries or things on your mind. That is the atmosphere we wanted our orphans to grow up in. The orphanage consisted of seven acres of land, lots of buildings, and suspect infrastructure such as power lines and water pipes. There was never a shortage of things going "kaput".

A lady acquaintance was coming to Mwanza to teach at a Bible seminar. Her husband, William, was coming along as well, but she felt he would be bored sitting around on his own all day. She asked if there was anything he could do to help us at the orphanage. Given that he was approximately 80 years old, I tried to find work that was not too strenuous and could be done over a period of time—tasks that were "no rush." Those kinds of jobs always seemed to end up last on the list and never got done. William's help was much appreciated.

Each morning I picked him up, and we drove the 10 kilometers out to the orphanage. I would give him some jobs to keep him occupied and then take him back into town in the evening. He was well into retirement, so the heat of the day pretty well sapped his strength.

On the way back into town, he would slump down in the car seat and not show any interest in anything going on around us. In Africa there is always something going on. A few minutes into the ride, he had nodded off and stayed that way until we arrived at the place where his wife was staying.

One job I had not gotten around to doing was erecting some clothesline posts. Forty feet was the total length of the clothesline, which necessitated one sturdy middle post. Even then it would be a stretch to expect the 6 inch steel posts not to eventually bend. Together with William, I measured out the placement and left him to dig the holes while I returned to town to pick up some things. Returning, I found William sitting near the workshop. Hardly giving me a chance to get out of the vehicle, he said, "Guess what?"

I said, "You punctured a water line."

"Yes! How did you know?" As I was in town, I began thinking of the area, figured in Murphy's law, and knew what the result would be. It was easy to shut off water flow through that part of the line. Repairs were easy too. Locate the muddy spot, dig up the pipe, then as tightly as possible, wind binder twine—the original made from sisal—then wrap some strips of bicycle tire rubber around that, and presto! The leak is fixed. Sisal absorbs water and expands, producing a watertight seal. Similarly, it works well when used on a pipe thread.

As we all know, railway crossings never have a train on them until you are in a hurry or very tired and wanting nothing more than to get home. One evening we encountered a train shunting cars back and forth. As usual, William was nodding off and beginning to breathe deeply. The train whistle managed to get his attention. Groggily, he opened his eyes and looked at the train locomotives, which were parallel to the road. Suddenly he shot up as though he had just sat on a sharp needle. Now wide-awake, he took a piece of paper out of his pocket and began jotting something down.

I was very intrigued about what had so abruptly captured his attention and radically transformed his demeanor. Suddenly, his fountain of youth had returned. He hadn't said a word, so I asked him about it. Excitedly he explained the numbers on the locomotives. He recognized them as having been made in Canada, and he knew exactly how many had been produced—20 of that vintage were still in operation. Seeing these locomotives, he knew exactly where they fit into the scheme of railway memorabilia.

He had worked on the railway as an engineer, and this brought him back to a time when he still felt useful and able to contribute something. This communication opened a whole new relationship between us. I now knew what really interested him, and I learned a lot about locomotives and trains in general.

Life after "work" can make us feel that we are redundant, if not a downright burden, to society or even our family. Early in life it is good for us to realize that our worth does not lie in what we do, but in who we are. Early years we spend in 'doing' and 'being'. Later years are focused more on 'being'. That is the way it should be. Any feeling of guilt about not 'doing' so much should be soundly rebuked. When well meaning people ask me what I am doing now, I enthusiastically tell them that I am 'being' retired.

Don't Point That Thing at Me!

Picture taking for us is a great pastime. We like to record things and file them away. Eventually, they end up on the dreaded to-do list, which never gets done. Years later, children and grandchildren wonder who those people in the picture are or where that picture was taken. If we really want to irritate our friends, we show them slides or movies of our exotic safari. Only their soft snoring reveals any hint of boredom.

People in strange lands are especially interesting for us, and we continually point our cameras at them. Their way of life is so different, and we try to capture that. We seldom stop to think what our lives would be like if, every time we walked down the road, carloads of chattering people were pointing some metal object at us. Sometimes that object even sends out lightning.

No wonder, then, that we can have some objects resembling rocks coming our way. We can count ourselves fortunate if it is not a spear. One tribe believes that if a picture is taken of them, it steals their spirit. When they see a camera pointed at them, they prostrate themselves on the ground and do not move.

Others see it as a moneymaking adventure and encourage you to take a picture of them. It is wise to negotiate a price beforehand or it will end up being a very expensive memento. You may even end up with a smashed camera if your payment is withheld or is seen as being despicable. Admiring another culture is great, but it must be done in a respectful way. Sometimes it is best to simply store the picture in your mind.

Reflections on a Signboard

Some churches in America don't even sport as an impressive signboard indicating the location of their church, as Pastor Samuel's does. He is a church planter living in the coastal city of Dar es Salaam. His face has pitch-black engravings that stand out even against his already black face. They go around his eyes, on his forehead, and down the side of his face. You take one look at him, and his tribal identification is immediately evident.

With little education and not much more in assets than his fierce-looking face, he left his tiny village and came to Dar es Salaam. He was accustomed to hard work, especially gardening, so finding work was relatively easy. With Christian nurturing he slowly sensed that God wanted him to work in His garden. Upon the departure of his employers, Samuel launched out—stumbled is a more accurate description—to plant a brand-new church. Eventually, airline executives, businessmen, bankers, schoolteachers, and ordinary laborers were attending the church. Samuel never felt intimidated by any of them; they were coming to hear the gospel, not him. Many people responded and became members of the church, which met in an unfinished home.

Samuel realized that his God-given talent was to plant churches rather than live off the fruits of hard labor (many church members), as well deserved as those might be. So off he went and planted another church a few miles away.

Back to the signboard ... It stands on the main road leading out of the city to luxurious beach hotels. Those advertise physical pampering and comfort; Samuel's church sign leads them to spiritual rest. Don't let the sign fool you; it is the best physical part of the church. Off the main road and down a roller-coaster road with more craters than you'll find on the moon, around a few dozen sharp corners, through sludge-filled potholes, and more than once thinking the road has come to an end, you see another sign. This is small and unimpressive compared to the road sign. You have arrived at the church. It's a good thing the sign is there because, if not, you would not know you had arrived at a church. Looking around for the church steeple, all you see is tall grass, ditches filled with oozing mud, a shabby house back on the lot, a few nondescripts leaning against a tree, and an open-sided lean-to building on the right.

Eventually, a woman emerges from the house, weaves her way along the footpath through the grass, jumps over muddy potholes and approaches the car. "Good afternoon! How are you?" She speaks perfect English.

Judging by the surroundings and with the impressive road sign only a dim memory now, we expected imperfect Kiswahili at best. After the usual greetings, she motions us to the side building indicating it as the church—a good thing because we would never have guessed it. Under the low tin roof it was hot. A hodgepodge of chairs, benches, seats and pulpit authenticated her statement—it was the church. It was afternoon, and Samuel was at the local market handing out tracts and witnessing to people. Here before us was evidence again of Samuel's heart for church planting. The congregation is a group as keen and sharp as the signboard on the main road. My motto has always been that we need to do things to the highest standard. Even a signboard can be a witness that draws people. After all, we are children of the King, and we are inviting others to join the best family in the world. Eventually the meeting place would resemble the impressive road signboard.

Lumbering

With no machines available to lift or move large trees, people living in rural areas devised their own appropriate technology to circumvent the problem. The tree, measuring three to four feet in diameter, is felled with a large crosscut saw. The log is cut into lengths of 10 feet. Just beside the log, a hole measuring seven feet deep and 14 feet long is dug. Supports are built in the hole so the log can be rolled onto them. Once in place, lines are marked on the top and bottom of the log. A man stands on top, and a man in the hole. In unison they pull and push, sawing the log along the lines. That is how boards are made out of the log. In populated centers large band saws are used, but each log still has to be manually pushed through the saw. Because of the weight, it takes up to four strong men to manhandle these green logs. With that weight and with that many men pushing, it is a marvel to see straight boards falling off the log. It is also a marvel that they can united push the log at a steady pace that the band saw can handle without binding or breaking the blade.

Leisure—A Long Way Around to Get to the Beginning

On a trip in a six-seater plane, I overheard two European businessmen talking. That is not a great feat in a six-seater plane. The bigger feat may be that you don't hear them speaking.

These two men had been sent by their aid organization to bring help to Tanzania in the form of selling fertilizer at an affordable price. From the conversation, it was obvious that each man had gone into a separate area. One of the men recounted his experience to his buddy. The conversation went like this:

"I hired a vehicle and drove out into a rural area. I found a man sitting under a tree and, going up to him, engaged in conversation.

After greeting him, I shared the good news that if he uses our fertilizer, his harvest yield will be double.

"He asked me, 'Why would I want to do that?'

"I answered, 'You will have more extra money.'

"He asked me, 'Why would I need more money?'

"I said, 'So you can buy more things to help you raise more corn.'

"He asked, 'Why would I want more corn?'

"I answered, 'So you can have more money. More money would mean you could have more things and more leisure time.'

"Then he asked me a question: 'Why would I go through all of that to have more leisure time? Don't you see me now? I already have leisure time without fertilizer. I am sitting under this tree.'"

The man said that he was defeated by the farmer's logic. Not being familiar with "on the ground" realities, they did not understand the finer points of rural living.

Having "more things" translate into greater leisure or happiness is a Western concept that is foreign to many cultures. I have personally seen people accept fertilizer to increase their crop yield. It was actually a great blessing. Without fertilizer the man had to dig 20 acres to get 20 bags of corn. Now, with fertilizer, he only had to dig 10 acres to get 20 bags of corn. You see, the important thing to him was to get 20 bags of corn, which is all he needed to feed his family. Anything beyond what he needed was a bother. It was prone to theft and a lot more work.

Governments and organizations spend millions of dollars to find ways to increase yields. They also focus on developing new varieties of plants which are resistant to certain pests or diseases. This misses the mark. Crop storage is the number one cause contributing to starvation when drought hits. Most areas rely on two rainy seasons to sustain them. There are years where bumper crops are harvested. However, where corn and beans are the main crops, weevils have eaten it all within six months of storage. I have lifted a burlap sack holding 220 pounds with one hand. It used to be that heavy, but now weevils have hollowed out every kernel, leaving only the husk.

The man under the tree does not want more than 20 bags of corn or beans because that is all he can store. It is tremendously intensive-intensive to store the staples of corn and beans. Different methods are used. One method used is that, every 45 days or so, the corn is poured out of the sacks onto the ground. Incidentally, that is most likely how it was gathered together in the first place. Corn is left to ripen rock hard on the cob. Then it is either shelled by hand, which is a huge job that requires callused fingers, or it is placed on the swept ground and repeatedly run over with a tractor to dislodge the kernels. It is then swept up and put into burlap sacks. The sacks are then stacked in a storage room where rats and weevils can have a feast.

Back to the storage ... As the corn is put back into sacks, it is mixed with wood ash. This again only lasts roughly 45 days before the operation has to be repeated. Another method is to build a relatively airtight mud storage hut with a small opening at the top. Corn is poured in, and the hole is sealed. If the whole structure is airtight, weevils are deprived of oxygen and die. However, eventually the thing has to be opened to begin using the corn, so weevils are given a new lease on life. Other methods exist such as pouring corn or beans into 45-gallon oil drums, leaving a bit of space and lighting a candle in it before sealing. When all of the oxygen is gone, theoretically the candle goes out and the weevils do not survive. All of this entails a huge amount of work. In the case of the oil barrels, seldom are they ever clean inside. There are no steam washers in rural areas, so the stored grain can become contaminated.

Chemicals may also be used, but they are mostly unaffordable and dangerous to boot. In the dry season, when the crop is consumed, water is at a premium, so having sufficient water to wash chemicals off the crop before consumption is not feasible. Our pastor, whose home was a literal hotel for people passing through, each year harvested 150 bags of corn, plus beans. I have seen his storage problems first-hand and can attest to the labor involved in storage.

Over and over, in years of bumper crops, we heard reports of large amounts of produce and harvest rotting on the field due to a lack of storage or, at best, no way to transport it to market. Dirt roads are no match for the monsoon rains.

One crop that is immune from weevils is rice. If left in the husk, it can be stored for any number of years.

Testimonies

If there is one thing I loved to hear, it was personal accounts of what God had done for people. Justus has been a cripple for as long as he can remember. In a country where even healthy people can find it a challenge to negotiate dirt walkways and uneven ground, Justus manages well. When his specially built tricycle becomes too cumbersome, he simply maneuvers his way off it, slips his shoes onto his hands and hops along on all fours. It can be dusty and muddy down there, but he is no quitter or complainer. I can't even begin to comprehend the effort it takes to maneuver around in those conditions.

He lives in a rural area of dirt, rocks, sand, paths and uneven ground. Washrooms are outside, if they exist at all. It would seem better just to be carried from place to place by friends. Upon finishing his fourth year at Bible college he said, "Even though my legs are crippled, I did well in serving Satan, so there is no problem in serving Jesus even better." A healthy spirit is to be coveted more than a healthy body.

Winnie stood up in church and testified that her young son drank some kerosene. He became violently ill, but she never took him to the hospital. She said, "The hospital might not have been able to help him, but I know the Great Physician. I turned to Him to help my son, and He did.

My Name Is Musa

In another publication I will tell more about the time when we operated Starehe Children's Home, our orphanage in Mwanza. One particular child always comes to my mind when I think of desperation.

Social services had a young boy of four in their care. They called us to see if we had room to accept him. We did. When the social worker dropped him off at the orphanage, she was near to tears as she related what had just happened.

Vehicles are not only expensive to buy; they are also very expensive to operate. Taking a small bus is much more reasonable as far as cost goes. On the way to the orphanage, out of the blue, little Mohamed looked at the social worker and said, "My name is Musa (Moses)."

"But your name is Mohamed!"

"No, my name is Musa."

"I know your name is Mohamed. Why are you telling me your name is Musa?"

"Because when my dad comes to the school (he didn't know what an orphanage was), he is going to ask for Mohamed and he won't find me."

It still breaks my heart that a four-year-old boy was so badly treated that he didn't want his father to find him. I also marvel at the lad's thinking with such adult logic. No wonder LouDell and I never considered working with the children at the orphanage as a burden. On the contrary, we were blessed to have the opportunity to do it. Musa has grown into a fine young man, and his father never did find him.

Flying Suitcase

Life in Africa was filled with opportunities to be flexible. Not that I always wanted to be, nor was I always in the frame of mind to be so. It is a situation that must be accepted, just as the burning tropical sun is accepted. It may be hot and uncomfortable, but there is no escape.

We were on our way from Arusha to Nairobi to drop our children off at the Rift Valley Academy boarding school. Living in Arusha, it was only a touristy jaunt of four hours for us. After having lived in Bukoba, which was a tiring, long two-day drive, this was a true picnic. Seeing the occasional policeman on the road was common. Due to lack of police vehicles, they were dropped off at some point on the road early in the morning.

Traffic could be very light, so they made themselves comfortable under a tree to make use of the shade. If they felt like it, they would get up, walk up onto the road and flag you down, ask to see your driver's license, chitchat a bit, and say goodbye.

Having our driver's license with us was not always a given. In Tanzania we carried only a photocopy in the vehicle. Going on a safari, it was a good idea to have the real thing, especially since we would be driving in the neighboring country of Kenya. "Take Driver's License" figured prominently on our permanent list of things to take along on safari. In Nairobi, a policeman stopped us. In response to his request, I gave him my driver's license. In the yellow light of his dying flashlight, he examined it.

"This has expired!" he said.

"What?"

Turning to my wife, I asked her for her driver's license — that is acceptable. The policeman examined it and made the same remark as he had made about mine.

After some contrite remarks, made in deep humility to acknowledge a serious misdemeanor, we were allowed to proceed with the promise that we would have our licenses renewed as soon as we returned to Tanzania. We would still be spending a week in Kenya, but the license issue could again be explained if necessary. I don't know what the big deal was as they had only been expired for six months. In Tanzania, one never received any notices of anything expiring. You had to remember the expiry of your driver's license, vehicle insurance, house taxes, lease terms, land taxes, work permits, gun licenses, and a host of other things.

Back to our Nairobi trip ... Moving with more agility and purpose than was usually the case, I surmised that the policeman waving for us to stop had more than the normal checking of documents on his mind. Coming to a stop as quickly as possible, I rolled down my window to see what was happening. Pointing to the back of the vehicle, he said, "Your car is on fire!" I looked back, and all I saw was a cloud of smoke — no road. It had happened again. The motor had given up. The first set of rings had packed it in at just over 40,000 kilometers, and here I was at around 80,000 — so it was no big surprise that it was giving up again. It was a big inconvenience, though. We were only halfway to Nairobi, and there were no garages in between. The only thing to do was pour in a gallon of oil and let it belch smoke. I always carried extra oil. Things were banging, but we were moving. We received many strange looks as we met oncoming vehicles. About the time I thought a gallon had burned through, I added another gallon — our last.

It took us to within 10 kilometers of Nairobi, when the motor finally said, "The end." A passing motorist gave me a ride to the Volkswagen dealership. The vehicle was towed into the garage and assessed. It would take three weeks to repair. Arrangements were made to rent a mission vehicle so we could take Rhonda and Stephen to boarding school. A flight on an Africa Inland Mission plane was arranged, and we flew to Arusha. As usual, we had things planned and didn't want to cancel anything just because of a slight hiccup like a blown motor. Within three days we were scheduled to be teaching at our Bible college in Mwanza.

Planes were not exactly plentiful, but we managed to get a booking. That was the reason we were sitting in front of the booking office in a taxi waiting to head to the airport. Taxi owners are not known for spending recklessly. Maintenance is done only when absolutely necessary—like when the vehicle refuses to run any longer. Grinding brakes are a minor detail, as are doors randomly flying open and windows not rolling up or down. It was hot, and three of us were squeezed into the narrow back seat of the Peugeot 504. Our fellow passenger was not exactly a person who had the physique of someone who had been on a diet recently—if so, maybe a diet of sweets.

The two front bucket seats already had the driver and a passenger, so why we were not leaving didn't make sense. Besides, the moments were quickly ticking toward our check-in time. With so few planes, one did not have to be too late before another eager passenger on standby, with money in hand, grabbed the opportunity. Money under the table was also an avenue to get a seat.

"We are waiting for one more passenger" was the driver's reply when we asked about it. Where were they going to sit when the vehicle had a stick shift console in the middle of the front seat? Desperation was beginning to surface when a rather substantial Somalian lady showed up and actually squeezed herself in beside the driver. This meant he was pushing pretty hard against his door, with his arm and half his shoulder sticking out the window.

The trunk had just enough space for the other passengers' bags. We had a large suitcase, so it went onto the roof rack. Every vehicle in Tanzania has a roof rack. The driver showed no hesitancy as we lurched out onto the main road. The airport was a 45-minute drive if no slow-moving trucks or animals were encountered. Small minivans used as buses had to be added to the equation. Without any regard for the speed limit or safe driving conditions, they sped along the road. Often they followed close so that any slight miscalculation or misjudgment of the lead driver's intention resulted in a bad scene. These narrow roads were built when vehicles were few and speed was no issue. Road shoulders were no more than a meter wide, maximum. They proved to be a greater hazard than the road itself. Gushing rainwater eroded gravel right beside the pavement, leaving sharp drop-offs next to the pavement. Drifting off the road into one of these erosion ditches was a certain rollover situation. Cruising along, these buses would without warning, and seldom with brake lights, suddenly come to a stop on the road to pick up a passenger who had just emerged running out of a banana plantation, waving his hands. These were accepted bus stops. Oncoming traffic usually made it impossible to pass.

We were beginning to have hopes of just squeaking in by the arrival time deadline when African reality struck. When things are going well, hold your breath—because things may return to normal at any time. The taxi began to sputter and then died as we coasted as far as possible. Without a lot of fanfare, the driver peeled himself out of the door—remember the well-built Somalian middle passenger. He opened the hood and pulled off a hose. He began sucking on it like crazy, and pretty soon he spit gasoline out of his mouth. Quickly the hose was reconnected, the bonnet slammed, the engine started, and we were off again. Apparently, the fuel pump had notions of taking a rest once in a while. By this time we were again in "angst" mode about arriving on time for our flight.

Not to worry! We were approaching a nice long hill on which, with a lot of rattling and vibrating, the car hit 130 kilometers an hour. Not exactly a comforting feeling. Suddenly I had an all-too-real vision. That sound was our suitcase, in spite of its weight, flying off the roof rack. Wrenching my neck to look back, sure enough, there was our suitcase sliding down the road vainly trying to catch us. Remarkably, it never began to tumble. Our taxi was coming to an excruciatingly slow stop and beginning to reverse when I saw a huge truck coming down the road. The Lord's hand was upon us again as the suitcase came to rest smack-dab in the middle of the road, and the truck passed within a few inches of it. We made the plane but were already exhausted before the trip started.

Stay Away From Water

I'm not sure if it was meant to be shock therapy or teaching, but before heading to Africa we attended sessions held by the PAOC School of Missions in Peterborough, Ontario. It was mandatory for 'first-termers' and returning missionaries. Dr. Lenczner, head of the Tropical Diseases Unit at the Toronto Hospital, did not show us pictures of serene landscapes, beautiful flowers, birds or animals. They were enlarged pictures taken from under the microscope. Worms under the skin, in the gut and the intestines are what we were shown, up close! They were of all shapes, colors and lengths. They had names such as hookworm, mango worm, roundworm, Guinea worm and foot worm. All were gross, but the tapeworm took the cake. It could not be killed by merely taking a heavy and sometimes prolonged dose of some pills; it had to be rolled up on a stick as it made its exit. That meant only a twist or two a day. If the worm was torn apart, it just continued to grow—yes, inside of you. Ugh! "Slow and steady" was the motto. Comforting was the fact that these hitchhikers could reach a meter in length. No wonder the people carrying these parasites around were extremely lethargic.

Together with the detailed description of all of the terrible things they could do in your body, a brief instruction on how to avoid them in the first place was included. The instruction about water was repeated in a way similar to our Western advice on buying property: "Location, location, location." The avoidance maxim was: "Stay away from water, stay away from water, stay away from water!" Water in Africa seems to harbor most of these little critters looking for a warm place to have babies and pay no rent. Water is wet and warm and has all kinds of potential hosts coming to it.

Being a pastor and adhering to our biblical interpretation of water baptism by total immersion, water was the place where my responsibilities took me. I am thinking that Africa was the birthplace of the mainline church's practice of interpreting baptism as sprinkling. I was constantly baptizing people in Lake Victoria, which has the dubious distinction of being a beautiful lake straddling the equator, yet one whose waters are to be avoided at all costs. It harbors the slow-killing parasite known to cause bilharzia. This parasite has the rather mystical medical name of schistosomiasis. It lodges in the liver and ensures an eventual albeit slow death.

More times than not, baptisms took place in the dry season and away from the plentiful water of Lake Victoria. On those occasions I waded into a mud hole with a probing stick to measure any unforeseen holes that may have been waiting to surprise me. These were true mud holes, where mud far surpassed any water volume. Fish also called these places home, and they could add a little excitement when they hit your leg. In the beginning, I wore a pair of hip waders under my pants to keep water from touching my skin. It didn't take long before they were too cumbersome and

I took my chances without the waders in all kinds of places that passed themselves off as water holes.

Coming back to Canada after four years on the field, we were scheduled to see our friendly tropical doctor Lenczner. He was rather aghast when he surveyed my pre-examination medical questionnaire. Every question that pertained to nearness to water I had answered, "Yes, and in it." As he ran blood tests and examined stool and urine culture samples, he was not too confident that I would have a clean bill of health. Obviously, I had not taken anything to heart that he had so passionately taught us four years earlier.

Results in hand, he simply shook his head. I did not carry a single parasite. But my wife and children all had little hitchhikers.

CHAPTER 22

African Time

"African time," in most cases, is an expression used

to convey the meaning of lateness. Well, it is that if understood from our Western mindset. However, "time" in Africa has its own meaning. Important people show up when the less important have already gathered and gone through the preliminaries, as it were. If a governmental official is invited to participate in an official opening, you can be sure his arrival will be much later than you expected. They do not want to waste their time sitting through preliminary parts of the meeting that are not important. Besides, if the important person arrives early, who will see him or her? By the time the important person arrives late, all of the latecomers have also arrived, so he has a full audience. The motto is "Last not late." That was hard for this punctual Western boy to swallow.

An elderly pastor especially stretched my promptness tendencies. When we were in his area, we slept in his yard in our Volkswagen Camper. This meant that his five children did not have to huddle together in a corner all night under a blanket while we occupied their bed. From his hut we drove two hours to visit a church. Arriving there, the people would already be singing and

414

well into the service. I was chomping at the bit because I felt bad for the people and we were late!

Hearing our vehicle arrive, the local pastor would come to greet us and welcome us to the church. Pastor Paulo would tell him to bring two chairs outside to place under a tree, where he should bring us tea. He regularly explained that as pastors we need energy going into a service. It was not good to travel a long distance and then enter the service tired. A few cups of hot, sweet African tea would encourage rest and a fresh injection of energy. After all, it was not going to be a typical 60-minute Western service. Four to five hours was the norm, not counting food after the service.

African time places more stress on personal engagement than on an appointed time of meeting. If, while on the way to an appointment, a friend or acquaintance is met, spending time chatting at that unexpected meeting is more important than the more distant anticipated encounter in the future. Eventually it will take place, but this "happening upon" a friend will not happen again in the same way. Besides, it is a wonderful opportunity to catch up on the latest personal news and information about other relatives, acquaintances, tribal members and friends.

Irksome to me, but of utmost importance to them, was the attendance at a meeting. I never did understand this concept, but then I did not understand a lot of things which were important in the culture. Not understanding never translated into despising the culture. Simply showing up at an event is the important thing. It doesn't matter that one is not present when the opening bell is sounded. Even arriving as the closing bell tolls meant that the event had been attended. Many people arrived within the first 30 minutes. This meant that although, at first, some empty chairs stared you in the face, at least there were enough occupied to dampen the hollow echo in the place.

Watching a bus pull away from a station was always intriguing. In reality, even though the schedule says that the bus leaves at 11:15 doesn't mean it has left at 11:14. Therefore, it is not necessary to get on the bus early. When it begins to actually get those wheels in motion and pull away, that is the time for serious boarding. If I can run faster than the speed with which the bus is moving toward the exit, I still have time to say goodbye to a friend or grab a basket of fruit. That is what I call making good use of time, the attribute that so few of us have come to appreciate. We waste so much time being early for everything. There is too much "army" in us: run to stand in line and wait. That is me, at least; my wife has another view of arrival time. In 50 years, we have not yet managed to synchronize our idea of being on time.

There were times, during the height of shortages for everything in the country, that when we did see a lineup, we immediately joined the line and then asked what was being sold up front. If we had no need of that item, we kept on to wherever we were going. If it happened to be sugar, rice or propane gas, we crept along with the rest, always hoping that stock would still exist by the time we reached the front. "Line jumpers" added excitement as they were either loudly berated or even set upon. There's nothing like an interesting sideshow while waiting for the line to move.

Sometimes lineups had a slightly funny side to them as well. There was a huge shortage of gasoline, but every once in a while it was available. On and off, this offered fertile ground for the rumor mill. Rumors, sometimes true and sometimes false, circulated that a certain petrol station was receiving some fuel. Lineups of vehicles formed. Blocking the main road was an accepted byproduct of it all. No one minded. Lines of up to six vehicles wide were common, as were sharp disagreements about who was cutting whom off. Naturally, at the pump, everything had to filter into one line. The desperation of one taxi driver gave him the idea of coming in from the opposite direction, where vehicles waited impatiently. Seeing one accomplish the goal of getting nearer the pumps encouraged others, and soon there was a complete gridlock. Vehicles faced each other headlight to headlight.

I joined the lineup waiting for gasoline to arrive. Nothing was moving, so I walked home to have lunch. There was no use in wasting my whole day. When I returned an hour and a half later, my driverless vehicle stood all by itself in the middle of a busy main street. The rumor of petrol had been false. Without any objection or annoyance of any kind, other vehicles were merrily going around mine. This again underlines the people's commendable acceptance that things don't always work out, and getting too agitated about it is needless. Life can be a bit unregulated, but the world will not come to a standstill because of it.

Tanzania uses two different ways of telling time: Western time and their time. Western time uses the Western clock: 7 p.m. is 7 p.m. In African time, 7 p.m. is one o'clock at nighttime. Twelve noon Western is six o'clock African. It is the same time as that used in the Bible. When making arrangements, it always had to be determined which "time'" was going to be used as the standard. "Are you talking your time or our time?" is one of the first questions to ask.

The standard of many things in our day and age is something that seems to be movable. Many cults adopt the use of Christian expressions, but they have vastly different meanings. When it comes to morals, character, commitment or salvation, the Bible is the only standard. It is the only true standard that promotes the well-being of people.

Another area of vigilance is the use of calendars. Many calendars have Saturday as the first day of the week. At the same time, there are calendars where Sunday is the first day. Paying attention to this is critical when making a booking. Look both at the date and the day.

Although not directly related to time, there is a factor connected to time and ranking the importance of a person. When attending a gathering with a large number of various leaders present, the importance of a guest in the minds of people can be gauged. Thankfully, status of position is not something that the common person sees as being overly important. Usually, the person designated to introduce guests is the "top man" and needs no introduction himself. He makes a short speech and then begins to introduce guests. Seldom has a list been compiled beforehand. He turns around to survey the guests: there may be up to nine or ten. Then, beginning with the least important—at least in his mind and that of most people—he begins his introductions. The last person to be introduced is the most important guest. This can be disconcerting to guests from our country as they are used to being first on the list of introductions.

What Is That Garbage Truck Disgorging?

First impressions sear into your mind like a hot branding iron. With vividness they unexpectedly come back in full colour and motion. The first thing LouDell remembers seeing is a man—who happened to be wearing a suit—operating a caterpillar bulldozer.

It was our second day in Africa, and we were in the office of our East Africa overseer in Nairobi. His office was right downtown on the third floor overlooking a back alley jutting off into a triangle of alleys. We were curious and, while we waited, gazed out of the window to take in the sights. Naturally, everywhere we looked we saw something new.

Even a city garbage truck with a dome covering attracted our curiosity. That curiosity then turned to disbelief. The truck stopped, and suddenly men dressed in suits and women dressed in their finest began pouring out of the opening. They were office workers heading to work. Over the years we witnessed people putting convenience over "being proper," according to our Western ways. That was refreshing. I have a feeling that the city garbage truck was washed out overnight and then served a dual purpose in the morning as a bus before "going to work." I also have a feeling that the driver was moonlighting with a city truck—just pocketing a little extra cash to start his day right. The workers were probably also happy as they didn't have to contend with small, overcrowded minivans that were speeding while being used as buses. Locally, these are called *Dala Dala*. It was a good arrangement for everyone.

Africans are pure, unadulterated, unconvertible capitalists. They are very generous and giving when the situation demands it, but if there is a slight hint of monetary advantage, it is immediately pounced upon and put into motion. It all stems from having to engage survival skills right from childhood.

Rains are always a welcome sight as they carry with them the promise of an abundant harvest. Unfortunately, they come in big amounts in a short period of time, causing all kinds of havoc with roads.

We were on our way to Nairobi from Mwanza, driving along a recently constructed, newly paved highway. Too many times we had driven that old road and were pounded to pieces by the potholes in it. Heavy rain or not, this was an enjoyable trip until we came upon a lineup of vehicles. The gushing waters eroding the embankments had washed out a very short bridge. The embankment was probably 15 feet deep, or I surely would have been tempted to test my four-wheel driving prowess. It wasn't long before I noticed some vehicles turning around and, a short distance back, heading across country to skirt the high embankment. I followed suit, and we soon found ourselves in a very muddy field that did test the four-wheel drive. The soil there consisted of a good percentage of clay, which becomes like grease in rain. Slipping and sliding, with the engine screaming and all four wheels spewing mud, we bounced around in deep ruts left by our predecessors. Coming around a corner in the cornfield, we encountered a crudely erected barrier. This was private land, even though the government owns all land in Tanzania. We had to pay a toll to cross it. The sisal pole was pulled aside, and we were on our way again. "Yup," we smiled—yet another good opportunity was seized to capitalize on a circumstance. After not more than a hundred yards of slipping and sliding, we encountered yet another barrier: we had to pay to *leave* the field! The road was such that stopping would more than likely see the vehicle stuck. That would result in people rushing out of the bush to push, and that would translate into more money.

Keeping my foot on the pedal, I smashed through the barrier. Out of the corner of my eye, I saw the "gatekeeper" hit the deck to keep from getting hit by the sisal pole barrier. Paying once was tolerable, but not paying twice. I am a keen admirer of ingenuity and opportunity, but admiration does have its limits.

This desire for money sits upon a very thin wedge. It can be a help to cover the living expenses of a struggling family, yet there yawns the steep precipice of greed. A pastor friend was riding in a minivan bus when it was involved in a serious accident near the small village of Kikatiti in Arusha. People from the over-crowded bus were lying all over the pavement. Some were dead, and others were badly injured. My friend was still able to walk. He observed a man wildly running from one dead person to another, and also to those who were injured, stealing their possessions. He did not even realize that sticking out of his own chest was a piece of steel. Suddenly he fell over, joining the dead.

By accident or sheer ingenuity I do not know, but a hotel in Arusha became very modern and installed new "pay as you go" washroom doors. They were of European design where deposited money opened the door. These doors were installed backwards. Horror only registered after the door clicked shut and your eyes took in the scene of a coin slot with instructions. Fortune indeed smiled on you if you had the required coinage to extricate yourself. Unless you were of beanpole shape and size, squeezing under the door or under the partition to the adjacent stall did not lend itself to a rescue. At this point, any preservation of dignity was set aside. If the faintest footsteps were heard, cries of help were in order. Even begging for the right coinage was not seen as an act of begging. Again, the biblical teaching that it is best to go with a friend is advice well taken—unless your friend also finds himself or herself in the same predicament right beside you.

Don't Come Just Yet

When making a life-changing move, a person does not just go down to the airport and buy a ticket at the counter and take off. An enormous number of things must be arranged, both at home and also at your intended destination; for example, such things as designating a guardian for your children if you should die. Passports must be obtained; banking is a huge aspect spanning both worlds. When furnishing the house, is it more economical to ship your old furniture or buy new over there? On and on it goes. The to-do list can run into several pages. Some things cannot be done until something else has already been set in place.

We were at the point where everything had been cared for. We were on our third assignment to Africa. Arrangements at home and on the field were in place. Our arrival in Dar es Salaam went as planned. Once there, we contacted the people at our final destination in Mwanza. It was a reconfirmation of our correspondence regarding our expected date of arrival, which was scheduled for two days later. They were in charge of getting the house where we were to stay all cleaned up and ready for us.

The next day, they called us back and asked if we could postpone our arrival by at least two days as they had not done anything to prepare the house where we were to stay. By this time, we had been living out of suitcases for well over a week. We did arrive a few days later, well into the afternoon. Still suffering from jet lag, tiredness, lonesomeness, and now frustration, we walked into our house at around 4:30 p.m. We felt like turning around and heading home, but the die had been cast, so we had to grit our teeth and face the situation.

The house reeked of the pesticide used to fumigate the dwelling for bats, cockroaches and other bugs. Everything from the kitchen was piled in a heap on the dining room table and floor. The beds were unmade, and we had to hunt around for linens, towels and soap. The floors were gritty. The hot water tank had not even been turned on so we could take a warm shower. Everything would have been somewhat excusable except that this couple had at least a few months' notice of our coming, and our house was no more than 50 yards from theirs. They did not have to do any of the work themselves; other people were available to do the work. It was a severe case of procrastination. Later we found out that they were not too happy about our coming, perhaps they were putting off our inevitable arrival. They felt threatened that we would take over their job. It's amazing how our minds can mislead us and cause unnecessary stress. We were there to do a job, not to usurp another's job. Having said that, I observed that things on the field could become a big issue, whereas at home the very same thing would hardly be mentioned. Did it have something to do with having only very few people of like culture around? I am certain that more times than we wish to admit, it had something to do with the spirit of darkness endeavoring to bring dissension, thereby shunting to the sidelines our major work.

You Light the Bier

The father of a friend of ours was visiting, and each evening he walked down the road to a local pub. He had just retired and was waiting for his wife to join him from overseas after she retired in a few months' time. One morning his daughter awoke and did not find him at home. A search was begun, and he was found in a ditch along the road. Dead. It seems that while coming home in the dark, he had fallen and hit his head on a rock.

The family had no church connection and it was decided that cremation would take place. The only crematorium in Mwanza was the Hindu temple. They agreed to allow the cremation to happen there. It took time for his wife to arrive and make arrangements for his burial. In fact, a week had gone by before she arrived.

Erase any picture you may have of a huge, ornate Hindu temple such as we find in Eastern countries. This building was low, about 50 feet in width and 30 feet from front to back. The back wall and sides were open. On the front wall were painted pictures of a number of Hindu deities. People sat on backless cement benches. The temple was located just off the main thoroughfare down a steep road and situated about 50 yards from Lake Victoria.

The coffin arrived on the back of an open Land Rover. It was carried in and placed at the front. The man's daughter, wife and friends were seated at the front, a few feet from the coffin. As it was carried past us, the distinct reek of death washed over us. I have smelled this on a number of occasions; it is unmistakable and over-whelming. This odor is common when a body has not been properly refrigerated, as we understand refrigeration. Morgues in Mwanza are not known for cool temperatures. Even while the family and friends shared remembrances of his life, the smell intensified and far too often wafted over us. The temple had a low ceiling, it was hot outside, and there was no breeze to offer the least bit of air movement.

As the remembrances carried on, through the open wall we could hear the *boing, boing* of dried logs being thrown into a pile.

When the time came to exit, the wife, family and close friends followed the crude, handmade, wooden coffin to the bier. It was an open area with a covered roof. The body, wrapped in a shroud, was lifted out and placed on a cement slab with logs surrounding it. Those in attendance stood around. Very close to the foot of the bier stood some men near a large kettle. In it was melted *ghee*, a rancid-smelling liquid similar to butter. Large pieces of clothes were dipped in the ghee. This was sloshed over the body. What remained in the large kettle was thrown over the body.

The wife, standing very near the head of her deceased husband, was then given a match to light the ghee. When it began to burn and ignite the ghee and the wood, a large fire erupted. Popping noises began, and an accumulation of odors filled the air. It was at this time that my wife grabbed my arm and we headed to our vehicle. There were too many things going on for the mind and senses to compute.

Stranger Than Fiction

Examples of strange things happening in Africa are as numerous as at home. An 80-year-old American came to Africa after his wife died. He had the name Esther somehow impressed on his mind. Therefore, the first Esther he met was destined to be his wife. They got married. We happened to know Esther. She was a wonderful lady whose husband, a number of years ago, had been found dead alongside a road. The cause of his death was never determined. She left the coastal town and moved to Mwanza with her nine daughters. Miraculously, she raised them on her own. She lived a typical lifestyle, eating only the local food she grew up with. Possibly once a month she was able to supplement her diet with goat or chicken meat. Her clothes were washed by hand in a metal tub. A daily trip to the local market provided fresh vegetables. Cooking was done outside over rocks, and charcoal was used as fuel. Church services were attended by walking two kilometers.

Her clothes were sewn by private seamstresses, parked with their sewing machines on the sidewalk that was already too narrow. If they were progressive, they would erect a torn tarp on some sisal poles to shelter themselves from the hot tropical sun. Cloth dragging on the dirt as a dress was sewn was not a big deal as the dress could always be washed once finished.

These shops sold a huge variety of household supplies, including cooking pots, matches, kerosene, kerosene cookers, charcoal, mosquito nets, stirring ladles, cooking oil, sugar, soap, mosquito coils, and a host of other things. Goods were crammed into a tiny space. This left no room for browsing. You could only walk up to the counter, look into the shop, and ask for things from there. This carried the advantage of reducing theft. Imagine Esther travelling to the USA. All that met her there is beyond anything I could envision. Esther from a poor third world country suddenly finds herself on a plane and in civilized America.

Food in one country is not the same as in another country. People from a certain country came to work in Tanzania to build a series of water canals based on the Egyptian plan, where water is channeled from a body of water out onto fields. Where these men came from, dogs were a delicacy. Dogs in Tanzania were plentiful and lethargic due to lack of food or, most likely, worms and parasites. Crossing the road, they were unable to judge the speed of a vehicle or were simply too weak to move quickly enough to avoid being struck. Because of road conditions it is next to impossible to swerve to avoid a dog—or any animal, for that matter. If an attempt is made, many times a rollover is the result, along with serious injuries or even the death of the vehicle's occupants. It was a regular occurrence for the men to shout to their driver to stop the vehicle so they could inspect the dead dog to see if it was still edible. Drivers of the Muslim faith complained, and eventually the practice had to be abandoned.

Resourcefulness

African people continually astounded me by their resourcefulness. For me, it would be thinking outside the box. For them, it was assessing what was at hand and how it could be used in the present situation. Here are some examples.

Paper matchboxes are used to bring stool samples to a hospital.

Bicycle tire tubes, having accomplished their life's work, are cut into thin strips and make excellent tie-down straps. They are extremely stretchy, so they can be cinched down very tightly.

To seal a leaking vehicle radiator, a couple handfuls of cornmeal flour are put into the radiator. Another method used is to add a few raw eggs into the rad. A handful of pepper also works.

Tough grass is tied together to make a broom.

Sisal (binder twine) strings are tied onto the end of a stick that is 10 feet long to remove lake flies or other insects from walls.

To catch a monkey, put something into a glass jar. He will grab the item but will not release it, thereby not being able to withdraw his hand from the jar.

Catching bugs alive: turn a matchbox upside down and put the bugs in from the open bottom. Bugs always fly upwards so do not escape from the bottom.

To bend a PVC pipe without developing a kink, place sand in it and heat it over charcoal until a bit soft, then bend it.

To remove a flat tire, rocks are placed under the axle and then a hole is dug under the tire to give it the clearance to be removed. It is taken somewhere to be fixed. Few people see the necessity of driving around with a spare tire that is not used every day. Better to sell it and enjoy the money. African people are practical, looking at the here and now and not worrying much about what may happen tomorrow.

Many people in the rural areas cannot sign their names. Thumbprints are the usual seal used instead of a signature. In the cities, ink stamp pads are a part of the stationery equipment at every desk. The thumb is pressed onto it and then pressed onto the area reserved for a signature.

When wanting to tear off a piece of paper from another, the paper is folded and the tongue licked along the fold. Presto! It comes apart as nicely as if it were cut with a pair of scissors.

Have a tire that is too large for the rim? Take the inner part of a larger tire and place it over the rim first, then put your smaller tire into that. I have seen it done on large truck tires and am amazed that the whole thing did not fly apart.

Children's bicycles are made out of wood.

Sandals are produced out of old worn-out tires. They still have a few thousand miles left in them.

AA radio batteries are hooked up in sequence to produce enough volts to operate a speaker.

A sprinkler is made by punching small holes into a plastic bottle and then tying it to a stick that is stuck in the ground. A water hose is taped to the neck of the bottle and presto—you have a sprinkler!

Wooden guns are made for army practice.

Children make soccer balls by rolling up plastic and tying it into a somewhat round shape of a ball.

Sisal string is used to stop a leaking faucet.

Old tires are cut so that the remaining part becomes a swing seat.

By weaving the fine wire out of old tires, traps for mice are made.

Drums are made out of used cooking oilcans.

Guitar strings are made from fine wire.

Flat tires are a regular occurrence on bad roads. This possibility is heightened when sharp rocks are part of the road. I was travelling on the gravel road from the tiny rural village of Mto wa Mbu (River of Mosquitoes) to the small village of Mkuyuni. It was always a relief to reach Mkuyuni, as that was where the pavement started. Arusha was only a tantalizing two hours away. By this time I was hot, sweaty and tired. Suddenly the sickening, if not unexpected, rumbling sound came to my ears above the banging of the vehicle on this bone-jarring, rocky road. Coming to a stop on the dusty road, I was prepared to grit my teeth and get the job of changing the tire done. This had never happened to me before. Both a front and a back tire were flat. The quick appearance of some Maasai herdsmen immediately brought the picture into focus. They had hidden a board with nails just under the powdery dust. Now they were going to charge me for rolling my tire back to the village to have it repaired. Ha! I had outsmarted them because I always carried two spare tires. Disappointment showed on their faces.

There were only a few bakeries in our city. When most commodities were extremely scarce, we could always count on the Greek baker to hide a few loaves in the back for his friends. During the war with Uganda, he informed us that the army had commandeered his total output of bread. He told them how many loaves he produced each day, and they wanted it all. A few days later, we passed by to see him, and he asked us if we needed bread. We were amazed that he was still selling bread. We asked him how that was possible in light of the army's taking his total output. He replied, "Oh, that is easily done. I just reduced the loaf size. I am able to give them the number of loaves I promised and still have others to sell."

Seeing large trucks and vehicles broken down on the road is a common sight. In the case of large trucks, where towing is not a viable option, they are repaired right on the side of the road. That includes everything from flat tires to complete motor jobs. A tarpaulin is laid out and the parts, as they are removed, are nicely arranged on it. There is a special technique for cleaning the parts. A mouthful of diesel or petrol is taken and expertly directed at the part in small spurts. I have been amazed at how much the mouth can actually hold.

Africa does not have many straight roads. This can result in some very unexpected situations when coming around a sharp bend and being suddenly confronted with a large truck being repaired on half of your already narrow lane. More often than not, oncoming traffic hugs the centre line because only a foot from the edge of their lane is a washed-out shoulder. Emergency flares, African style, are used to warn others of a broken-down vehicle ahead or around the bend. Piles of branches, brush, thorn bushes or grass are laid on the side of the road just before the bend in the road. Every African traveler immediately knows to slow down.

Stomach cleanser formula: half a bottle of Coke is mixed with super gasoline and guzzled down. It is referred to as "the Supa Klensa." The recommendation is to take this every six months. We stuck to more traditional Western methods.

Every culture has its unspoken language. Attending our first service at a rural church, our cultural training began instantly. The local Luo tribe has many customs. The one revolving around tea we never forgot. First of all, they boil their tea over an open fire so it is piping hot when it comes to you. Their love of tea is reflected in their huge cups. The lady of the house, who is not sitting with you but flitting in and out, keeps a close eye on your cup. As soon as it is empty, it is filled again. Chai was new to us, so finishing a large cup was a victory. We motioned that we did not care for a refill, but the lady insisted. Eventually, we were told that the silent language indicating "no more" was to turn the handle of your cup facing the centre of the table.

Some good-humored comments were directed to us by a pastor of this tribe. As I said, their cups are huge—about the size of our soup cups. When we poured tea for this pastor he would say, "Posein, our noses are not as long as yours. You can fill the cup to the top." I didn't think my nose was particularly long, nor theirs particularly short. Neither did I think the cup could hold much more. It was nice to have friends with whom you could share a joke.

CHAPTER 23

Flights

Due to the long distances and the bad roads, we flew by small plane whenever it was convenient. Not only was driving time-consuming; it was also expensive in a number of ways. Fuel prices were high: over $3.50 a liter. Extra fuel had to be carried since the next gas station might not have a supply. Ask me how many times I sucked gas and didn't quite get the timing right, thereby ending up with a mouthful of gasoline or diesel. Secondly, the roads were torture on the vehicle. Shock absorbers went through a major workout on every trip. With our Volkswagen Kombi, I always carried an extra set of shock absorbers. It was a regular thing to have fatigue overcome them along the way. To remove the shock absorbers, a pair of good gloves was essential. They would be so hot that it was impossible to touch them with bare hands. Tanzanian roads and Volkswagen torsion bars were not friends. Sand had a tendency to pile up in the middle of the road. As the vehicle went over a bump and came down, the torsion bar softened the bump by flexing downwards. They then connected with the sand and were bent. Often I had to drive up to a tree, hook a cable around the bar, and pull it straight—a fine mechanical art for sure.

Flying meant that we missed seeing all of the activity going on in the countryside and villages. We did drive enough times to make up for the occasions when we flew.

Commercial airlines were few and mostly unreliable as far as a schedule went. Mission planes, such as those flown by Mission Aviation Fellowship (MAF) and Africa Inland Mission (AIM), were the most reliable. We could book them and be assured we had a seat. These planes were normally six-seaters, which had their own quirks. When booking, some standard questions were asked: "How many pieces of luggage do you have?" "What is the approximate weight?" "How much does each passenger weigh?" Now that can get a bit tricky when answering for your wife.

There is something exhilarating about walking on the tarmac where planes are parked. You can actually touch them and be there amid all the action of getting things ready for flight.

All luggage was weighed before takeoff. All of these things affected how much fuel could be carried as well as the safe operation of the plane. Turbulence was magnified on these small planes, and the pilot's "before takeoff" instructions always included a clear explanation of where airsickness bags were located. In order to avoid the turbulence, a pilot could sometimes get above it. Not having a pressurized cabin, the pilot had to become the barometer of what was high enough. Sitting beside him, he would grab your hand and check to see if your fingernails were blue. If they were, the maximum height had probably been reached.

On one of our flights, we were one of two missionary couples heading home on furlough. We left Mwanza Airport heading for Nairobi, where we would board an international flight. Pilots are not only pilots; they are masters at packing and knowledgeable about weight distribution. We had large suitcases, but not that much weight. Space became the issue. The last suitcase was shoved in against my feet, and the door forcefully slammed shut. By the time the two-hour flight ended, my legs were numb from not having moved the whole time.

Small planes meant that seldom did you sit beside your wife since weight distribution was such an important aspect of plane handling. Six-seater planes are not known for being roomy. It was pretty much shoulder to shoulder. In addition, there were no washrooms on board. That could prove to be interesting if you had a weak bladder or a turbulent stomach. Typically, the seating arrangement was that the heaviest person sat next to the pilot. Immediately behind were two seats, and behind them were two more things called seats. That was just to fool you. Reality set in when you crawled into the plane. They were just slightly elevated platforms on the floor. By the second time around, you were praying that you would not end up in that section. By then you knew better than to call them seats; they were just spaces.

The order of seating was organized. First, those sitting in the "back spaces" crawled past the two middle seats and usually plopped into their space unceremoniously. The second person, to some extent, fell onto the person who was already there. No matter how hard you tried, getting in could not be done gracefully. Then the "middle seaters" entered. After that, the front passenger was seated, and last of all, the pilot. He made sure that everyone's seatbelts were properly adjusted and that all the doors were secured. Before clambering aboard, an external, detailed inspection took place. Cables running from the rudder to the flaps are important, so they are carefully inspected. It's always good to check that the fuel cap has been properly fastened. I was at the airport when a plane took off and just barely made it back when it was discovered that a fuel cap had not been properly closed. Air coming up over the wing formed a vacuum which sucked out fuel at an alarming rate.

I really believe that the designers of these planes had a sadistic streak in them. Seats are so placed that when you open the door to get in, you have to step up onto the wheel frame, then hoist yourself up through the narrow door and maneuver your body past the seat to actually sit down. By the time the pilot gets in, he has checked the single propeller for any rock chips. If there are any, he has filed them down smooth. Apparently, if left unfixed, it greatly increases the "weak factor" of the propeller. Having only one, it is probably a good idea that it be in the best possible condition. Rock chips occur when the plane is taxiing on a dirt runway and the propeller sucks up a rock. Once inside the plane, the pilot straps himself in with over-the-shoulder safety straps. As each person has contorted their body and managed to get in, the pilot has helped them into these straps. Somewhere between comforting and arousing nervousness are one's emotions when the pilot says, "Let's pray." He now gets all of his headgear on and speaks with the airport tower for permission to leave. "Prop clear" are his final words before he coaxes the engine into life. This may seem redundant, but it has happened that an engine has cranked over and the propeller has hit a person. No loading tunnels exist; people walk from the tiny terminal over the tarmac to their planes. People are walking around among the closely parked airplanes with their attention on different things. Some are carrying luggage to another plane, some are disembarking and walking to the terminal, some are going to do final inspections before their flights, and there are always those passengers who are "excited to be going somewhere in Africa" and are not too careful.

The passengers are all quiet, and finally there is the slow, dull *thump, thump, thump* sound as the engine builds up momentum in turning the propeller. The whining of the prop relaxes everyone. A great delight about flying in these machines is that they fly relatively low, so huts, cows, people and wild animals are visible. For maximum fuel efficiency it is strongly suggested that the plane get up to cruising altitude as quickly as possible so the engine doesn't have to work so hard. Being that they are mission planes, keeping costs down is always a priority.

On a flight over the world-famous Serengeti Park, the pilot was less concerned with the financial bottom line as with experiencing life. Flying from Arusha toward Mwanza, we climbed high enough to clear the hills surrounding the Ngorongoro Crater. Actually, just a bit north, an extinct volcano still pushes out smoke. Right after clearing the hills, the ground falls away onto the Serengeti plains. This is one sight a person never forgets or tires of seeing. It is a panoramic view of the days of creation. It takes in isolated *kopjes* (small hills)—it remains a mystery as to how they got there—as well as the large ravine known as Olduvai Gorge. Close to it, pressed into the hardening lava spewed out of the Ngorongoro volcano, are the footprints of earliest man. Those who know such things tell us they are the footprints of a man, a woman, and a child. The woman exhibits signs of fear as her footprints suggest that she half turned to look back. Could it have been the biblical Lot's wife?

Zeppo, the pilot, was a man after my own heart. His philosophy was that we only live once, and surely it would be committing a sin if we did not take advantage of seeing God's creation up close. As we cleared the walls of the Ngorongoro Crater, he said, "Get your cameras ready." We were heading down to see animals up close. It didn't take long before we spotted a herd of giraffes. They are well known for their graceful lope. What is not well known is that this graceful lope is a result of their moving both legs on one side of their body at the same time. This is very rare in animals. Due to the length of their legs, their kick is known to have decapitated a lion more focused on a meal than on protecting his life. The herd consisted of 18 giraffes. Hearing the airplane, they began to move as in slow motion. The pilot asked me to take the rudder as he banked the plane to get a better shot of the loping giraffes.

On our way to Nairobi, the plane delivered some supplies to the Lutheran Mission Hospital in Haydom. Mission hospitals are a great blessing to the local populace as they are exactly what their name suggests. They are there to serve the rural population who are far removed from a city and access to medical care. Professional staff, made up mostly of missionaries, are competent and caring. For some reason that does not readily come to mind right now, we were loaded to our maximum weight capacity. It was rainy and soggy, which did not exactly make for a friendly dirt runway. We would attempt a takeoff, but if weight proved to be a factor in the ability to take off, we would abort, return to the small one-room building, and drop off a suitcase. Who, when travelling, can afford to leave a suitcase in the bush? The fellow beside you, hopefully!

Having said a prayer, the engine revved to its maximum allowable, and we began to sluggishly head down the muddy runway. A saving factor that the pilot had pointed out, and which we were not too interested in exploring, was that at the end of the runway there was an extension of the Rift Valley escarpment. It is a world phenomenon running from Egypt to South Africa. If we were within acceptable chances of successful takeoff, we would shoot over the end of the runway, over the escarpment, and have 2,000 feet of depth of air to help us stay airborne.

On the spur of the moment we sometimes do not sufficiently think these things through. Once the plane is three-quarters of the way to the escarpment abyss, reality explodes in your head. *What was I thinking to agree to this?* Suddenly, one less suitcase, even though it contained the essentials of life, may have been the best choice. If we didn't get the escarpment calculation just right, the suitcase wouldn't matter anyway. It wasn't by much, but we did make it, for which our children and families at home were very grateful. I am too; otherwise, I would not be writing this story.

Being shoulder to shoulder is great if you are trying to push over a wall. However, sitting on a plane with a nervous stomach and no visible washroom sign at the front of the plane—and sitting that close to a stranger—fringes on the uncomfortable. As every person knows, no bathroom break is necessary until it is not available. Staring at the escarpment edge rapidly coming toward you, even though you are the one heading toward it, doesn't do much for settling the stomach or other parts of the system that are prone to overreacting. My wife's weight dictated a centre row seat. She was sitting next to a man whom we casually knew. What we didn't know was that, even in good times, his bladder had its own unalterable schedule. Abruptly, he turned to my wife and said, "Look the other way!" At least he was prepared with his reserve tank in the form of a bottle. That is more than can be said for the dear Catholic sister on another flight where a bottle was not sufficient. She was crossing herself at such a rate that her arms were a blur. No more details are necessary on this subject.

Does Your Machete Have Blood On It?

International organizations, as well as local governments, were taken by surprise when the Rwandan genocide, as it is referred to by many, suddenly erupted. Within a few days, thousands of "refugees" fled the country. Local newspapers reported that within one 24-hour period alone, 250,000 poured into Tanzania. Within a matter of months, the total reached 2.5 million.

From Rwanda they fled across the bridge at Rusumo Falls into the Ngara region of Tanzania. Border immigration officials to whom I spoke said that machetes were piled high, resembling an old-time prairie straw stack. Tanzania allowed them in, but at the border no one could take in their preferred combat weapon, which was a machete. It had to be abandoned.

Water is always an essential, so a small lake 10 kilometers inside of Tanzania became the area for settlement. In the providence of God we had an established church a half-kilometer across the road from the camp. It was rather a nice setting as it sat on top of a hill giving an excellent, if not surreal, view of the camp. One day before, the area consisted of long grass and trees, and now 250,000 people called it home. At night, clearly visible from our church, a thousand twinkling fires dotted the camp. Had it not been for the reality behind these fires, it would have been a beautiful sight. In the next few days tiny grass huts were built.

News of the genocide—some even called it ethnic cleansing!— spread quickly. The United Nations labeled it as "the largest and fastest refugee exodus in modern times." I quickly began to see it as an obligation and opportunity for the mission community to do something. Mwanza was only a six-hour drive from the camp, and I knew we already had churches in the area. Within a short while, leaders of mission groups that were active in Mwanza were contacted. Independently, each felt as I did.

A day later, leaders from The Pentecostal Assemblies of Canada (my denomination), the Assemblies of God U.S.A., the Baptist Mission, the Pentecostal Evangelistic Fellowship of Africa, and the African Inland Mission met. Since I was seen as the original catalyst, I was chosen as chairman. Immediately we organized an organization known as Evangelical Mission for Rwanda Refugees (EMRR).

Our next step was to fly into the area to survey the situation. AIM Air donated the expenses of the plane. Paul was our pilot and did an excellent job of flying over the camp and low over the Kagera River, which is a natural shared border of Tanzania, Uganda, Rwanda and Burundi. Even from the air we could see that things did not look good. Landing at a muddy airstrip, we were able to find a vehicle to take us to a United Nations field office, which had hurriedly been set up in a house. When asked what the most urgent need was, we were told that food was not among them. Burundi had been boiling politically, and the United Nations had already stockpiled massive amounts of food in anticipation of the need. Therefore, when Rwanda erupted, food was already stored in a neighboring country, ready for shipment.

The United Nations' involvement in the crisis, unfortunately, had some negative side effects for the local economy. In the interest of getting things done quickly, the United Nations pays exorbitant prices. In the Ngara rural area, they paid $1,000 per month for a small house that was far from European standards. Even homes in the city did not fetch half that much rent. Whenever possible, they purchased local food, so that can drive the price up 300 per cent or more. That is great for the seller in the short term, but the local population that needs to purchase food is left in dire straights. Greed creeps in quickly, and farmers even sell their seed stock and then have to pay huge sums to get more when the next rains appear. We were caught up in the problem as well. I was building some churches in that area and could not get a truck to deliver material for a reasonable sum. The drivers said it is better for them to just wait for the UN, who pays them many times more than what their normal charges would be to me.

Even though not officially registered with the government, we succeeded in opening a bank account. Letterheads were printed, and we were in business. Each organization was asked to immediately request funds from their headquarters so we could begin work. All groups responded, and shortly we had nearly $35,000. Without delay, a steel container was purchased and filled with beans, rice, blankets and small kerosene cooking stoves. From the local Bible Society we also found 25 Rwandese Bibles. Once the container arrived, its contents had to be inspected by the United Nations personnel. Because of the 25 Bibles they insisted on rejecting the whole donation. We finally agreed to remove them, but we did keep them at our church and later passed them out to refugees.

One contact led to another, and more containers of blankets, beans and cooking utensils were delivered. However, a small but significant change took place. Aimo's mission released him to work full time with the refugees. He became our point man. Aimo was a good carpenter, and with the help of competent assistants, a camp was developed on the grounds of our church just across from the refugee camp known as Benaco Camp. We were able to purchase two used metal rondavels in Mwanza, which we dismantled and reassembled at our camp. One became a cookhouse, and the other served as sleeping quarters for visitors and workers. Slowly we acquired steel containers; Aimo fashioned these into nice sleeping quarters and a sitting area. We also built a motel style building with six rooms to house increasing staff and volunteers at the camp. What all of this meant was that we distributed things from our camp, and not from within the camp that the United Nations sponsored. Eventually, even the United Nations realized that religion was a big part of people's lives. Therefore, we were at liberty to distribute Christian materials along with aid for the refugees.

Over a period of a few years, eight camps were established by the United Nations. In total they housed over 570,000 people. Because of our good name we had the liberty to go throughout the camps and hold church services. These were led by pastors from Rwanda. A rough estimate: more than 25,000 people were baptized during the four years the camps were operational. Earlier on, I mentioned the word "refugee." In our understanding of refugees, the first 250,000 who came into Tanzania at the Benaco Camp were not refugees as we understand the term *refugees*. They were murderers fleeing to avoid arrest. Speaking with a United Nations official, she said that in their estimate, on average, every woman in the Benaco Camp had killed 25 people, and every man 50. Shocking and sobering statistics!

War correspondents from all over the world set up broadcasting stations within a few days of the genocide. Considering the United Nations statistics, little wonder then that they said, in all of their coverage of war throughout the world, they had never seen such a massacre nor felt such a presence of darkness. To highlight this, let me share an account.

A certain pastor came to visit me on his way to help the refugees through preaching to them. I warned him that the camp was a dangerous place, but that we had established a good rapport with people in the camp, and they did not see us as a threat. I strongly advised him to go to our camp. A Rwandan national would go into the camp with him and his co-worker. He scoffed at the idea. He did not need our help. Within a few days I received the report that they had indeed gone in to distribute tracts. They had been viciously attacked. The co-worker was killed, and the pastor barely escaped with his life. The vehicle was badly damaged but drivable. He then had to pay people who knew where the body of his co-worker was buried so he could get the keys for the vehicle. The co-worker's body was never recovered. It was during those early days that I personally stood on the Rusumo Falls Bridge and saw bodies floating down the river like so many logs. Some were swirling around in eddies but were eventually released to join others on their journey to Lake Victoria. For people around the world, proper burial of the deceased is of extreme importance. The emotional and spiritual toll on Rwandans, not being able to bury their dead, was unimaginable.

Early on we found a young Rwandan Christian man to be the cook and general overseer of the EMRR Camp. He had his own horror story to tell. As he was leaving his hut in Rwanda, militiamen came along and asked him, "Do you have any blood on your machete?" Meaning, had he done his share of killing? When he said no, he had not killed anyone, they pointed to five people coming down the road and instructed him to kill them. This was beyond him, and he fled for his life. Making his way through Burundi, he eventually made it into Tanzania. That is where he found our church.

Within two days of the refugee's arrival, and before we had set up our own camp, we went into the Benaco camp and located a pastor. He was to locate and inform all the pastors in the refugee camp to come to a meeting the next day. One frustration of working in a crisis situation and soliciting a lot of donations is the duplication. Some get multiple help, and others get nothing. To mitigate this, we registered all of the pastors, no matter which denomination they were affiliated with. Each pastor then registered orphans and widows in his denomination. In that way, when we distributed goods, we knew exactly to whom they were going, the last time when they had received something, and exactly what that was. It worked very well and, also forced, the pastors to work together. A little later, the Catholic church operated their own distribution; but they shared their information with us, so no duplication took place.

I do not recall exactly how it transpired, but we got in touch with Haven of Rest Ministries in the United States. You may recall they operate the radio broadcast to sailors. As a young lad I heard their radio broadcast and still, to this day, vividly recall their opening, "Ship Ahoy". They raised money to purchase, airfreight and erect eight huge circus type tents. Instructors were sent with the first tent and they taught us how to erect them. These tents, with their bright, striped, circus colors on the roof, could be seen from miles away. They were literally used 24 hours of the day. Each church group was given a time in which they could meet to have services. When that wasn't happening, the tents were used for schoolrooms. I believe God's hand was upon those tents. Even after four years, when the refugees were forced back into Rwanda, these tents had not deteriorated sitting in the hot African sun. Over time, other organizations had erected smaller tents; these had long ago succumbed to the sun's strong rays. In fact, when Benaco and other camps closed, we dismantled the tents and moved them to the Kigoma area, where they were further used for refugees coming into Tanzania from Burundi.

Fundraising by Letting Go

Fundraising is becoming more and more common for all of us. Not that we are necessarily fundraising for ourselves, but for someone or for some organization that we think is doing something worthwhile to help the less fortunate. Websites spring up spontaneously when a person or family has been hit by an unexpected tragedy. People feel good about being able to share. It is also the Christian thing to do.

In Africa, fundraising is mostly connected to a church or school. The events are elaborate, with much forethought given in planning them. The invited and honored guest knows that he is expected to give a substantial donation as well as act the part of auctioneer fundraiser. If you are attending such an event, you had better set your dinner slow cooker timer for late, late afternoon.

The shotgun approach is used to "guilt spray" everyone in the room. "Come up to donate if: you are wearing socks, you showered last night, you have children, you are a pastor, you are wearing a tie, you had breakfast, you arrived on foot, you have a bicycle, you have more than one child, you were not born in Tanzania, you have a garden, you have chickens, you have goats, you planted corn ..."

Don't think that all designations are mentioned in one breath, as I have done above. Each one is mentioned and is accompanied by long cajoling. Just when it seems that the special designation has been milked to the extreme, another person slowly comes up with a donation. That reignites the hope that there are more if the fundraiser can just be a little more persuasive or tarry a bit longer. It takes time for the ladies to be persuaded and begin to undo the tight knot that has been tied in the tip of their wraparounds, which hold the money. Similarly, although taking less time, the men have to dig around in their socks or take off their shoes. No wonder that before such an event people collect all of the five-cent pieces they can find in the neighborhood. They will contribute under many different categories so five-cents at a time will work.

Fundraisers were a community event, so to accommodate everyone, they were held outside. People really get into the spirit of it, and donations go far beyond money. Donations can range from a pair of pants (which someone may quickly run home to get) to shoes, jackets, shirts, colorful wraparounds, neckties, soda pop, roofing sheets, blouses, pens, goats, rice, corn, chickens, watches, and literally all kinds of things that are unimaginable to us. All of that takes a long time.

Just as things are beginning to wind down after three hours of this and your mind is turning toward food, stark reality jumps out of the bushes. In actual fact, there is another segment. Few places that sell building materials accept non-monetary items for payment. Now an auction must begin to convert goods that have been piled high into money. It is best to put your mind into neutral for another two hours.

I was invited to attend a school fundraiser as a guest and ended up as the main fundraiser. I was only made aware of this as the fundraising part was ready to start. One can never be sure what is going to happen at these events. It is best not to get too comfortable until the event is over.

One Falling Star

Looking into the dark African night sky, we see a falling star. Living within 15 degrees of the equator, that is not unusual, mainly because we could see stars in both the Northern and Southern Hemispheres. That is a lot of stars. But does it really matter that there is one less star in the universe? Among the millions of others, will it be missed? We grappled with that same question as we were operating Starehe Children's Home orphanage. If another child in Africa died, did it make much difference? Or if we saved another child from death, did it really matter? It had been a year and a half since the Tanzanian government, through social services, had asked us to operate the orphanage. The 41 orphan children we had were young and so few, compared to the thousands beyond our boundaries. In these early months, we wondered if we really could make a difference in their future lives. Then the Lord encouraged us in our work through meeting two young people working with the Swedish Mission. They were Tommi and Saara, a couple from Finland. Saara was born in Bangladesh, a country that regularly is ravaged by the forces of nature resulting in thousands of orphans. Saara was one of them. Let her tell you the story as she shared it with me.

It was 1971, during the coldest season on the Bhola Island in the Bengali Gulf, when I was born. My father was the chief of our village. I was often told that we were perhaps the richest of all the villagers because we owned domestic animals, good pasture, and a nice hut to live in. When my father married my mother, she was a beautiful girl who could not read but excelled in housekeeping. My sister preceded me by four years.

Bangladesh was fighting for independence from West Pakistan at the time I was born. The war made it difficult for all. Even my father, as the chief of the village, was struggling financially since he could not cultivate his lands in due time. As if the war and financial struggles were not enough, a major flood hit Bangladesh during the monsoon rains. Overnight the floods swept away an estimated 500,000 people. The water rose so high and the current so powerful that it took huts, houses, flocks and people as it rampaged through the villages and towns in its path.

When the flood hit our village, Mother tied my sister and me to her hips with an Indian dress known as a sari. She held onto a tree branch, and we barely managed to cross the strong, swollen stream. The disasters were not over after the flood. The worst was yet to come. All sorts of epidemic diseases afflicted the people, and our home was not spared. Sadly, my father also died in this epidemic. My sister and I eventually had to run to my uncle to find refuge in an area that had been somewhat spared from the worst of the flood. Being Islamic, it was the duty of the father's eldest brother to take care of us. Mother had to return to her own family.

Already having six children of his own, my uncle could not always supply us all with enough food. One day our uncle said that we needed to find food for ourselves. I was about two and half years old at the time. That was the beginning of our lives on the street. During the day we begged, and at night we slept at the post office.

Around this time some Swedish people opened an emergency food distribution centre. All the children on the streets got food and were provided with some health services. We were also standing in line for food when I, for the first time, saw a white person. She looked so strange to me that I got scared. However, she was just smiling kindly at the children. The best moment of the day was to receive a portion of rice and marvel at the white woman. Later on, while other children were watching her from a distance, I clung to her.

One week, the white woman noticed that we were not in line for food. She could not sleep until she sent a night guard to search for us in the town. It was a seemingly impossible task for the guard to find two little girls among the many people sleeping beside the streets and close to nearby houses. Miraculously, he found us in our place at the post office. We had occupied it since we began living on the streets.

The white woman took us to a house. She bathed us and gave us clean clothes. She put us to sleep in a soft bed. Despite the clean clothes and soft mattress, I could not sleep. Curious to find out where the white woman had disappeared, I took my pillow and knocked on the door through which I had seen her leave. She came to the door and wondered why I was not sleeping. I said that I would like to come to her room and stay the night with her. From that moment we started to develop a very special relationship. Soon I began to call her Mother.

When I was five years old, I visited Finland for the first time. Later, at the age of 13, I moved permanently to Finland. I studied

and got my degree in Medical Surgical Nursing. Soon after grad-
uation, I married Tommi, and together we were sent to Tanzania. I
do not believe it was mere coincidence that I had this kind of oppor-
tunity. I am grateful to God for His guidance and to missionaries
who cared for one orphan—me. It feels great that now it is my turn
to help AIDS orphans in Tanzania.

With God's help and the sacrificial support of many people,
LouDell and I saw hundreds of orphans helped through Starehe
Children's Home. Yes, saving, loving and teaching one child is
definitely worth it. These children will go on to touch many lives.

CHAPTER 24

Typical Italian Reaction

Over the years one meets many people through various circumstances, and some of them actually become good friends. Mario was one of those people whom I met in Arusha. Trying to find competent trades people of any sort in a Third World country is a challenge. This is especially true in a society where "yes" is the answer, no matter what the question may be. For example, if you ask a person on the road, "Does this road lead to Nairobi?" the answer is always "yes," even though the road is heading in the opposite direction. You must ask, "Where does this road lead?" They are then forced to give the right answer. You see, Africans do not like to disappoint, so if you want the road to lead to Nairobi, why would they disappoint you and tell you it does not lead there?

It was while I was searching for an electrician that I met Mario. He was an Italian prisoner of war sent to Tanzania. After the war he stayed on and ran his business. As the years passed, Mario's liver began to break down from too much alcohol indulgence. In order to get proper treatment, he usually took a bus to the city of Moshi, where there was a good hospital run by the Lutheran

Mission. Doctors from Germany staffed it. The bus ride took about two hours and consisted of many, many stops on the way.

That was OK if you had the time and if your medical condition was not serious. However, one day Mario, in desperation, called. His pains were severe, and he needed to get to Moshi as quickly as possible. I hurried over to pick him up, and we were soon on our way. We weren't even out of town when he hollered at me to stop at a local clinic so he could get a shot of morphine to tide him over during the trip. One can do that in Africa – without a prescription. I wheeled into the place, and within a short time Mario had received his shot and was back in the vehicle. No doubt it had cost him a few shillings, but those shortcuts had to be taken at times. Off we headed, with Mario regularly having outbursts of loud, lamenting cries of "Mamma Mia!" The first time, it caught me so off guard that I nearly drove off the road.

Arriving at the hospital, I had a hard time keeping up with Mario as he headed straight for the doctor's office. He was a regular, so he knew exactly where to go. Being that this was Africa, I was able to follow him right into the doctor's office and examining room. The loud cries of "Mamma Mia" continued. Before long Mario was given another shot of morphine, and it took effect.

I sat by the doctor's desk as he filled out the report on Mario's card. Judging by Mario's outbursts of "Mamma Mia," in my mind he was in great distress. Casually the doctor wrote, "Typical Italian reaction." I smiled to myself and tucked that information away for another day. An emergency is gauged by more than loud outbursts.

What Kind Of Soda Do You Want?

LouDell has a very soft and compassionate heart for suffering people. Children especially grip her heart, so it was no wonder that she came up with the idea of starting an orphanage. However, going beyond that, she also became the co-coordinator of ChildCARE Plus for Tanzania. That is part of the humanitarian arm of The Pentecostal Assemblies of Canada. The overall program is called ERDO (Emergency Relief and Development Overseas). Within the city of Mwanza and the surrounding area, LouDell helped to connect 228 sponsors with needy children.

Agatha was one of those children. When her father died of typhoid fever, Agatha, at age 13, became a lone and defenseless child. She was abandoned and left with her uncle. Abandonment simply took on a new face as he did not look after her very well.

Normal household furniture is not taken for granted by an orphan in Africa. Agatha was appreciative of a home, and the fact that she did not have a bed, mattress, pillow, or even sheets, did not embitter her spirit. A burlap sack laid on cardboard was her nightly mattress. It did not offer much protection from the stinging scorpions scurrying around on the dirt floor at night.

Agatha's diet consisted of very normal, basic Tanzanian food. A measly $6 per year is what school fees amounted to; however, even that amount is beyond the reach of thousands of parents. Schooling for an African child is number one on their list of desires. Many times Agatha was sent home from school because she did not have a school uniform, shoes, a scribbler or sufficient school fees. When she finally got a school uniform, she had to wash it by hand each night.

In moments of despair, Agatha would wonder out loud why she was born to suffer. Sickness was especially trying since no money was available for even the most basic medicine for malaria. Family chores such as helping with the cooking, sweeping floors, washing clothes by hand, gathering firewood, and carrying water from the nearest public water spigot were part of her daily routine. Her hobbies included drawing, singing Christian songs, playing games with her friends, and attending church.

After six years of being cared for by her relative, LouDell, through the local pastor, became aware of Agatha's need for help. Through the ChildCARE Plus sponsorship program, LouDell was able to show her love and supply her with sufficient funds to attend school. If any medical issues arose, those could also be covered as well as essential food items. The acceptance of Agatha in that program literally shone a bright light through the hitherto black clouds, which had been part of her life. She could now attend school without interruption.

As time went on, Agatha excelled in school and eventually graduated. Her sponsor very kindly agreed to continue to support Agatha for a further year so she could take courses in English and computer. That extra year was a catapult that propelled her into areas of unbelievable opportunity. She excelled in both of those courses, which qualified her to become an employable person. That was borne out a few days after she completed the courses.

I remember the day Agatha came to our house. Sitting on the deck, she lamented that her school time was coming to an end, but she needed a job so she could support her younger sister. Because we were operating the orphanage, she intimated that possibly we could employ her. At that time we did not need anyone and were up front in telling her that. We sensed this was not God's plan for her. A job is a job, but the Lord always has a specific job for us where we feel fulfilled and fill a unique place. In prayer, we agreed with her that the Lord would provide the right employment for her. He wanted her to experience a miracle from Him, not from us. She left rather disappointed. It reminded me of a pastor who wanted to build quite a large church and called me for help. I was excited that he would undertake such a project. He was very pleased when I assured him that I was willing to help. The pleasure in his tone changed dramatically, though, when I said, "I promise to help you by praying for you." I was serious. They successfully completed construction without my contributing any money. Our Western all-inclusive remedy for problems in the Third World is to throw money at it. Measured support and good advice are more effective.

For a reason, which now escapes me, I went to a local private hospital—most probably to get a blood test for malaria. It was not the place where I usually went. In casual conversation with the receptionist, I commented on the fact that they were still recording patient information by hand. She excitedly told me that they were getting computers within three weeks. I happened to know the Dutch doctor in the hospital and asked to see her. I mentioned my conversation with the receptionist and suggested that I knew a girl whom she might want to interview. Early the next morning found me at the hospital with Agatha.

She was taken into the office to be interviewed by the doctor. It went on for some time before the doctor and Agatha emerged. What a shock, albeit a pleasant surprise to me, when the doctor said that Agatha was hired and she was to start work immediately! She couldn't even go home to change her clothes. She was an orphan girl one hour ago, and now she was working and earning her own way! She had gone from no family, no money, and no future to suddenly being thrust into the room where all of these things were possible.

Think of it: now she can pay tithes, place something in the offering plate, buy her own clothes, contribute to food costs in the household, buy her own bed and mattress, and no doubt help others in need. What a loving Father we serve! ChildCARE Plus meant the difference between hope and depression. Jesus is Agatha's hope. We are so thankful for the ChildCARE Plus sponsorship program of the PAOC that offers hope to desperate, needy children who don't know where to turn. Thank you, ChildCARE Plus and those who sponsor children through them. You have made a tremendous difference in Agatha's life! You have given her the first chance in life. She is only one example of many.

I frequently returned to the hospital to see how she was doing, as well as to undergo tests for malaria and other little hitchhikers one picks up in the tropics. Agatha was stationed at the reception desk while they waited for the computers to arrive. She was so efficient that she was able to do work that they had previously needed two people to do.

While sitting in the waiting room one morning, Agatha came to me and asked, "What kind of soda would you like?" Culturally, it is an indication of friendship and respect. Mentally, my first reaction was *Hey, I am the "rich" person! Why are you wanting to purchase a soda for me? I should be buying you one*. Just as quickly, I picked up the true meaning of her question. It was an expression of thankfulness for our help, but way beyond that, it was an act of pride that now she could be the giver instead of the receiver. Every sip of that soda was taken with gratitude that God had allowed us to be a part of His plan for Agatha's life. She is now married to a wonderful pastor, living in a nice home, and has children of her own.

In conjunction with the government, a study was carried out in our area involving orphans and other vulnerable children. The study was done by visiting these children individually at their places of residence. To the question "What do you see as your greatest need?" the overwhelming answer was, "I need love." A very telling priority for those who don't receive it. The need for love superseded money, food or a bed.

Nature

In Africa one has close contact with nature. An occasion that still stands out in my mind happened while heading home through Serengeti Park. It also constantly reignites my regrets that I am not artistic. I would love to paint the picture that resides in my brain.

I was driving through scrub brush on something called a road. Actually, it was two narrow, sandy tracks with grass growing in the middle. The track headed down a bit and ahead I saw a corner, which concealed my long-range view. Coming around the corner and in front of me, sloping down from the road, stretched a grass savannah as far as I could see. Zebra, wildebeest, Thomson's gazelle, Grant's gazelle, warthog, giraffe, hartebeest, and topi were all grazing together. It looked like the dawn of creation.

How can the land sustain all of these animals? is the question that came to my mind. It certainly did not look overgrazed. Sometime later I spoke with a game warden about it. He explained that the concentration of so many animals did not pose any problem at all. Each one has its own plate of food. The gazelles eat one type of grass, the wildebeest another, and the topi yet another. The animals that do eat the same type of grass eat it at different stages of growth. Zebras, for example, like only the taller, tougher grass. Warthogs dig around with their tusks to get at roots.

Gazelles only very rarely need water. Grazing in the early morning, any dew on the grass is enough to supply the moisture their body needs. Their digestive system is very efficient at drawing out every bit of moisture. Their droppings are totally dry pebbles. Elephants, on the other hand, have one of the least efficient stomachs. They only process about 40 per cent of the grass they eat. Their large droppings consisting of mostly undigested grass, attest to that. They have a keen sense for detecting water. I have crossed a dry riverbed and have seen places where elephants have dug holes in it to get to underground water. Adding to the amazement is that they could actually dig a hole in the first place with their large feet, which are certainly not made for digging. They are also very destructive. In areas where herds of elephant roam, trees are knocked over. The bark of a certain tree is a delicacy, so they reach up with their trunk and pull branches down. If that fails, they push the whole tree over. The areas get this natural pruning from time to time, that is for sure. A period of up to seven years may pass before a cow produces an offspring. A certain grass governs the time frame. When the grass is plentiful, it triggers their reproductive system. This actually happens with a few other animals as well.

Wildebeest are known as the clowns of the savannah. They can contort their bodies into all kinds of shapes as they run and jump around. They need water every day. They are only one of a few animals that bear young each year. That is why they number in the tens of thousands on the African savannah. Large herds naturally have numerous sick animals, and these become the mainstay meals for lions and hyenas.

Phones Are There to Be Answered

Well, sometimes. Other times they are a nuisance and may represent work requested by whoever is calling. There are few things as annoying as hearing the phone ringing incessantly on a secretary's desk while she ignores it in an attempt to outlast the patience of the person on the other end. Crowded government office waiting rooms are the norm; so are vacant desks. I discovered that many officials spent a lot of time out of the office due to their low pay. They had to have second jobs to survive. If these happened to fall within the hours of a person's "first job," that was regrettable but necessary. Knowing that fact never lessened my annoyance. If the person I wanted to see sat in another room, I became knowledgeable in understanding what I call the second language to Kiswahili — the language of deception used by his secretary.

"He has left for a short period of time." Actual meaning: he never did show up at the office today or he *just* left.

"Just now he will return." Oh, sure!

"Just wait a while." Remember, their time is not our time!

"He has not come today." Actually, he is sitting in his office and doesn't want to see you. Seeing the secretary carrying tea into his office is a dead giveaway.

"Come back tomorrow." That is a delaying tactic for certain.

If you are a polite and blissfully naive Westerner, you even say thank you and politely sit down and wait and wait and wait or even go home to come back tomorrow. If you do guiltily ask when this man may return, the same soothing answer is given: "Soon." After a few hours, you may actually get the hint and decide to come back tomorrow. Once I out-waited the person who was "coming back soon"—the person whose signature I needed. Returning at eight o'clock at night, the person in question was surprised to see me still in her office. It took another hour, but I did leave with her signature on the document.

Once you have acclimatized to the language, you begin to ask more direct questions like "What time is he coming back?" or "What time did he leave?" or " Will he be back today?" You may even coax the truth out of her by saying, "He isn't coming back today, is he?" Obviously just giving a simple yes to your question is easier than her providing an explanation.

It was during one of these waits in an office whose secretary was absent that the phone began to ring. After about nine rings, a fellow "waiter" got up and casually answered the phone. He gave the appropriate "Oh, yes. OK. Fine. That is good. Yes" (etc.) for a good two minutes. Again, calmly, he hung up the phone and rejoined the rest of us on our hard chairs.

Speaking of which, be careful of chairs. Do not take anything for granted. We have the saying, "Don't judge a book by its cover." In Africa that is so true, except don't judge a chair by its cover. On one occasion I was in the Dar es Salaam Airport. Our flight from Mwanza had been delayed a number of times, but eventually we did arrive late evening into Dar es Salaam. Everyone was tired as we waited for our luggage to be off-loaded manually and brought into the terminal. A weary male passenger, who would shortly have other problems to worry about, dropped himself into a comfortable looking chair. Within a split second, he was wide-eyed and in deep trouble. Had things been what they were not, he would have leaped right out again. But this was different and required some calculation of measured extradition. You see, right in the middle of the chair, lengthwise and standing on edge, smugly hiding under what used to be the cushion, lurked a strong, thin board. Who knows how many weary passengers had been impaled on that thing? The moral of the story: when you see an empty chair in a crowded public place, sit down gingerly. As with so many other things in a strange culture, this can be learned only by painful experience or the eagle-eyed observation of others.

Back to the empty chair of the secretary ... She eventually returned to her perch. After settling in, she began writing down the names of newcomers who had come into the office during her absence. Everyone must take his or her turn. Naively, I was waiting for the "wannabe" secretary sitting beside me to inform the real secretary about the phone call. But it never happened. This reinforced what I already knew: never think you have actually talked to the person in charge via a phone conversation. It doesn't work. Face-to-face encounters are time-consuming; however, Africans are "people" people. Face-to-face contact is very important to them. A telephone is inanimate and cold. Therefore, it is not an accepted instrument of communication. Important issues are discussed face to face.

You Were Trying To Steal That!

Immanuel and I were in the city of Dar es Salaam on business. An ocean-going container with mission supplies had arrived on the ship and needed to be released from the port. We were there to go through the laborious process of getting it cleared through customs and getting it on its way up country. Payment was in cold, hard cash. Port charges, customs clearing charges, and clearing agent charges could add up to a tidy sum. Immanuel was nonchalantly carrying that money in his briefcase. Don't look nervous, and suspicions will not be raised about anything important being in the briefcase.

It would be much too convenient to have all of these offices together in one building. We had successfully navigated a few offices, but dinnertime had overtaken us, so we had to wait until offices re-opened after lunch. To waste a bit of time we popped into a used bookstore. The briefcase was heavy, so Immanuel placed it on a chair while we browsed. However, with that much cash, one is like a nervous new mother: one eye is constantly on the crib. There was a boy in the store, and he bent over to pick up the briefcase. We were hot, tired and sweaty in the coastal humidity, but Immanuel made a leap for that briefcase. This frightened the daylights out of the poor boy. In fact, Immanuel's reaction was so swift that the boy's hand hadn't even touched the case yet.

"You were going to steal the briefcase!" Immanuel suggested, not exactly in a pastor's voice or tone. Stammering with fright, the boy said he was hired to clean up and he just wanted to wipe the dust off the chair. After that, we figured it was best to keep that thing in our clammy palms.

I'm Catholic

A fellow missionary and I were on our way to deal with a serious church matter in a rural area out of Arusha. The church where we were meeting was four hours away over very bad roads. Knowing that the meeting would be going well into the night, we decided to stop at a local eating-place. Having a good grasp on culture does pay dividends. Normally a meeting concludes and then food is brought. In this case, I was anticipating that might not happen until 1 or 2 in the morning. As it turned out, my calculation was right. I felt sorry for the cooks who were on call.

Local rural eating places have their own character, which one learns to accept and actually appreciate. Chicken and rice is always a pretty safe bet when ordering food. Given that lighting in most places is pretty dim, the presentation of food is enhanced. One tends not to notice so many of the pinfeathers and small feathers still protruding from the chicken skin. No matter, though, rice is always eaten with care— more like gently mouthing it rather than actual hard chewing. It is common, from across the table, to hear the crunch in your friend's mouth as he chomps on a tiny piece of gravel. The crunching sound sends shivers up your own spine and slows your chewing considerably. Harvesting rice is very labor intensive. Apart from planting it by hand into water, it must also be thinned and weeded in the water. Once it gets close to ripening, bird chasers with slingshots must be stationed in the field to chase the ever-present birds. They come in swarms of thousands. Harvesting is done by hand with a small sickle. The stalks are tied into bundles to transport to the area where it is flailed to dislodge the kernels. Here is where unwelcome bits and pieces get mixed in. Cleaning is painstaking. These small rocks are the same colour as the rice, so it is impossible to spot every one of them.

Just as we were finishing our strong ginger soda, the food arrived. We had been on the road for three hours and were hungry, so the food looked great. Before eating starts, the waitress brings a pitcher of warm water and a basin. Turning your hands to the side of the table and over the basin, she slowly pours warm water over your hands as you wash them. This is a ritual we always enjoyed and appreciated. Sometimes a towel was offered to dry your hands. It was best to do an air dry since the towel may have undone all that the washing was supposed to do. Our hands now washed, the young lady smiles and half turns to leave. We catch her off guard by telling her she has to pray for the meal. The conversation goes like this:

We say, "You have to pray for the food."

She says, "No, I don't pray."

We say, "Oh, yes, you do. We know your culture: the lady who brings the food prays for it."

She then says, "I don't know how to pray."

We reply, "That is too bad. You have to. It's your responsibility. We can't eat until you do."

She then informs us, "I'm Catholic!"

We respond, "We don't care what you are. You must pray."

Finally she prays and does a commendable job in spite of using every excuse she could think of to get out of it.

At the end of the meal, she cleared the table and again returned with warm water, a piece of soap, and the basin. This washing of hands happened every time we ate anywhere. It was very welcome because our hands grew sweaty from the heat, and also because people are handshakers. They cannot understand our custom of blowing our noses into a cloth, putting it into our pockets, and taking that home. They use the "thumb to nose" method a lot – in your sight just before extending the hand for the warm shake.

In some tribes, after the meal people take a deep sip of the water, swish it around in their mouth, and spit it into the corner. Rooms are small, so it is not necessary to leave the table. Just turn a bit and *whoosh*! This made us squirm every time it happened! Mostly the floors are dirt; nonetheless, it was a practice we never became comfortable with.

Watchmen

If you don't have them, you are a bit nervous; if you do have them, you are asking yourself why. More and more, it is expected of Westerners that they employ local help. That is a fairly good idea in itself. Unfortunately, the Western understanding of work ethic differs substantially from that of the local population. In their eyes, all Westerners are "rich"—and we are, compared to them. When comparison is analyzed, they are not that bad off either. Income tax for the rank and file is not heard of. No such thing as house insurance exists, nor does life insurance or employment insurance. Fruit and vegetables are priced within their means. Working for aid organizations, of which there are many, puts them far above the national average income. Some organizations actually pay local help on the same scale as they would pay anyone at home. To me this is absurd because living conditions are not even close to being equal. I don't see these same organizations at home basing the salary of their workers on the cost of living in Japan, for instance, where costs are sky-high. So why this inequity in Africa? I have always believed in paying our workers above the accepted national average income while still trying to maintain their dignity in being able to live within their means.

For many years we never had a watchman; we relied only on our German shepherd dogs. Whenever we came to the gate at night, they were awake. This could not be said of a watchman. After honking the car horn and getting no response, we opened the gate ourselves. Once I drove into our yard and actually shone the car lights onto the watchman. He had wedged himself into a drainage ditch and was obviously enjoying the comfort of it. I went out and, grabbing him by the shirt collar, started shaking him and making loud roaring noises. He came to in a fright, bellowing as though he were severely being beaten. Don't even waste your breath asking him to admit that he was sleeping—because he was not sleeping! He will tell you that himself.

Even stealing his spear or club and hiding it will not solicit an admittance of sleep. The last watchman we employed was done solely for the purpose of keeping our dogs from barking all night. We could always tell where he was sleeping by where the dogs were. That worked OK because if something was happening, the dogs would wake the watchman. It seemed to be a good arrangement. Unfortunately, there were also times when the watchman took advantage of his position and stole things. Once we were out visiting churches on the weekend. When we came back, the watchman had disappeared, together with a brand new grinding mill that we had purchased for a village. It never was recovered. One other thing that happened, and for which one could not fault them, was that the watchmen helped themselves to food that was meant for the dogs. Our dogs were fed food that was totally edible by anyone. Each day they received a portion of a tiny fish called *dagaa*. It was like a small dried sardine. That was boiled together with cornmeal. Twice a week, meat was added to the menu. Many times people lived from hand to mouth, so we could not become overly strict even if we noticed that the food meant for the dogs was disappearing more quickly than normal. Partly to occupy the watchman, it was his responsibility each night to cook the food for the dogs—so he had easy access to it.

It is important to cultivate a good sense of humor. While still living in Bukoba, we went to Mwanza twice a year to teach at the Bible school for three weeks. Around 2 a.m., there was a huge ruckus outside, and we heard the watchman talking angrily to someone. I went out and discovered that he had apprehended a thief. I do not like to be unduly disturbed at night, so I helped the watchman tie up the thief, with the promise that I would deliver him to the police station next morning. Most probably, if I had taken him that night, I would have been there until morning. Finding anyone awake and able to coherently write up a police report by hand in the light of a kerosene lamp takes time, a lot of time.

Rendering the thief immobile seemed to be the most efficient use of time. However, after I got back into bed, the thief kept mournfully shouting, "Help, help, help!" Even taking him to the police station would be futile as he had not stolen anything yet, so no charges could be laid. We finally untied him, led him down to the main road, and warned him not to come back. I don't think he did, unless he was the one who stole our car battery and car tires in the midst of a tropical downpour a month or so later. Thieves never seem to mind the rain. It actually is the best camouflage you can have if you are in that business: it provides lots of noise, and certainly no one is taking a casual midnight stroll. Oh yes, they also pray that they will return home alive. A friend assured us that thieves do pray! Considering, that if caught, they are often stoned to death, prior prayer may not be a bad idea.

There are many good and faithful watchmen who go far beyond even what is required of them. They become like family and are very loyal. Many are in danger because they are the ones to be hacked to death by serious criminals intent on stealing.

CHAPTER 25

You Are Quarantined

When travelling abroad it is always best if you have a person on the ground at the destination you are heading toward. It also helps if you are not too proud to ask questions and don't rely on the home country to know things that the "travelling book" at home does not show.

A lady was coming to visit us and had all of the current information about inoculations that were needed in the destination country. At least, that is what she was given to believe at home. Home said that a yellow fever shot was not needed anywhere in the world anymore. The trouble was, they did not know local regulations, which supersede World Health suggestions.

The visitor arrived at the small local airport late one Friday afternoon. When I say "airport," think of an oversized fruit stand. Health authorities requested to see her yellow health card. No yellow fever shot was listed, so she would have to be quarantined at the airport. Monday morning was the earliest any medical personnel would be available to administer the shot. The airport had no regulation quarantine room, causing a big scare for the traveler. She would have to spend the weekend in a small makeshift room that had no windows and only one naked, dim, light bulb dangling from the ceiling. She would be alone in a strange country with a language she didn't know.

Being confronted with this possibility was not something that calmed her heart. Frantically we made phone calls home to receive any official document to indicate clearly the World Health travel regulations. Her father, now in a panic himself, assured us that they had checked at home, and no such shot was needed. The trouble was, they weren't dealing with home when travelling.

Through much negotiation and attempts to use logic with a bit of common sense thrown in, she was released from the airport with the promise that she would return on Monday morning for the shot. In spite of our letters concerning "what to do before you leave home," some people would just ignore them with the rationale, "At home we were told ..."

Convincing people to wear a hat was a constant challenge. The tropical sun is not the same as in North America. Heat stroke caused by dehydration can hit. It can ruin a person's holiday pretty fast, and for a good number of days. When we insisted that people wear hats, the typical response was "I never wear a hat at home!" *Wakey, wakey! Are you at home right now? No, you are in the tropics, very near the equator.*

Some humorous things do happen. A visitor was coming to Tanzania and was in a discussion with the immigration officer about a visa requirement. They were going back and forth and just didn't seem to be connecting. Finally, the immigration officer asked our friend if he spoke English. Coming from Canada, he thought he was speaking English the whole time. He was about to ask the immigration official the same question. Both had been speaking English and were not understanding each other.

You Will Die of Starvation

Faraja was a happy eight-year-old boy living near Arusha. His father was an animist who did not want anything to do with a religion that followed Jesus. Faraja snuck into a small church one day. There, for the first time, he heard about this man called Jesus. This man loved everyone and invited people to follow Him by giving their hearts to Him. He felt warmness in his heart and made the decision to follow Jesus.

Only a few days passed before his father heard about it. He began a regimen of whipping Faraja to force him to leave his new religion. When that did not produce any results, he took him outside to the edge of their property near the forest. There he was tied to a tree. He was to be left there to starve or until he would renounce his new religion. His mother could not bear to see her son suffering. Every night, after the father was asleep, she took food to him. Faraja was also badly beaten each day. After two weeks, his father relented and let him go. Faraja escaped to live with his grandparents. He eventually followed his dream and became a pilot.

Snakes

My hope is that this small portion of the book does not scare people away from visiting Tanzania and thus missing a wonderful country and people. When I talk about snake encounters, it must also be remembered that I am covering a period of 35 years. Months can go by without your ever seeing a snake.

As intriguing as the wildlife of Tanzania is, the subject of snakes gets the heart pumping just a bit faster. Snakes are part of African life. They show up in the most unexpected places. More likely, we show up unexpectedly in their territory. Tanzania has all kinds of snakes, 90 per cent of which are poisonous. This is reassuring in that it takes most of the guesswork out of assessing whether a snake is poisonous. Many people do not get a fuzzy feeling when it comes to reptiles of any sort, and certainly not when it comes to snakes. There are the odd few (odd in number, maybe even in mentality) who see them as creatures that are as likable as any puppy or kitten. A young student we knew always had a pet snake in his shirt pocket or pant pocket, even when travelling on a plane. These days, I am sure some scanner would quickly pick it up and send people scurrying to find the nearest bench to jump onto.

In North America we think little of very cold temperatures, the reason being that we make preparations to deal with it. Now, if we insisted on going out in -40 degrees in a T-shirt and shorts, there would be a hazard. However, when wearing the right clothes, people can go outside and enjoy whatever they like to do, no matter how cold it is. The same holds true for snakes. There are ways to minimize the risks they pose. Some fundamental precautions exist.

How many of us in North America carry an emergency kit in our vehicles in case our vehicle develops problems on a freezing cold day or night? I am sure the answer is not enough of us. The same goes for the following precautions concerning snakes. Never walk outside in the dark with open-toed shoes. Always use a flashlight. Avoid walking through tall grass; if you must, then walk slowly. Snakes pick up motion from vibrations in the ground. Therefore, if you walk slowly, most snakes slither off without your ever seeing them. The times when I have seen the most snakes was when I was following another vehicle, which was probably 50 yards ahead of me. This happened especially if we were in a bushy area and driving slowly. Apparently, after feeling the vibrations the first vehicle produced, they sensed it was safe to cross the road. When picking up a board or anything lying on the ground, always pick it up so the open side faces away from you. That way, if there is a snake or a scorpion under it, their escape route is away from you, not toward you. If you are entering a crawl space or a room under a building that has not been opened for a while, do it cautiously. Even if you were in it yesterday, exercise caution as snakes do move around.

Similar to bears in North America, it is the surprise encounter that triggers a defense mode, sometimes resulting in an attack. An exception exists: the puff adder. At maturity they reach four feet in length and are about the size of a small grapefruit in diameter. Puff adders are sluggish, so they do not move when they feel a vibration on the ground. Warning to the squeamish: please skip these next few sentences. Puff adders also have a very slow venom injection system. When they strike, they must hang on to their prey for about a minute until the venom is injected.

Welcome back to those who quit reading a few sentences back! A friend of ours was bitten by a puff adder and survived. More than likely, the snake had recently used its venom on prey and it had not yet had time to build up another lethal dose. Experts maintain that no snake is sufficiently venomous to kill a person. The procedure is simple. Don't panic, breathe slowly and deeply and, if possible, lie down and relax. It is the sudden rush of the poison into the bloodstream that the body cannot tolerate. Now, if you are able to do these things, you are a very disciplined person indeed and are an expert in mind over matter. The book says you should take a good look at the snake so you can tell the doctor what type it was — and oh, yes, don't get too close to it in case it comes back to bite you again! I am sure that if I were bitten by a snake, I would be following it to get all the details for the doctor's report. Not quite! I would be heading in the opposite direction — and fast! Oops! I realize that I am supposed to slow everything down. You could also take the advice connected to every medical ailment in books such as *When Not Near Medical Services*. It states, "Get to a hospital as quickly as possible." I thought I was reading the book to see what to do when services were not near at hand.

Some methods have been developed to deal with snakebites. Quickly getting to a hospital is not one of them. Anti-snake venom is very difficult to store, so it is best not to rely on that possibility. Apparently, an electrical shock from a vehicle alternator can neutralize the poison. A stun gun has been developed, but where would you find that in a hurry when you have been carrying it around in your vehicle for the last 10 years? There is a much better and more effective method, however. The Catholic White Fathers in Belgium have developed what is called "Snake Stone" or "Black Stone." It effectively works like a poultice to draw out venom. The area of the bite is first made to bleed; the stone is then placed on the area. Within a very short period of time — say, a minute — the stone attaches to the skin and begins to absorb the venom. In fact, it is not possible to pull it off. Once the poison is out, the stone releases. It is reusable after boiling it in milk and rinsing it with water. Once dry, it is ready to go again.

I always carried a stone in every vehicle, on the tractor, and in a building that was in close proximity to where I worked. Fortunately, I never had to use it, but I did have a friend who had to—many times. He worked close to the Mara River, where snakes liked to lie in wait for frogs. Children going to fetch water would occasionally be bitten. Once a young boy was bitten in the hand, and the pain had progressed up his arm to the shoulder. After an hour had passed since my friend applied the Snake Stone, the pain began to move back down the boy's arm. It was the only time where he had to use two stones to completely draw out all of the venom. Amazing to us was the fact that children at our orphanage would come across poisonous snakes, yet they were never bitten.

I came uncomfortably close to collecting first-hand knowledge of snakebites. I was walking through tall grass following a path at Nyasirori, which was our first and only mission station in Tanzania. Mission stations are great, but they do not normally lend themselves to neighborliness. Seeing your one and only neighbor seven days a week and observing each other's comings and goings can lead to complications. Ours complemented the complications as the houses were only 50 meters apart. A shared driveway did not help. It was four miles of bad road to the main bad road and another 21 miles to the small town of Musoma. Getting out of the station was not an everyday occurrence. Pulling the shutdown cord of the generator every night at exactly 10 p.m. served as a form of caste system hierarchy. The senior missionary pulled the cord. Add to that chickens scratching in a garden where they were not supposed to; this led to some flying feathers. By the way, did you know that chickens become hypnotized if you lay them on their backs and stroke them? Seeing their chickens lying as if dead gives any chicken owner a severe case of heart palpitations. Barking dogs at two o'clock in the morning add a bit more spice to life.

My path took a bit of a turn and, as I rounded it, a puff adder was lying there, no doubt catching some tropical sunrays. These snakes are coldblooded, so they need some external heat source. Had I not been alert, I would have stepped right on it. While my wife was helping me skin a specimen (which she dropped at every quiver of the nerves before bolting), I discovered that a snake's heart is about one foot back from its head. Again from experience, I found out that it is very difficult to immobilize a snake. This was reinforced time and again as I witnessed other people trying to do it. Snakes just seem to be able to survive a tremendous amount of punishment without throwing in the towel. When you see them limp and stretched out straight, you think it is game over. Come back in the morning, and they have moved some distance or have totally disappeared. Seldom have I seen a person encounter a snake without their doing everything to kill it. There are always more of them. It seems that snakes never will be on the endangered species list. They are actually the bad weeds of the reptile kingdom. No matter how many are destroyed, another one or two pop up.

Snakes have a variety of venoms. Some attack the nervous system. They strike and then move away, waiting for their prey to die. When startled, cobras can be very menacing in appearance in that they will rear up to three feet and flare their neck ribs to form a flattened, widened head. As with the black mamba, even in that position they can move with lightning speed.

Two types of cobras exist. One is known as the spitting cobra. It accurately shoots venom into the eyes of its prey to blind it. If washed out with water, the blindness is temporary. There are a few tree snakes that are harmless but can get your heart pounding just by unexpectedly encountering them. LouDell suddenly opened the door leading from the kitchen to the outside, and a snake that had been sunning itself on the ledge dropped onto her arm. There was lots of shrieking and jumping! At our house in Mwanza, we had mango trees growing very close. A certain thin green snake loved to live in them. Every once in a while, they would crawl from a branch onto the roof and into our attic. From there it was an easy road down into the house. Our granddaughter, Marlyse, and her friend from Canada were visiting when a snake suddenly came out from under the bathtub enclosure in their guest bathroom. They were leaving for home the next day, so we did not tell them about it until they had left the house. Otherwise, I am sure they would have insisted on sharing our bathroom.

Another close encounter I had was early on during our time in Africa. In Musoma we purchased cooking propane gas in 70 lb. cylinders. Availability was haphazard, so we always had six or eight on hand. Full cylinders and empties were kept outside, beside the house. Over time we would forget which were full or empty. Doing an inventory, I would tip them to see which ones were empty. One particular day I rolled one around on the palm of my hand in a merry-go-round fashion. Suddenly, out of the corner of my eye I saw something and instantly felt a knick on my toe. I was wearing open-toed sandals, which is a no-no in Africa. I immediately let go of the tank. Most of the cobra snake was pinned under the tank, so it was not able to strike at full length. It had found a nice warm place in the open base of the tank. No doubt mice or small rats liked to dart in and out of the space as well, creating a ready meal for the snake. That was an early lesson I never forgot.

An amazing thing with snakes is that you can spot them in the grass and somehow never be able to follow them with your eye. Your eye reaches their tail and cannot pick up their body again. If they are in an open area, you can follow them, but even then it is a challenge.

On one occasion we were travelling from Bukoba to Mwanza, which was a 14-hour drive on a good day. Flat tires or encountering some sudden road construction could make it a lot longer. Torrential tropical rain did the same thing. That length of road runs through forest for a long stretch. Certainly, there were no eating establishments along it either. Because of the activity of thieves, driving at night was not a good idea. They would place large boulders or a large log across the road so there was no way to avoid stopping. Daytime was safer, and we always gauged our time to arrive at that section during midday. We were having our lunch, and a large snake appeared out of the ditch heading toward the vehicle, where my wife and children were eating their lunch. I attempted to head it off, but it just kept on coming. Seeing that it would not be deterred, I jumped aside. It quickly slithered under the vehicle and across the road, disappearing into the grass. Wherever it was going, it had a plan—and nothing was getting in the way of it. I didn't mind that.

While I was driving near a river, I came across the largest snake I ever saw. Coming around a bend, I was unable to quickly slow down. The snake was on the road and unable to make a quick exit. It reared up and slapped across the hood of my vehicle, which was at least four feet from the ground. That snake must have been 11 or 13 feet long and reared up well over six feet. Just being hit a glancing blow in no way injured it. When I reversed the vehicle, there was no snake to be seen. Another time I saw one that stretched across a road that was eight feet wide. The head was in one ditch, and the tail was just coming out of the other. Those are the times when it is good to be inside the vehicle with the windows rolled up. More than likely these were black mambas, which can grow to 15 feet in length.

Another snake that is common, especially around water, is the python. They grow to 21 feet and are as round as a melon. As with most snakes, it dislodges its jaws to swallow the prey whole. Pythons are easily able to swallow a goat, a small wild animal or dog. When this happens, the python lies in the shade for a month or so, digesting. If seen right after it has swallowed something, a huge bulge can be observed where the prey is in its body.

A friend shared an experience he had in Central Tanzania. He observed a python on a path, meeting a wild animal coming the other way. He said that when the animal saw the python, it totally froze. The python easily approached the animal and captured it. A python gets its leverage by anchoring its tail. When people try to capture a python, up to four men rush to grab the tail and wrestle it off the ground. Another person grabs its head to keep it from turning around. They do bite but have no venom.

In the light of people's fears of snakes, many possibilities of mischief present themselves. A number of ladies who were staying in another house very close to ours volunteered at our orphanage. Missionaries have two favorite subjects that always seem to mate-rialize as topics of current interest. In North America we begin to talk about the weather. Missionaries begin to talk about stomach problems and snakes. Stomach trouble seems to be a topic that is discussed after dinner around the table. Visitors squirm a bit, but for missionaries it is like recounting a trip to the market.

We had gotten past the stomach part and were into the snake encounter part. By the time we had finished relating our stories, darkness had settled in. The girls were thoroughly hypersensitive to the possibility of seeing a snake. Having to make their way in the dark from our house to theirs added to their jumpiness. Always wanting to lessen anxiety whenever I could, I escorted them to their house, shining my flashlight around to help dispel their acute paranoia.

However, when a wonderful opportunity to exploit a shaky, unstable thought pattern presents itself, it cannot be passed up. The ladies reached the door and one girl, in her angst, was fumbling around with the key to unlock the door. Having seen a small snake curled up on that very doorstep the day before did not help. Impatiently, and looking over her shoulder, the other girl crowded right up to her, further impeding the unlocking process. The door could not open fast enough. I reached down and pulled out a piece of grass. Just as the first girl managed to unlock the door and begin wrestling it open, I touched the leg of the second girl with the grass. A shrieking, screaming, clawing stampede is the closest I can come to describing what happened next. Had the door not opened, it would have been knocked down, I am sure. Before the ladies could regain their composure and sanity and decide to wail on me, I headed home.

Give Me a Ride

In a country that seldom sees temperatures fall below 70 degrees Fahrenheit, bugs of all kinds thrive. Added to their wonderful life is the fact that they like to hitch a ride on other warm bodies; still other bugs are actually hitching a ride on them in the first place. Malaria is a hitchhiker. I am sure studies have been done to figure out why the mosquito provides a comfortable home for the malaria parasite.

Numerous studies, in my estimation, are absolutely worthless. Of further annoyance to me is the fact that someone is actually paying people to come up with this worthless information. I suppose the annoyance comes because I am actually paying for it through my taxes. Why has no one asked me what studies would benefit me as a taxpayer?

Regularly prisoners, some after spending decades looking out through bars, are being exonerated. They have been the wrong person tagged in a crime. Similarly, we are vilifying the poor mosquito and doing everything in our power to wipe them out. Why not go after the malaria parasite and leave the mosquito to live a happy, buzzing life?

That not being the case, we have been left to spray, swat, mangle, electrocute, fumigate, smoke out, and net out the mosquito. We try to kill them before birth, in fact. We pour all kinds of things on their living room floor to suffocate them. Let me share my personal preference. It is to "net out" the little buzzer. Few things are sweeter to my ear, and bring on a more peaceful sleep, than hearing the high-pitched drone of a mosquito outside of my mosquito net. I don't want to leave you hanging: the other aid to sleep is to put on a movie. Actually, it is not such a bad thing because I really milk the worth out of a movie. I watch it five times, between slumbering, to get the whole story. The trick is to begin slumbering at different moments each time I watch it.

Pardon me for pricking a hole in the wonderful, philanthropic act of supplying mosquito nets to my dear rural African friends. If you are eating inside and cooking inside and chatting inside and have an indoor bathroom, nets are great. Oh, yes—I forgot to mention ... Nets are also great if you have doors on the house and wire netting on the openings where windows should normally be. Doors for most of the day and evening consist of fabric hangings. However, the African lifestyle mostly utilizes the outdoors. Add to that the preference of mosquitoes to emerge after dark. That happens before the cooking outside, before the chatting outside, before the last jaunt to "the long drop." The prick in the net issue comes because it is highly uncomfortable to be walking around and cooking on rocks while enclosed in a net cocoon. Children playing outside also do not do well with that costume.

I cannot count the number of times I have observed the angelic glow on Western faces when they come, as long-sought-after saviors, to ward off malaria. It usually goes something like this. A vehicle drives up to the edge of a small local fruit and vegetable market that carries much more than vegetables. It also harbors a good number of pickpockets. A crowd of people has already gathered near the car before it has come to a stop. The trunk is popped open, and everyone in sight gets a net. "How many in your family? OK, 10 nets." In America everyone has his or her own bed! They are oblivious to the fact that 10 people in Africa represent three beds. Finally, the world is being saved! Hopefully, the driver has kept his head and firmly locked his vehicle because, if not, not only will his trunk be empty, but also his vehicle. Everything movable in his car will have disappeared.

The dust of their good deeds has not yet settled behind them, and African boys have already repurposed the nets for soccer goalposts or fishing nets. When they are actually used for their intended purpose around the bed, all they do is trap the mosquitoes inside for the whole night. To be effective, the net has to be put down around the bed before dark. It has to reach to the floor, and there can be no opening at all for mosquitoes to find their way through. This has to be done every day. In Africa it is impossible not to be bitten by a mosquito. To be fair, I need to say that people do see the need for netting; however, the logistics of implementing their use on a regular basis is difficult.

Back to mosquito land ... It is when I stop hearing the drone that I get nervous. Did she did find a way into my net and is right at this moment drawing blood? If you are wondering why I wouldn't feel its sting, I am wondering the same thing. My wife goes crazy, violently swatting everything in her path when she feels the bite. I keep telling her that all of the motion encourages them to gather, but she does not believe me. I never feel a thing, and unless I see the little blighter already bloated with my lifeline, I have no idea she is there. Apparently, only the female mosquito carries the parasite, but with the deception rampant in the world, I don't even trust the mosquitoes anymore. Maybe even the males are carriers.

Since the mosquito will always be with us, we need to look at their hitchhiker, malaria. A question we have been asked over and over by people is whether we have had malaria. Some have a look of awe as they ask, "And you survived?" Yes, yes—to both questions. Few ask, "How does it affect you?" Let me give my side of the story. Everyone, it seems, has different telltale signs that malaria is attacking. For me, I get an instantaneous hit in the stomach. The motto? When you feel you are getting sick, treat for malaria first and then look for other causes.

Following the clandestine hit in the stomach comes the high fever. Later, or in combination with that, come cramps, diarrhea, sweating, chills, a splitting headache, and weakness. No doctor has to tell you to go to bed and rest. I have left the best for the last. Malaria can produce acute depression. You want to literally die, but not in the sense of committing suicide. That is left for those who are trying to look after you.

There are prophylactics that lessen the impact. Doctors prescribe the most commonly used pills. A problem arises when you are not common. Some of these pills are taken as a daily dose or once a week. It doesn't take a brain surgeon to figure some things out. It only takes a little brain. The pill, if taken once a week, has lost its potency by the time days six and seven roll around. Taking one pill a day is best.

Arriving in Tanzania, I was prescribed a prophylactic that only partially did what it was supposed to do. Each time I got malaria, the treatment of a five-day pill regimen worked only for a short period of time. Eventually, the treatment did not work at all, and I became very sick. A doctor in Kenya prescribed another prophylactic, and that did work. It took me more than six months to fully recover my strength from that bout of malaria. There is an interesting fact about malaria that many people attest to. The last food you eat before your first serious bout of malaria is something you seldom have a craving for in the future. My "last meal" was a pomegranate picked fresh off the bush. In 35 years I never ate a pomegranate again.

I was hit with malaria while far out near the Uganda border. It struck on a Saturday, and by Sunday morning I knew it was severe. A water baptismal service had been planned. Many people from outlying churches would be coming, so to cancel was not an option. Our denominational regulation stipulated that only pastors who were ordained could do the baptizing. No other pastors in the region held that qualification. That morning I informed two of the local pastors that they were the baptizers. "We are not ordained" was their surprised reply. "You are now" was my reply. Had I gone into the water, someone would probably have had to fish me out. Watching from the shore, I did experience a twinge of guilt as I saw the freshly ordained pastors baptizing. First, you must know that people in this tribe were not swimmers or friends of bodies of water larger than a foot basin. They were lowered very slowly into the water. I could see the wide whites of their eyes as they rolled them upon seeing the water coming closer and closer. No one drowned, so it was successful in that sense. Later, when the real ordination came around, we joked that this was actually their second ordination. First they were ordained by Posein, and now by the national church. After the service I had a long 11-hour drive ahead of me to reach home.

Ticks were plentiful, and many carried tick fever and Lyme disease. I had been in tall grass during the day, and when I sensed an itching, I realized my whole body was covered in thousands of little red ticks. Apparently, a mother carrying her babies had attached herself to me and, when brushed, they all scattered. Or, more likely, they smelled food and scrambled for the feast. Getting tick fever is a very real possibility, but through all of my times in the bush, I never contracted it. Among animals, warthogs were favorite hosts for ticks. When skinning them, you had to be very careful that the ticks did not migrate to your skin. Removing a tick is tricky. Leaving any part of it attached can cause an infection. Nationals heat the point of a needle and touch the tick; it immediately releases itself from the skin.

Tsetse flies carry sleeping sickness, so it is not a good idea to mess with them. They are similar in size to a horsefly and produce the same sharp pain when they bite. Apparently, they do not bite so much as saw into your flesh. Their stingers are serrated. Bushy areas are their favorite habitat. These infested areas become uninhabitable for livestock. If you get into an area with tsetse flies, it is possible that elephants are around. They love elephant blood. Another of their peculiarities is that they can fly about 40 kilometers per hour. Drive anything below that speed, and they stick onto the window of the vehicle as though they have suction cups on their legs. Through experiments it has been discovered that they will not cross an open area of more than 40 meters. If they do get into the vehicle, they tend to head for the dark and pretty soon are biting your ankle. In spite of getting a number of bites, we never did contract sleeping sickness.

Typhoid is a waterborne disease that we were not so fortunate in evading. Raw water, no matter how clear it looked, was always avoided when outside of our home. Even bottled water was suspect because the "opportunists" would repurpose used water bottles with water from who knows where. Even if the water used was from a dirty source, that could be remedied. Crushing moringa seeds and mixing them into the water settles solids to the bottom, and clear water remains. "Clear" does not translate into "clean," as your stomach will announce to you within a very short period of time. We were never sure how we contracted typhoid, since we were certainly aware of its source through water. It is highly contagious and can even be spread through food contamination. It is another tropical disease that is not pleasant. Recovery takes a long time.

Although not fitting into the disease category, food poisoning is a possibility when eating meat. A friend who had worked in butchery in Canada visited us. When he saw how meat was handled, he became an avowed vegetarian for his entire three-week visit. He hadn't even seen the small meat pickup trucks. Every morning they are out on the highway heading into the countryside to find cows, goats or sheep to butcher. Butchering is done under the hot sun. The meat is unceremoniously thrown into the tin box on the back of the pickup truck. When full, they head back into the city. I have never seen any semblance of refrigeration attached to the trucks. The tin box does not exactly keep cool for too long, I am sure. The positive side to this procedure is that the meat does not need to hang long to cure. It is probably is well on its way to being cured by the time the truck returns to the city.

My wife and I ate at a reputable restaurant in Nairobi, Kenya. Within an hour my wife suffered severe abdominal cramps, vomiting and diarrhea. An hour later, she was totally dehydrated and in the nearest hospital on a drip getting fluids and minerals back into her body. Surprisingly, she was the first person who ever complained about getting sick from that restaurant's food. Don't believe me? I asked the manager; he said so himself.

Doctors say that once you have had food poisoning, you are very susceptible. We discovered the truth of that rather quickly. Three days after LouDell's episode, we attended a potluck dinner back in Arusha. Within two hours we were in the car making the four-hour trip back to Nairobi. Arusha at that time did not have a reputable hospital. A little secret after that was to eat only the food we brought or that of someone whose cooking and hygiene we knew and trusted.

CHAPTER 26

On Top of the African Continent

I have visited a number of countries on the African continent. Tanzania seems to hold more history and uniqueness than most. Africa is huge, so to boast having the highest mountain, the largest game reserve, and sharing the largest part of the second largest fresh water lake in world is phenomenal. I have not mentioned Lake Tanganyika as the deepest and coldest lake in Africa nor the historical story surrounding Tanzania (formerly Tanganyika and, before that, German East Africa) as the central area to be searched for the source of the mighty Nile River. Famous explorers such as Speke, Livingstone and Stanley crisscrossed the country. Villages mentioned in that epic search still exist today. As John Anonby relates, "… [Mount] Kilimanjaro is … higher than any peak in Europe, Australia, Antarctica and 'all the islands of the seas.' "[6]

The quest to stand atop Africa's highest peak began as a promise that was not well thought through. Our daughter, Rhonda, in Grade 3 at the ripe old age of nine, said that she wanted to climb Mount Kilimanjaro. Young minds create and discard plans on a regular basis. How many young children mention wanting to be a fireman or a policeman, only to end up being a cook, a brain surgeon, or a locomotive driver?

To see that promise never needing to be fulfilled seemed a sure thing. I readily agreed that I would go with her. Many times we make promises on the assumption that the circumstance will not materialize and that we will never have to carry through on them.

My promise to Rhonda was simple: "I promise to climb Mount Kilimanjaro before you graduate from high school at Rift Valley Academy." Nine years seemed like a pretty safe bet that the memory bank would have many files piled on top of her wish to climb. By the time it resurfaced, a new file, "I wish I would have …" would already have been started.

But it was not to be. In the middle of Grade 11 while on school leave, Rhonda said, "Dad, remember you promised we would climb Mount Kilimanjaro before I graduate and return to Canada?" Exposed! There was no way out; I was a rat in a trap. By this time we were living in Arusha, just 60 kilometers from Kilimanjaro. I had driven past it many, many times, always hoping that the cloud-shrouded, snow-capped peak would be visible. On a number of occasions that did happen, and I had wonderful photos of the majestic, proud sentinel rising up out of the African plain. Standing alone, it commands a respect which would seem to be dishonoring if other mountains stood in its presence. It is a scene of towering power. Standing only a few degrees from the equator, it seems impossible that snow could be seen here. My desire for the mountain never reached beyond the side of the main paved road and my camera lens.

Preparations began. Injecting excitement into the climb was the fact that a number of missionaries from Tanzania and Kenya were enthused to attempt it as well. On December 10, 1984, our group of 26 set out.

The first part of the trip consisted of staying at the Marangu Hotel situated at the base of the mountain and snuggled in the rain forest part of lower Kilimanjaro. It was rustic and served to align our minds to the unknown harsh realities that lay ahead. A German lady ran the hotel, and she gave us the terse, no-nonsense orientation of what to do and what not to do. "If you are a smoker," which one visiting girl was, "just stay at the hotel because you will never make it!" The girl hadn't come this far to stay in a dark, damp hotel.

Not being experienced climbers, we looked a mess as far as clothing was concerned. Being that we lived in a tropical country and were heading into some cold atmosphere, the majority of our coverings were scrounged from used clothing parcels. What we couldn't find in them we were able to get from the hotel's moldy storeroom. Such items were collected from previous climbers' castoffs, which had never even been in the shadow of a washing machine. These included parkas, climbing boots, gloves, welding goggles to shield our eyes and prevent snow blindness, and toques. The next morning, in the light of day, we were not a pretty sight. The night before, we had also been introduced to our porters and guides.

We had the option of beginning our climb from the hotel or driving up the steep ascent to the collection point and park gate. Sanity ruled, and we drove the seven kilometers. Our porters and guides were already waiting for us. With our fees of $500 paid, we milled around like a herd of nervous cattle eyeing a wolf. Walking sticks in hand, and our small backpacks holding a favorite energy snack secured, we headed for the starting gate. Each porter lowered his pack so it could be weighed to meet the regulations of humane weight. Unscrupulous guiding companies regularly overloaded the porters to reduce the number needed on the five-day climb. Porters, needing the employment, would agree to the overweight. We were relieved to see that this weigh-in precaution was in place. The need of employment was brought home to us two days later when we reached snow level. One porter wore only a pair of thin, torn shoes—not even a pair of socks.

The first day was not kind to us. In the hotel, the night before our climb, we met climbers who had successfully reached the summit. They told of beautiful weather and clear vistas. We began our climb in pouring rain. Only a few in our group had managed to find some semblance of rain gear. Others became inventive and slit head holes in black garbage bags. The bags had been brought along for another purpose, but this was the urgent need right now. Passing us on their way down were people with top-of-the-line, brightly colored outerwear. This made the contrast with our clothing all the starker. On the way to the first camp at 9,000 feet, we ate our sandwiches while standing in the rain. Already five days looked like a long way off, and at least one adventurer was open to heading back down with any companion also so inclined. The anticipation of a nice, warm, dry spot with a crackling fireplace at our first camp kept our spirits buoyed. Reality snapped us back to the fact that we were on an adventure.

Oxygen—or, better said, the lack of it—is the great sifter. Everyone's body reacts differently as the oxygen level decreases. This is a crucial part of the great adventure of climbing "Kili," as it is affectionately referred to. Professional mountain climbers begin using oxygen at around 13,000 feet. Summiting Kilimanjaro at 19,341 feet without oxygen is the challenge. Some people's stomachs rebel, some begin to feel light-headed, vomiting is common, and fatigue is a given. Along with the excitement and challenge, there is the sobering aspect that every year up to 15 people die on the mountain. Anyone lagging on the trail has the guide shining a flashlight into their eyes to check for altitude sickness. It can be fatal and, when detected, the only immediate option is to get the person down to a lower altitude at a dead run.

Arriving at our first "hut" (the word is used loosely in order to soften the blow), we are wet, cold, and unbelievably tired. Tour company advertisements are best left to tabletop books. Their description boasts "Full service camping." These three words have a thousand ways of being interpreted. The reality is that you are taking your food in a cold, windy tent. Reality continues to reinforce the adventure side of things when it comes to sleeping accommodations. "Accommodations" is too kind a word. Sleeping quarters are located in a long, narrow A-frame building. It is nine feet wide and is divided evenly into three sections. On one side are three-foot-wide continuous beds—head to neighbor's toes. Luxury is a two-inch by four-inch board between you. The highest part of the A-frame is a three-foot-wide walkway; on the other side, it is a replica of the first side. There is no heat and no light. Men and women use the same facility. By that time, who cares? Maneuvering through the narrow corridor with soaking wet backpacks and sleeping bags, dozens of people are trying to find a bed. Wet clothes are strung up in the vain hope that they will magically dry overnight. Sleeping bags are damp. Food is prepared outside by the porters and is heavy on cabbage—great energy food, but also a full production factory for bloating gas. Not the most exciting when single file trekking the next day. The toilet is outside in the cold.

Imagine the reality check of two bankers who did not attend church, one from New York and the other from Chicago, joining our group. They were friends and wanted to do something exotic on their holidays. Kilimanjaro was the answer. In their minds, they imagined a leisurely climb. At the end of the first day, there would be chalet accommodations, beer, women, and robust nightlife. Everything except the beer was there, but not as they had conjured it up in their minds. The nightlife in Africa was slightly different from that found in New York. Exhausted, wet climbers barely had the energy to unroll their sleeping bags and plop them, along with the sleeping foam, onto the hard plywood. Life, as portrayed in the slick, colorful tourist brochures, was radically different from what we had imagined. Real life too can be far worse than imagined, but it can also be much better. As a result of experiencing Christian witness for five days, one of the bankers became a follower of Jesus when he returned home.

The final push to reach the summit of Uhuru (meaning *freedom* in Kiswahili) Peak begins at midnight. There is a long, exhausting day ahead! We are not the only group climbing. Headlamps create a ribbon of spotlights as people slowly make their way up the steep path. By now it has become tough sledding in cold and snow. To keep the stomach from rebelling with overwork, breakfast consisted of two cups of sweet tea. People progress at their own speed. The usual is to take three very slow-motion steps, stop to rest, and then do a slow count of one, two, three again. Constantly reminding yourself that it is the day when you will reach the top helps one to take the next steps. At the side of the path people are throwing up, moaning, and in general stages of distress. Staying focused is the key. This is complicated by the below-freezing temperatures and the freezing rain. Gillman's Point is counted as reaching the top, but there is still a long way to go to Uhuru Peak. The snow has piled up deep, and from Gillman's Point to Uhuru the path follows a narrow ridge around a deep ice crater. Slide into that, and there is no rescue process available. Occasionally it does happen. Our guide was not keen on taking the chance.

The night before we had met a German who can best be described as a "mad German." He was alone with his guide. He said he was going to Uhuru Peak, no matter how much snow had fallen. He departed Gillman's Point earlier than we did, so in fact broke trail for us. By this time daylight had arrived. The snow was waist-deep, with an avalanche being a real possibility. Our guide had his eyes glued to the snow to detect any crack indicating the slightest movement of the snow and a possible slide.

The stretch from Gillman's Point to Uhuru is possibly the most difficult—physically and psychologically. To this point, the path has been a steep climb, but the terrain now begins to descend into slight depressions and then out again. Muscles, after having experienced the luxury of going down instead of up, rebel against this chore of having to climb up again. Around 10 o'clock, four of us out of the original 26 stood by the summit sign and flags. John, Kathy (the smoker), Bob and I shook hands with our guide, Emmanueli. The wind was howling in the bitter cold, and blowing snow pelted against our jackets, so we did not waste a lot of time savoring the moment. It was another two days down. Glaring sunlight met us on the second day's descent.

Here is a checklist for success in conquering the mountain:

Take some food to snack on, peanuts or granola bars or raisins. These are great anyway as they pack a lot of energy in a compact space.

Every evening, when arriving at your sleeping place, climb a few hundred meters higher and spend some time before descending again. This helps to combat altitude sickness.

On the night before the summit attempt, take only liquid food like consommé soup and drink tea.

For breakfast on the early morning start, drink only sweet tea. At this stage your stomach cannot tolerate the exertion of digesting anything beyond that. At home, in preparation for the climb, trim your toenails as short as possible. Coming down the steep grades, your toes push into the toes of your shoes. If your toenails are protruding even a bit, you will lose them in a month or two. The obvious needs to be repeated: wear climbing boots that you have broken in. Blisters make aggravating walking companions.

Another thing that's obvious: bring clothes for weather ranging from hot to bone-chilling cold. Take warm gloves, a toque or a balaclava. Eye protection is a must as the sun reflecting off the brilliant snow will soon cause snow blindness.

Take anti-inflammatory medication; your joints need it. The joints endure a lot of jarring, especially on the way down. The anti-inflammatory should be the main ingredient, not just a small part of the product. Any rock steps that do exist are just a bit further apart than what a normal step so at every step the knee joints get a wicked jolt.

You Can't Deny Me Entry

Border crossings seemed to be part of our lives. Kenya was our most frequent destination. This was necessitated by a number of things. Our children attended boarding school at Rift Valley Academy. Along with about 1,000 other organizations, our East

Africa Mission was headquartered in Nairobi. Twice yearly, missionary retreats took place in Kenya because it had the easiest access from Uganda and Tanzania. Supplies of any kind, ranging from groceries to stationery to vehicle parts, were either non-existent or very expensive in Tanzania. Kenya had modern medical facilities, good labs and doctors. For normal tropical diseases such as malaria, infections, stomach ailments, and bone fractures, Tanzania did have appropriate facilities.

It was on the occasion of an East Africa missionary retreat, coupled with our children having midterm break, that found four of our missionary couples from Tanzania in Kenya. The weekend now over, we were all getting ready to return home. From the mission guesthouse I popped over to a store to pick up some last-minute supplies. Storeowners in the small strip mall across the street from the guesthouse knew us well. Some had formerly lived in Tanzania, so were interested in hearing the latest happenings. Going into a store, the owner expressed his condolences that the borders between Kenya and Tanzania had yet again been sealed. Taken aback by the news, I said, "What?" He assured me it was true and showed me a headline in the local newspaper.

Returning to the guesthouse, I gave the others the "good" news. We huddled and strategized as to what we were going to do. One thing was certain: no one wanted to remain in Nairobi to wait out the politics. We all had differing opinions. In our favor was the fact that if we acted quickly, the border posts might not yet have received an official directive.

One missionary decided to head to the coast and try the border from Mombasa going toward Tanga. Another headed off to the Taveta border in the direction of Arusha. Another headed toward the Tarime border leading to Musoma. My decision was to try the park border between Maasai Mara and the Serengeti. It was primarily a tourist crossing with only a sign saying, "Leaving Maasai Mara Game Park, Kenya," then going through about 100 meters of no man's land and coming to another sign which read, "Entering Tanzania. Report to Immigration at the Lobo Tourist Lodge."

Uncharacteristically, when I reached the small Kenya border exit, a policeman was guarding the gate barrier. The road is a single-lane sand road, so a pole is easily placed across it. Given the news of the border being closed, I acted surprised and began relating to him all of the reasons why I could not drive six hours back to Nairobi. I was only three hours from home. Anyway, the announcement did not say that I couldn't leave the country. Eventually, he said that even if he allowed me through, then I would be denied entry into Tanzania. My reply was that he should leave that to me since Tanzania could not deny me entry. I was a resident there and had a residence permit stamped in my passport to prove it. Unable to refute my logic, he relented, and I was off through no-mans land and up to the Tanzania barrier across the road.

Again there was the same message: the border is closed. Now I had an extra weapon in my argument chest. With an air of one who knows the law, I said that he could not deny me entry because I was a resident. How can a country deny a resident entry? Besides, I was in no man's land so was technically already in Tanzania. Using all the logic I could think of, I persevered and continued to press for entry into Tanzania. Finally, he, too, pulled the sisal pole away and allowed us to pass. We drove 10 kilometers to Lobo Lodge to have our passports stamped with the entry date by a surprised immigration official. "But the border is closed!"

"Yes, I know, but here we are." In three hours we were home.

All the other missionaries were turned back at the border posts they tried. Returning to Nairobi, they had to charter mission planes to get home. Within a few weeks, arrangements were worked out so each one could go back and get their vehicles. I loved that about Tanzania: many times there seemed to be a way to negotiate things. I had learned what to do when up against an official who had the authority to make a decision but was not seeing it my way. After intense but respectful head butting, you had to show respect for their authority and say, "I understand your point; you understand my point. You did not make the law, you have only been put here to enforce it. In my present situation it is not working. Now how can we resolve this issue?" In most cases they honor your arguments and allow things to go as you wish. Alternatively, they may suggest a compromise that is also generally acceptable.

Survival of the Fittest

I was constantly amazed by the adaptability of the African people. Extremes are tolerated far above anything that we as North Americans could comfortably endure. Ours is a pretty narrow window of survival. Their window of tolerating extremes is wide.

Many organizations dealing with aid send parcels of used clothing. As is imaginable, all kinds of things come in these parcels. Without showing any signs of perspiring in 90-degree heat, I have observed people wearing fur coats. Woolen toques are favorites—even one-piece winter underwear! Heat, especially when prolonged, does a number on us. Numerous times as we drove over the plains of Serengeti Park, we encountered temperatures nearing

95 degrees. Running across the parched ground toward us, leaving his herd of cows, sheep and goats, would be a young Maasai boy not even wearing a hat. We would usually drain a number of water bottles within a short period of time.

I would be working outside in oppressive heat and be looking to see how I could drag my work into the shade. When I suggested moving to a shady area, my helpers would shrug and comment that they were just fine. Apparently, the dark pigment in their skin reflects the sun's rays. What I witnessed with them and the sun would bear that out.

When faced with new challenges such as eating with Western table utensils, they cautiously observe how others are using them. Within a short period of time they, too, have mastered the skill. Even after 35 years of practice, I was still unable to daintily eat rice with my fingers.

The metal work industry is huge. Everywhere men are welding all kinds of things, from window security grills to truck chassis. Often welding takes place at night. This may be because they have been able to borrow a welder after hours from their employer or a friend. For the most part, welding glasses are not used; if they are, only the glass part is held in one hand. The co-ordination of starting the weld and getting the glass in place over the eye is critical. Naturally, in the dark or semi-dark they have to line things up with the naked eye. It is scary to see people welding without any eye protection, especially while wearing flip-flops. Workers' Compensation inspectors from America would have a heart attack. This manufacturing industry needs to have enforced safety guidelines. Handling chemicals is another scary industry.

On a regular basis, food consumption perplexed me. At breakfast I have sat beside someone who has obviously bitten into a soft-boiled egg that was rotten. The putrid smell certainly quickly dampens my appetite. Without flinching, he finishes it. It was normal to see a man pedaling his bicycle with stacks of trays of eggs on the back. Somehow, for me, hot sun and eggs don't go together. Then there is food intake. A person can eat a substantial meal once a day. This is normally taken around 9 or 10 at night. I have been with people over a period of five days where we have been offered food fit for important guests four times a day. Each time, my friends ate as though they hadn't eaten in a week. On the second day, I really didn't care if I saw food for the rest of the trip. Eating is a big part of their culture.

When the rains fail, people go hungry until the next rains arrive. There was a severe drought in an area where we worked. In travelling to churches, we would drive through fields of bean or corn that were nothing but dust bowls. People were literally boiling bark off the trees to survive. The drought lasted six months, and somehow the people survived. I can't see myself surviving that long on boiled bark or plant roots.

Water has long been the white man's killer in Africa. It carries all manner of sickness. Yet we have seen people drink rainwater that has collected in ditches. In the midst of a hot summer, a small pool will remain somewhere; this is used without any hesitation or fear of obvious harm. If the daylights have not been boiled out of such water before tea is made for us, our stomachs do not hesitate to rebel. It usually doesn't wait until you get home either. Water may still carry the smell of cows and goats, but any dangerous bugs have succumbed in the boiling process. Believe science, not your senses.

Children in Africa spend their time outside mingling with the goats, ducks and chickens, and all that they tend to leave behind in the dust. Yet they survive. Admittedly, our Western diseases are a challenge for them. Our Western obsession with extreme hygiene (disinfecting absolutely everything) has made us susceptible to things to which our bodies, over time, would normally build up immunity. A happy medium is relative to each society.

Hey, Who Is Tickling My Toes?

Returning to Africa from furlough we re-located from Bukoba to Arusha. Moving from the small town of Bukoba, where we were the only white couple for a year until another missionary family moved there, was a culture shock, even within the country. Arusha was a bustling place where farmers actually had tractors. There were large farms that grew beans and corn. Successful coffee planta-tions operated. Arusha was in the centre of world-renowned tourist attractions such as: Mount Kilimanjaro; Serengeti National Park; Ngorongoro Conservation Area; Lake Manyara (the only place where lions climbed trees); Olduvai Gorge (where the remains of earliest man were found by the Leakey family); Tarangire Park; and Arusha Park, which had colobus monkeys. A modern interna-tional airport brought thousands of tourists each year. It really felt like we were in a civilization somewhat resembling North America. Hotels existed to cater to tourists' wishes for a "real" African expe-rience. Fully explained, this meant getting up before dawn for a drive to see animals, back to the hotel for breakfast, then out for more "game drives." By evening time, it was back to the hotel for a hot shower, an actual menu at the hotel, and a comfortable sleep. The "real" African experience was restricted to daylight hours. The dusty and bone jarring roads were an 'experience'. At home the same set of circumstances would be met with loud complaints to the city mayor. After that, the comforts of home were demanded. To illustrate, 15 visitors came to see the church work being carried on by our mission in Tanzania. We picked them up at the airport in Arusha and headed toward Mwanza. The first 50 kilometers of the road are paved. Then brutal reality hit—dusty, rough road. After one kilometer, one of the visitors asked, "How long does this go?"

"Another 10 days," I answered.

Seeing so many tourists in the parks presented us with a few laughs along the way. While stopping on the road to view animals, it is best to take stock of your immediate surroundings. Many times animals were only a mere 50 yards off the road. In their excitement, few people ever looked down to see what was happening at foot level. Small red ants do not take kindly to having their territory invaded. They have a sense of humor and work with army precision. Without detection they swarm onto people's legs and, on command, begin to use their pinchers. Sophisticated people automatically bolt into defense mode and, oblivious to their surroundings or companions, drop their pants to get at these little blighters. The ants are not that easily intimidated and often have to be dislodged one by one. Small, stinging red welts will be a reminder of the experience for the rest of the day. One such experience was not soon forgotten. From then on, animal viewing was second on the list. First was checking the place you were standing.

Our work responsibility took us through the parks up to six times a year. Holding residence permits, we could get bed and breakfast for $45 per night. Tourists subsidized the rest. How nice! Six-thirty is the usual hour for tourists to check in after a long, hot, dusty day of nine people bouncing around in a minibus. The usual "touristy" things have been aggravating during the day. For instance … somebody abruptly standing up in front of you just as you were getting that once in a lifetime photo. Banging your head on the car roof a few times as a result of the driver's hitting a pothole doesn't help either. Most minibuses had portions of the tops cut out for animal viewing. The blistering tropical sun can dehydrate quickly or, at the very least, bring on a headache by evening time.

Add to that the little things the glossy brochures do not tell. One of them is the fact that washroom privacy begins and ends at the hotel. The local population knows that animals do not require washrooms, so why build any? In our travels through "tourist domain," we would often come upon Europeans taking a wash-room break – more accurately described as making a break for the washroom. The scene was known as 'Euro-peeings'. Their blad-ders dictated that they make a quick break. These are the plains of Africa we are speaking about—there are no bushes here. The rule is: men to the right, women to the left. On making one of these scheduled open-air pit stops, our lady visitor thought it beneath her dignity to squat out in the open. She insisted on waiting for the next bathroom that offered some privacy. Honoring her wishes, we stopped at one 20 kilometers down the road. Smugly, with an air of "See, you all could have waited," she entered the washroom. In a very short while she came out throwing up. Our warnings that the open-air facilities are preferable were taken seriously from then on.

There would be the odd "watering hole" which actually had washrooms. Dozens of tourist mini buses jammed the dusty parking lot as tourists see a bit of what resembles home and make a mad dash to savor it. An actual bathroom! My brother and sister-in-law were visiting when we arrived at this "relief" centre. Beyond the relief aspect of a hole in the stall or a continuous urinal, there is nothing—not even lights. Urgency does not offer the luxury of allowing time for the eyes to adjust from the brilliant tropical sun to darkness. My brother was lined up at the relief trough when he felt something warm on his leg. The fellow standing shoulder to shoulder with him hadn't gotten his bearings straight and was not exactly facing in the right direction. A good sense of humor is a plus in these situations.

Standing at the tourist hotel check-in counter, a humorous side-show could unfold. A tired, high-paying tourist had been to his room and was down at the desk loudly complaining. "The mirror is cracked (this is quite biblical: "Now we see through a glass darkly")! There is no hot water, the chairs have next to no padding, and the shower has no head and no curtain! I am paying a lot of money and want my amenities." Local people are amazingly patient. The check-in lady would sport a pleasant smile—not too broad as to indicate distain—and, in a sweet voice, explain that due to circumstances beyond anyone's control, things aren't as everyone wished they would be. We are, after all, a long way from a large city and the roads are bad, so supplies are difficult to get. In one government hotel, a bleached and weathered paper sign gave tourists the bad news that elephants had ripped up the water lines, so water would be brought to each room in a bucket. Tourists were so awed by this. It is a very clever piece of marketing to portray a bad situation as an adventure. People actually boast about it to their friends. It never entered their minds that possibly a water pipe could have been fixed since the time the sign had been posted many years ago. As seasoned travelers, we had long since learned to decipher the brochure lingo from reality. Few things surprised us when we got into our rooms. We were most surprised when the windows had glass and curtains; when the doorknob actually turned; when the shower yielded water beyond a dribble; when the toilet flushed or the single light bulb glowed.

Arusha city, being the hub of civilization, had many Westerners living there. These included missionaries, diplomats, people doing research on any given subject, and business people who had an idea of how to cater to tourists. I had been living and working in rural Africa for close to four years by this time. In that predominantly rural setting, 99 per cent of the population lived off the land. Most survived from hand to mouth. Watching the skies for rain was the norm as it literally dictated life or death for the next six months. Seeing the city of Arusha for the first time really set me back on my cultural heels. I couldn't even imagine that such a place existed in Tanzania.

Before the breakup of the East African Community, consisting of Tanzania, Kenya and Uganda, Arusha was the headquarters. Three impressive buildings, with 10 stories each, housed offices for each country. These were built in a semicircle facing onto a large, manicured, park-like courtyard. Inside they rivaled any executive building in America: massive conference rooms holding up to 1,500 people, with red carpet and mikes at every desk. Years later, the United Nations International Criminal Tribunal for Rwanda crimes was conducted from one of these buildings. The tribunal rented all ten floors of the building. I went to see a lawyer I knew who was working on the criminal cases. Security was very tight all around the building. Getting past the entrance gate was impossible. She had to come to the gate and escort me to her office. As modern as the buildings were, taking an elevator was always a sincere "Dear Lord" prayer time. Africa is not known for reliable power sources. Power outages just happen, either by accident or by design. It may also just 'happen' to come on some hours or days later, so one does not want to be stuck in an elevator.

With the diversity of the white population, some Western attractions also emerged. A dynamic little theatre operated. Regular productions were performed where comedy was a regular theme. The room was an amphitheatre, so everyone had a clear view of the stage. When the lights darkened, a mysterious atmosphere descended in the room. Tickling in the toes was a subconscious sensation brought on by the character of the place. This is the tropics, so sandals and open-toed shoes are the favored footwear. It was always a good idea to bring a flashlight because the tickling toes sensation could be some homey rat seeing if a bite of something might be about. A lady's shrieking from the audience was not usually part of the play script.

The little theatre also brought in the occasional film. They showed the five-day series of the film *Roots*. By the second night the room literally bristled with tension as the portrayal of black slaves and white masters unfolded. The scene showing a European beating a black man really electrified the crowd. Emotions can become uncontrollable at such times. We never did return to finish viewing the series. Having said that, during our 35 years we never encountered any animosity shown against Europeans. In fact, the opposite was true. Many times we were called to the front of the line in a bank or store. It was a sign of respect, but was also something we were uncomfortable with. In our eyes, we were no better than anyone else. Special treatment was not something we sought, but on many occasions it was respectfully offered and we received it as such.

CHAPTER 27

Miracles

At home, one question was asked more than any other: "Did you see any miracles?" Without the questioners elaborating, I am sure that what was on their mind were things like dead people coming to life or a shrivelled leg being made whole. Scripture tells us that the Jews demanded miraculous signs. Seems to me we all have some Jewish in us when it comes to miracles. To be very candid, if a dead person came back to life or if a leg was restored, of what importance are such things? People in the New Testament looked for such spectacles. Unfortunately, when miracles did happen, they wanted to see another one. Seldom are people satisfied to see their sports team score one goal. We want another and another. So it is with people who crave miracles: it becomes a desire for more. Their lives do not change because of one miracle. If they did, they would be satisfied with witnessing one miracle. Miracles can become spectator spiritual highs.

To answer the question, we did experience miracles. They may not fall into your general category of miracles, but they were miracles for us. Here are a few examples.

Being kept safe on the roads was a huge miracle for us. When you consider that we travelled in the vicinity of a million and a half kilometers without a major accident, that is a miracle. The underlying reasons are many. Drivers are poorly trained, if at all (driver's licenses can easily be purchased). Buses and trucks are large and minimally maintained. At any time they can have a tire blowout or a mechanical failure while you are nearby. Roads are narrow, one-lane affairs. Vehicles, at high speed, literally pass within inches of one another. Any small miscalculation spells disaster. People drive fast. Road signs warning of hazards such as a sharp corner or steep hill are few. Here is a true saying: "Few people are injured in a bus accident; they all die." Sometimes up to 60 people died in one accident. Our superintendent and general secretary were involved in a bus rollover. People came along, but instead of helping injured passengers, they stole all they could lay their hands on and ran off. Sadly, our general secretary passed away a few months later as result of complications associated with the accident.

We were kept from many of the fatal tropical sicknesses such as AIDS, yellow fever, and sleeping sickness. That is not to say we did not get sick with malaria, tuberculosis and typhoid. Other things affected us too, but the Lord always helped us get well. Nothing was so severe as to shorten our time on the field.

In a country where people live under very difficult conditions, their physical and spiritual needs are great as well. We witnessed many people being healed in a sense that their situations changed. Earlier on, I mentioned the orphan, Agatha. Stories similar to hers are many. Barren women bore children. People received direction for their lives. Paulo fell and broke his two permanent front teeth. Faith and prayer resulted in his growing two brand-new front teeth.

In my eyes, the greatest miracle we saw hundreds of times was people being given a new heart by God. Wicked people turned into good husbands and wives. Drunkards and drug takers were released from their addictions. Probably the thing that had the most impact was seeing people who were demon possessed set free. Evil forces weigh people down. It is not a matter of an occasional bad day. Every day they struggle with evil. To me, salvation was—and still is—the greatest miracle. Churches in large cities number in the tens of thousands. A church I am familiar with seats 8,000 people and conducts four services on a Sunday. The first time I sat in the original building (it had not yet been expanded), there were 25 people in the congregation. Twenty of them were missionaries.

On a personal level, it was a miracle for us that LouDell, in spite of her traumatic experience of being arrested on her first day in Africa, was healed of the constant fear she lived with. Another miracle was that, midway through our first furlough, she had a desire to return to Africa. Due to the constant bouncing on the bad roads, LouDell suffered from a painful back problem. While at home in Canada, God healed her back. Throughout our many subsequent years, she never suffered with any back issues again. The ability of both of us to learn the Kiswahili language was also a miracle. God's call was a miracle. The way Starehe Children's Home began was a miracle. The details of this will be described in another book dedicated exclusively to the story of the orphanage.

No less a miracle was our ability to weather storms of personal attack from fellow workers and leaders. Those had the very real possibility of overshadowing even God's call. Thankfully, they were not many in number, but they did occur. Our reliance upon God and upon His call is what kept us focused on the fact that we were His children, and He would fight for us. During one very trying time, a lady from home sent us pages and pages of promises from God's Word (70 in total). We never publicly shared our trials, but the Lord placed us on her heart. The Scriptures she chose were a great comfort to us. She never knew, nor will she ever know, why she was impressed to send those particular Scriptures to us. They were all lovingly handwritten.

Should I Go?

If God calls you to be a missionary, do not hesitate. Certainly there are things to consider, but do not overanalyze your present situation and what you think could happen in the future. Remember, nothing ever happens until we take the first step. This is a lesson we overlook in a baby's first step. Without that, the baby would forever be scooting around on his or her bum.

The same holds true for God's call. It will be confirmed in a number of ways that are totally random but directly related to your call. You will be the only one who connects the dots. Beware of prophetic utterances over you by others. Don't try to make things happen. Do not allow others to call you. This is a fine line to understand as there are times when God uses other people to nudge us in the right direction. Having said that, let me say that if another person's comment or invitation to missionary service ignites something genuine, expect other confirmations. These can come through the casual remark of a person who is oblivious to what is happening in your life. Uncharacteristic circumstances as well will confirm God's call. Being a missionary is a romantic lifestyle. However, all of the romance comes to a brutally sudden stop when the plane takes off and you are staring at four years without seeing friends or family. Lonesomeness can bore a huge hole in your heart. It relegates the romantic to the mission brochures.

The romantic anticipation ends when the main dish is cow stomach—poorly cleaned, at that. Forget about McDonald's; this is your new reality. The romantic anticipation also ends when a parent passes away while you are on the field. Intrigue and excitement properly occur in the planning stage of the trip. When that is over, realty hits—and it will be with you a lot longer than the romantic planning stage. Short-term visitors have this same experience of letdown. Possibly for years they have heard exciting missionary stories. On their three-week visit, they expect to experience what the missionary has experienced in his 20 or 30 years. It is the "show me the highlights" syndrome. The mundane and rocky parts are of no interest to them. To use an analogy, the visitor wants to see the building finished. They're not interested in the parts about lumber cutting, the hauling of materials to the construction site, the pouring of cement, or the erection of beams and rafters.

A missionary's victories are encased in disappointments, setbacks, opposition, and even disasters. To be very honest, if missionaries told their whole story, the victory part would make up a very small portion. The process of getting to that place of victory is where the majority of life is spent.

We know that accomplished people in all kinds of fields have spent hundreds and hundreds of monotonous hours practicing. Winning, or being at the top, is the payoff. My question is, what about the person coming in last? Have they, too, not spent hundreds and hundreds of hours practicing? Life can throw the same questions at us. A man or woman, after years of untiring service, may never even be mentioned in the denominational magazine, but I believe they are mentioned in God's list of heroes akin to what we find in Hebrews chapter 11. My favorite Scripture says that only one thing is required of a servant; that is, that he be found faithful (1 Corinthians 4:2). Sounds pretty basic to me. Too many times I have seen the so-called successful people drop by the wayside. There is inherent danger when one seeks to live in the glow of others' accolades. We can quickly forget who we really are and begin thinking we are the super "whatevers" that others say we are, such as: a great administrator, so loving, full of compassion, humble, godly, musician, and many more. True, we may be, or were, one of those things at some point in life, but life is made up of more than one thing and more than one day.

The pages of the calendar kept turning over. One beautiful day in Africa concluded, and another one began. Then suddenly, 35 short years—or so it seemed to us—had passed, and the calendar page was blank. It was time to retire. We had become the longest serving missionaries of our denomination in Tanzania. There were no more days crammed with so many things to do that the diary page couldn't hold all of the notations. Our time in Africa was over, but the experiences, both good and bad, accompanied us home. They will forever live in our hearts and minds.

You can leave Africa, but it will not leave you. Little wonder, then, that the heart of explorer David Livingstone was buried in Tanzania. That is where our hearts will always be.

Thank You, Lord, for giving us the opportunity to spend the best years of our lives in Africa.

NOTES

1. A similar account of this story has been published in the book *This Is My Story: Mission Stories from the Front Lines*, ed. Kathy Bousquet (Pickering, ON: Castle Quay Books, 2008), 158.

2. Ibid., 109.

3. "Breathe," Written by Marie Barnett, © 1995 Mercy/Vineyard Publishing.

4. A similar account of this story, titled "Only a Christmas Card," has been published in *This Is My Story,* ed. Kathy Bousquet, 70.

5. A similar account of this story, titled "He's a Witch Doctor," has been published in *This Is My Story*, ed. Kathy Bousquet, 114.

6. Rhody Lake, "Mountain-Climbing Professor," *The Pentecostal Testimony*, July 1989, Vol. 70, No. 7, 27-29.

CPSIA information can be obtained
at www.ICGtesting.com
Printed in the USA
LVOW07s0605110717
540884LV00011B/19/P